Embedded Machine Learning
with Microcontrollers

Cem Ünsalan • Berkan Höke • Eren Atmaca

Embedded Machine Learning with Microcontrollers

Applications on STM32 Development Boards

 Springer

Cem Ünsalan 🆔
Electrical and Electronics Engineering
Yeditepe University
Istanbul, Türkiye

Berkan Höke 🆔
Agsenze Ltd
Lancaster, UK

Eren Atmaca
Technical University of Munich
Munich, Germany

ISBN 978-3-031-70911-1   ISBN 978-3-031-70912-8   (eBook)
https://doi.org/10.1007/978-3-031-70912-8

Contents

Introduction

1

1.1 What Is Machine Learning?

Our brain is continuously exposed to data originating from our sensory organs as the eye, nose, ear, skin, and tongue. In other words, each sensory organ converts the world around us to data. Then, the brain performs two fundamental operations on this data: First, it learns what information the provided data is offering. Second, it applies inference based on the learned information. These operations are more or less the same for most living organisms. The aim here is interacting with the outer world in an intelligent manner.

We can also form intelligent systems interacting with the outside world. To do so, the system should gather data from its surrounding through its sensors. There should be a dedicated unit on the system to execute a learning algorithm and then apply inference to new data. As a result, the system can learn from data acquired from the environment and perform inference accordingly. Machine learning methods are introduced for this purpose.

We can group existing machine learning methods into two parts as traditional and neural network-based. Traditional machine learning methods are based on designing or forming a system via given data. On the other hand, neural network-based methods directly learn from the data. One approach may be more useful compared to the other for the problem at hand. Therefore, we will cover both methods in this book.

There are several ways in which machine learning methods can be used in an intelligent system. We can group this usage into three main groups as classification, regression, and clustering. The classifier assigns the incoming data to one of the known classes. Regression leads to predicting future values based on given data. Clustering groups the data based on its inherent characteristics. As a result, all these machine learning methods lead the system to interact with the outside world in an intelligent manner.

© The Author(s), under exclusive license to Springer Nature Switzerland AG 2025
C. Ünsalan et al., *Embedded Machine Learning with Microcontrollers*,
https://doi.org/10.1007/978-3-031-70912-8_1

Recent machine learning methods (especially neural network-based ones) require vast amount of data and huge computation power during training. Hence, a very powerful PC or workstation may be needed for operation. While doing so, data is collected from sensors. Then, it is stored in a local device or on cloud. Afterward, the collected data is used in training. Depending on the selected model and its parameters, this training operation can be done in minutes to days (depending on the computation power of the used device). The size of the trained model after training can be from few hundred kilobytes to hundred megabytes.

1.2   What Is an Embedded System?

Embedded system definition covers broad range of devices used in all parts of our lives. Since there are various embedded systems, it is unfair to give a strict and limiting description for them. However, we can make some general definitions about embedded systems as follows:

We can think of an embedded system as a computing device developed for solving a specific problem. To do so, it interacts with the environment by acquiring data, processes the acquired data, and produces a corresponding output. To perform all these operations, joint usage of hardware and software is mandatory. Hence, the embedded system designer should know how the hardware works and how the dedicated software should be formed for this hardware. More importantly, the designer should grasp the idea of joint usage of hardware and software to get the best from the embedded system.

The embedded system works in a stand-alone form most of the times. It may communicate with other nearby devices as with Internet-of-Things (IoT) applications. This way, it may share data to perform complex operations accordingly. Since the embedded system works in a stand-alone form most of the times, it depends on a battery or energy harvesting module to operate. Therefore, energy dissipation becomes one of the main concerns in embedded system development.

We can group embedded systems based on their hardware properties into five groups as follows: Field-programmable gate arrays (FPGA) form the first group. They provide the most flexible but hard-to-master hardware. They have been extensively used in forming custom embedded systems at gate level. Therefore, we do not program an FPGA. Instead, we describe the system to be constructed by a hardware description language such as Verilog and VHDL.

The second hardware group in forming embedded systems consists of microcontrollers. A microcontroller has limited memory and computation power. However, it can be programmed by a high-level language to perform operations on it. The microcontroller also offers a cheap and energy-efficient solution for embedded system development. The most well-known microcontroller families are Arduino- and Arm® Cortex™-M-based ones.

Microprocessors form the third embedded system group. In these, embedded Linux is used most of the times to control and organize operations. Microprocessors have fairly high memory and computation power compared to microcontrollers.

However, their weaknesses are high energy dissipation and price. The most well-known microprocessor-based embedded system is the Raspberry Pi family.

The fourth hardware group for embedded systems consists of system-on-chip (SoC) devices. They have a microcontroller and other modules (such as FPGA or wireless communication module) on them. Hence, they aim to benefit from both device properties. However, their programming and usage are still not as easy as a microcontroller (or microprocessor).

The fifth hardware group consists of graphical processing units (GPU). These devices allow parallel processing via high-level programming languages. Recent advances in deep learning and neural networks also led to devices consisting of neural processing units (NPU or TPU) dedicated to neural network implementation. Such units are becoming part of almost all embedded systems.

The mentioned five hardware groups have their dedicated development boards. Hence, they can be used easily in developing a prototype embedded system. They use cross compilers such that the code is written and debugged on PC. Then, it is embedded on the system.

1.3   Microcontroller as an Embedded System

As emphasized in the previous section, there are several hardware options which can be used in forming an embedded system. Although each group has its advantages, we pick the microcontroller as the embedded system in this book for three main reasons: First, microcontrollers are easy to program compared to other options. Second, they have a fairly wide usage area. Third, they are cheap. Hence, they can be used by a wide range of audience.

We pick the STMicroelectronics STM32F746NG microcontroller based on the Arm® Cortex™-M architecture in this book. This microcontroller cannot be used alone unless a dedicated development board is constructed for it. The best option for us, as early learners, is using the available 32F746GDISCOVERY kit for this purpose. Therefore, we will introduce the embedded system properties using both the microcontroller and the board. To note here, the selected microcontroller and its development board in this book have general characteristics of other STMicroelectronics microcontrollers and development boards. Hence, the software formed for them can be used in a wide variety of embedded systems. That is why, we set the subtitle of this book as "Applications on STM32 Development Boards."

We should also explain why the Arm® Cortex™-M architecture-based microcontroller is picked in this book. Arm® Cortex™-M architecture has gained dominance among microcontrollers. The main reason is that Arm® forms the IP and companies use it with their custom peripheral units to form a physical microcontroller. Hence, a large ecosystem has been formed. Besides, Arm® Cortex™-M architecture offers advanced properties in low-power usage. As explained before, this is extremely important for stand-alone embedded system development. Related to this, Arm® declared that its partners have shipped more than 180 billion Arm®-based chips [2]. Arm® also predicts that a trillion new IoT devices will be produced between now and 2035 [32]. Although this is a prediction, it shows the potential of growth in the microcontroller market. Besides, it indicates that we will be surrounded by more embedded systems in the near future.

1.4   Machine Learning on Microcontrollers

We mentioned in Sect. 1.1 that recent machine learning methods (especially neural network-based ones) require vast amount of data and huge computation power. The model formed after training may also be large. Therefore, it may seem counterintuitive to apply machine learning methods on resource-constrained microcontrollers. However, there are some available paths to do so: First, only inference can be applied on the microcontroller. Hence, the required computation power for training can be avoided. Second, fairly small models can be picked for operation such that they fit into the microcontroller memory. They may still be useful for solving the problem at hand. Third, microcontrollers are evolving and becoming more and more powerful in terms of computation power and memory day by day. Hence, they may provide sufficient power to implement a wider range of machine learning methods.

Although it may seem counterintuitive to apply machine learning methods on resource-constrained microcontrollers, the gain will be great if we follow one of the mentioned paths in the previous paragraph. As a result, machine learning methods can be implemented to run directly on the microcontroller. Hence, we can have more intelligent systems. Moreover, since data acquisition is possible in an embedded system, it can be acquired and processed on the edge (or microcontroller). Therefore, data can be kept securely within the device. Only the information extracted from it can be sent to a remote location. This allows a secure and low-cost data/information transmission.

1.5   About the Book

This book introduces machine learning concepts on microcontrollers. While doing so, we aim to form a complete structure. Therefore, we will introduce the hardware and software to be used in the book in Chaps. 2 and 3, respectively. To be more specific, we will introduce the STM32 board and STM32 microcontroller on it in Chap. 2. Then, we will explore STM32CubeIDE, Mbed Studio, and Keil Studio Cloud as software platforms to program the microcontroller in Chap. 3. Next, we will cover data acquisition methods from sensors in Chap. 4. We will cover the temperature and humidity sensor to acquire relative humidity and temperature data. We will benefit from the accelerometer, gyroscope, and magnetometer sensor to acquire motion data. We will use a digital microphone to acquire audio data. Finally, we will benefit from a digital camera to acquire digital images. We will also transfer the acquired data to PC for further processing.

As we form the foundations on the microcontroller and sensors side, Chap. 5 introduces fundamental machine learning concepts to be used throughout the book. Hence, we will cover the definition of feature and related topics. We will introduce traditional machine learning methods from Chaps. 6 to 8 as classification, regression, and clustering, respectively. In all these chapters, we follow the same strategy such that we will first provide the theoretical background for the method to be implemented. Then, we will provide the method to train it. Afterward, we will explain how to implement and test the trained method on the microcontroller. We will provide real-life examples as end-of-chapter applications.

We will start neural network-based methods by introducing the TensorFlow platform and Keras API in Chap. 9. These will be used extensively while handling neural networks on PC. Then, we will cover the fundamentals of neural networks in Chap. 10. In this chapter, we will explain important topics using a single neuron. The single neuron may not be sufficient to solve most real-life problems. Hence, we will introduce the multi-layer neural network concept in Chap. 11. We will emphasize training the neural network model on PC in these two chapters. We will next cover methods of embedding the trained neural network model to the microcontroller in Chap. 12. To do so, we will introduce methods based on TensorFlow Lite Micro and STM32Cube.AI. We will show how the multi-layer neural network can be used as a classifier and regressor as end-of-chapter applications in Chaps. 10 to 12.

Convolutional neural networks (CNN) are extensively used in computer vision applications. Hence, we will introduce them in Chap. 13. As in previous chapters, we will explain how to train the CNN model on PC and make inference by it on the microcontroller. We will cover recurrence in neural networks by introducing recurrent neural networks (RNN), gated recurrent units (GRU), and long-short memory (LSTM) in Chap. 14. We will also train these models on PC and make inference by them on the microcontroller. Finally, we will provide extra information on probability theory and advanced hardware properties of the STM32 board in Appendix.

There is another version of this book titled "Embedded Machine Learning with Microcontrollers: Applications on Arduino Boards." Both books aim to introduce machine learning methods on microcontrollers. The main difference of these two books is the embedded platform used. In this book, we pick STM32 boards (specifically the 32F746GDISCOVERY kit) as the target. These boards and their programming environment allow us to reach and modify low-level microcontroller hardware. Hence, the reader will have more control on the developed embedded system. The second book uses Arduino boards (specifically the Arduino Nano 33 BLE Sense Rev2 board) as the target. These boards and their programming environment allow us to form codes in high level. Hence, the reader can focus on the machine learning part of the project instead of dealing with low-level microcontroller hardware. Both approaches have their advantages. The audience for both books will also be different. One group will benefit from high-level abstraction property of the Arduino platform. The other group will benefit from the low-level hardware control provided by STM32 development boards. Therefore, we suggest the reader to pick the suitable book version.

Sample codes and libraries introduced in the book are available at https://github.com/EmbeddedML/sklearn2c. This GitHub repository also contains all

STM32CubeIDE and Mbed Studio projects for the sample codes and end-of-chapter applications in the book. We expect to have an evolving repository such that readers can improve our codes and libraries. Hence, we will have a more useful and living book for the audience.

Hardware to Be Used in the Book

2

2.1 The STM32 Board

Our codes for machine learning will run on a microcontroller. However, we cannot use this microcontroller alone since it needs extra hardware to operate. More specifically, there should be programming and debugging circuitry and power unit accompanying the microcontroller. Therefore, development boards emerged. These have all the necessary units to support the microcontroller on them. Hence, they provide a complete environment to use the microcontroller.

In this book, we pick the STMicroelectronics 32F746GDISCOVERY kit as the development board. For the sake of brevity, we will call it as the STM32 board from this point on. In this section, we will provide general information about the board including its external SDRAM. We will provide the pin layout in Appendix. We will also evaluate methods to program the microcontroller and power the board.

2.1.1 General Information

The STM32 board is as in Fig. 2.1. This board contains the listed items below. As can be seen in this list, the STM32 board offers a fairly wide range of options to be used in machine learning applications. For more information on the development board, please see [40]:

- STM32F746NG microcontroller
- Onboard ST-LINK/V2-B embedded debugging tool interface
- 4.3″ RGB 480 × 272 color LCD-TFT with capacitive touch screen
- User and reset push buttons
- User LED
- Ethernet compliant with IEEE-802.3-2002
- USB OTG HS/FS

© The Author(s), under exclusive license to Springer Nature Switzerland AG 2025
C. Ünsalan et al., *Embedded Machine Learning with Microcontrollers*,
https://doi.org/10.1007/978-3-031-70912-8_2

Fig. 2.1 The STM32 board

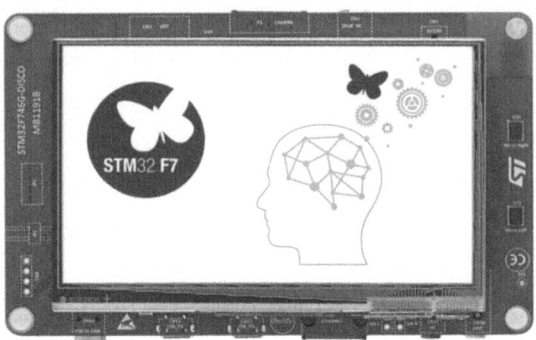

- SAI audio codec
- Two ST-MEMS digital microphones
- 128-Mbit Quad-SPI Flash memory
- 128-Mbit SDRAM (64 Mbits accessible)
- Board connectors: camera, microSD card, etc.

2.1.2 Pin Layout

The STM32 board has four headers (consisting of pins) such that a wire can be connected to them. Pins in these headers are directly connected to pins of the microcontroller. Hence, the user can reach the microcontroller pins through board headers. Pin layout of the STM32 board is as in Fig. 2.2.

As can be seen in Fig. 2.2, pins of the STM32 board are arranged in headers, named CN4, CN5, CN6, and CN7. Usage areas of these pins are tabulated in Appendix. The STM32 board has one user push button (B1), one reset button, one green LED (LD1) available to the user, and one red LED (LD2) indicating power. The GPIO pins for the user push button (B1) and green LED (LD1) are PI11 and PI1, respectively.

2.1.3 Powering the Board and Programming Its Microcontroller

The microcontroller on the STM32 board can be programmed easily by the onboard ST LINK/V2 B debugger/programmer. To do so, we should connect the board to PC via mini USB cable through its USB connector which is also indicated as USB ST-LINK on the board. Then, ST-LINK can be used for programming or debugging purposes. We will introduce methods to program the microcontroller this way in Chap. 3. The USB connection for debugging/programming purposes can also be used to power the board. Hence, whenever the board is connected to PC, it runs by the provided power. We can also use an external power supply to power the

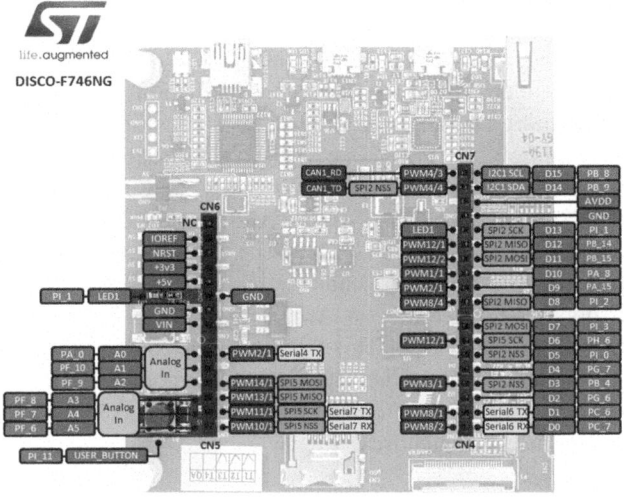

Fig. 2.2 Pin layout of the STM32 board [3]

board. However, STM32 microcontroller may not be programmed in these settings. Therefore, we suggest the reader to check the given reference for further information on this topic [38].

2.1.4 External SDRAM

There is an external 128-Mbit SDRAM available on the STM32 board. SDRAM is connected to the flexible memory controller (FMC) interface of the STM32 microcontroller. We can only access 64 Mbits of this SDRAM since its higher 16 data bits are not connected. This external SDRAM will be useful in the next chapters for storing images and allocating memory for neural network models.

2.2 Overview of the STM32 Microcontroller

As mentioned earlier, the STM32 board is equipped with the STM32F746NG microcontroller. For the sake of brevity, we will call it as the STM32 microcontroller from this point on. The microcontroller is based on Arm® Cortex™-M7 architecture. We will group the hardware and peripheral units of the microcontroller as CPU, memory, general purpose input and output ports, clock and timer modules, analog modules, digital communication modules, and other modules. We will briefly introduce each unit next.

2.2.1   Central Processing Unit

The central processing unit (CPU) is responsible for organizing all operations within the microcontroller. This is done by the instructions provided to it by the programmer. To be more specific, the programmer constructs the algorithm and forms the corresponding C or C++ code. This code is debugged and embedded into flash memory of the microcontroller by the help of an integrated development environment (IDE, such as STM32CubeIDE) running on PC. We will explain the usage of STM32CubeIDE in detail in Chap. 3. The CPU executes the commands by using its resources (such as peripheral units).

The STM32 microcontroller has an Arm® Cortex™-M7 CPU with a single precision floating-point unit (FPU). This CPU has nested vectored interrupt controller (NVIC) to handle interrupts, debug interface, memory protection unit (MPU) to protect memory regions of different tasks, digital signal processing (DSP) instructions support, and other features.

2.2.2   Memory

Data to be processed and code to be executed are stored in memory of the microcontroller. We provide introductory concepts here. Hence, the reader can grasp general characteristics of the memory and its usage.

The STM32 microcontroller has two memory types as static RAM (SRAM) and flash. SRAM is a special memory type which does not need periodic refreshing as RAM. It also provides faster access to data. The STM32 microcontroller has 320 kB SRAM of which first 64 kB is the data TCM (DTCM) RAM [33]. There is also instruction RAM (ITCM RAM) of 16 kB only reserved for the CPU in execution of critical real-time routines. More information on the TCM interface can be found in [4]. There is also an extra 4 kB backup SRAM within the microcontroller which keeps data written to it even if the microcontroller goes into standby mode. The STM32 microcontroller has 1 MB flash memory [35]. This memory is divided into four sectors of 32 Kbytes, one sector of 128 Kbytes, and three sectors of 256 Kbytes. This structure allows the programmer to erase a specific sector or the overall flash memory.

2.2.3   General Purpose Input and Output Ports

The microcontroller needs a medium to transfer data to and from the outside world. This is done by using ports of the microcontroller. A port is a collection of pins grouped together. Each pin can be taken as a wire with its electronic control circuitry. The STM32 microcontroller has 11 input and output ports called A, B, C, D, E, F, G, H, I, J, and K. The total number of pins in these ports is 168. Ports A to J have 16 pins each. Port K has eight pins. Each pin can be used for both input and output operations. Besides, a pin can be used for analog and digital operations. Hence, pins of the STM32 microcontroller are called general purpose input and output (GPIO).

Digital input and output values are processed in voltage levels as 0 V and V_{DD} (supply voltage). These correspond to logic level zero and one within the C or C++ code, respectively. Therefore, the reader should always remember that when the logic level one is fed to output from a pin of the microcontroller, the voltage there is V_{DD}. Similarly, when the logic level zero is fed to output from a pin of the microcontroller, the voltage there is 0 V.

2.2.4 Clock and Timer Modules

The CPU and peripheral units within the microcontroller are synchronous devices. This means that operations within them are synchronized by a common clock signal. Clock is a periodic square wave generated by an oscillator which can be located inside or outside the microcontroller. Frequency of the clock signal is measured in Hertz (Hz) which indicates how many periodic pulses occur in 1 second. The maximum clock frequency that can be reached by the STM32 microcontroller is 216 MHz. An operation requiring one clock cycle in the CPU actually requires $1/216 \times 10^{-6}$ sec. Hence, the higher the frequency of clock signal, the faster operations are performed within the CPU.

Due to the complexity of the microcontroller architecture and its peripheral units, the clock source for different units may not be the same. Therefore, the STM32 microcontroller has more than one internal and external clock source. A clock controller module (RCC) is available to manage all clocks and clock sources for this purpose.

The clock signal is used in timer modules as well. The timer module can be basically taken as a counter. It is responsible for counting specific events occurring in internal or external signals. For example, a timer can count rising edges of the internal clock signal. Thus, it can notify the CPU by requesting interrupts for periodic operations. The STM32 microcontroller has two advanced control timers, ten general purpose timers, two basic timers, one low power timer, one system timer (SysTick), and two watchdog timers. There is also one real-time clock (RTC) module for calendar-related operations.

2.2.5 Analog Modules

The STM32 microcontroller has analog-to-digital converter (ADC) and digital to analog converter (DAC) modules. Input and output from these modules are obtained from GPIO pins of the microcontroller. The ADC module generates digital representation of an analog voltage level fed to the microcontroller. The STM32 microcontroller has three 12-bit ADC modules. The DAC module converts a digital representation to analog voltage level and feeds it to output. The STM32 microcontroller has a 12-bit DAC module with two output channels.

2.2.6 Digital Communication Modules

Digital communication modules are used to communicate with external devices using dedicated communication protocols. The STM32 microcontroller has four inter-integrated circuit (I^2C) modules, four universal synchronous/asynchronous receiver/transmitter (USART) modules, four universal asynchronous receiver/transmitter (UART) modules, six serial peripheral interface (SPI) modules, two controller area network (CAN) modules, one universal serial bus on-the-go full-speed (USB OTG_FS) module, one universal serial bus on-the-go high-speed (USB OTG_HS) module with dedicated DMA, and one Ethernet media access control (MAC) module with dedicated DMA. These modules are extensively used when communicating with external devices or other microcontrollers.

2.2.7 Other Modules

The STM32 microcontroller has additional special purpose modules as well. These are the digital camera interface (DCMI) for receiving data from an external camera module; two serial audio interfaces (SAI) for controlling audio peripheral units using different audio protocols; LCD-TFT controller (LTDC) for driving LCD and TFT panels; flexible memory controller (FMC) for interfacing with memory blocks such as static RAM (SRAM) and synchronous DRAM (SDRAM); QUADSPI memory interface for communicating with external single-, dual-, or quad-SPI memory; secure digital multimedia card interface (SDMMC) for controlling multimedia, SD memory, and SDIO cards; and three inter-integrated sound (I2S) interfaces for audio data.

2.3   Sensors to Be Used in the Book

We will use sensors to acquire data from the outside world for machine learning applications. To do so, we will benefit from the STM32 onboard and external sensors. There are two MEMS digital microphones on the STM32 board. We will use them to acquire audio data. We will use the external B-CAMS-OMV camera module to acquire digital images. We pick the STEVAL-MKI141V2 sensor board to acquire relative humidity and temperature data. Finally, we will use the Adafruit BNO055 sensor board to acquire accelerometer data. These sensor boards can be purchased from an electronic supplier website. As we use these in specific machine learning applications in the following chapters, we will provide more information on them.

2.3.1 The Microphone

Our processing of the audio signal on the STM32 microcontroller starts with its acquisition. There are two IMP34DT05 MEMS audio sensors, U20 and U21, on the STM32 board. We will use them to acquire audio signals. More information on MEMS microphones can be found in [34].

Each audio sensor on the STM32 board converts detected acoustic waves into 1-bit pulse density modulated (PDM) signal. To process a PDM signal, it must be demodulated. In other words, the 1-bit PDM stream must be converted to a stream of digital values, namely, pulse code modulation (PCM) signal. The conversion is done by sampling the 1-bit PDM signal with a high sampling frequency, decimating, and filtering it. This operation is done by the audio codec available on the STM32 board in our case. More information on PDM to PCM conversion can be found in [36]. The acquired audio signal is transferred from the audio codec to STM32 microcontroller via SAI.

The IMP34DT05 audio sensor has five pins as given in [39]. These are VDD, GND, LR, CLK, and DOUT. The VDD and GND pins are required to power the sensor. CLK pin is to supply an input clock to the sensor. This clock frequency is equal to the PDM sampling rate. LR pin is used to sample the right or left channel using the rising and falling edges of the clock signal. DOUT pin is the PDM output signal of the microphone. DOUT and CLK pins are connected to the audio codec on the STM32 board.

2.3.2 B-CAMS-OMV Camera Module

We will use the B-CAMS-OMV camera module to acquire digital images in machine learning applications. This camera board has an OV5640 image sensor providing 5 Megapixels resolution and 8-bit color depth [37]. The front view of the B-CAMS-OMV camera module is as in Fig. 2.3.

The OV5640 camera module has a CMOS image sensor block produced by OmniVision. This sensor can provide 5 Megapixel images at 15 frames per second (fps). The camera module allows us to perform image processing operations such as exposure control, gamma, white balance, color saturation, hue control, defective pixel canceling, and noise canceling. More information on these operations can be found in [23].

The camera module can preprocess and send the acquired image to the STM32 microcontroller through its output pins. The camera module is configured through its serial camera control bus (SCCB) interface. SCCB has characteristics close to I^2C. However, the reader will not deal with this interface since we will handle its usage within our functions.

The B-CAMS-OMV camera board extends the required pins of the OV5640 image sensor to the CN5 connector. We will use this connector to connect the camera board to the P1 connector of the STM32 board. Pins of the camera module

Fig. 2.3 The
B-CAMS-OMV camera
module

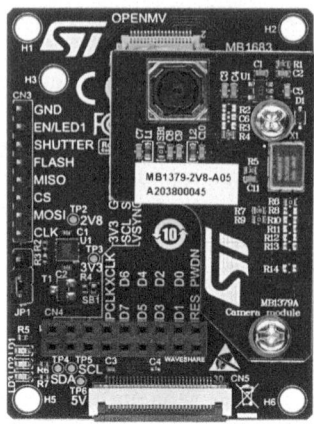

Table 2.1 Pin layout of the
B-CAMS-OMV camera
board

Pin	Description
SIOC	Serial interface clock
SIOD	Serial interface data I/O
VSYNC	Vertical sync output
HREF	Horizontal reference
PCLK	Pixel clock output
D0-D7	Digital data output
PWDN	Power down input
XCLK	System clock input
RESET	Reset input
3V3	Power supply
GND	Ground

used by the STM32 board are tabulated in Table 2.1. As can be seen in this table, there are three synchronization signals for the OV5640 camera module as VSYNC, HREF, and PCLK. VSYNC indicates a frame is being transmitted. HREF indicates a line being transmitted. PCLK indicates a new byte is available. Polarity of these signals can be adjusted via camera registers. Data output from the OV5640 camera is obtained from eight pins labeled as D0-D7. An external clock signal is supplied to the camera via its XCLK pin. SIOC and SIOD pins belong to the SCCB interface of the camera.

The OV5640 image sensor supports several output formats. These can be listed as grayscale, RAW RGB, RGB565/555/444, CCIR656, YUV422/420, YCbCr422, and compressed JPEG format. We will benefit from the grayscale, RGB888, and RGB565 formats in the following chapters.

Fig. 2.4 Adafruit BNO055
sensor board

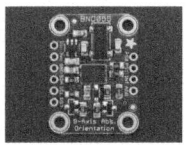

2.3.3 Accelerometer, Gyroscope, and Magnetometer

We will acquire and process inertial data in machine learning applications. Therefore, we picked the Adafruit BNO055 sensor board given in Fig. 2.4. This board has the BNO055 motion sensor containing an accelerometer, gyroscope, magnetometer, and Cortex M0+ microcontroller for sensor fusion operations. The accelerometer is a triaxial 14-bit sensor measuring the acceleration which can be configured to sense acceleration ranges ±2, ±4, ±8, and ±16 g. Gyroscope is a triaxial 16-bit sensor measuring the angular velocity which can be configured to sense from ±125°/s to ±200°/s. Magnetometer is a triaxial geomagnetic sensor measuring the magnetic field range ±1300 μT in x and y axes and ±2500 μT in z axis with a resolution of approximately ±0.3 μT.

The BNO055 motion sensor also provides sensor fusion in different modes to calculate the current orientation angles. These are roll, pitch, and yaw, also known as Euler angles. Due to the nature of accelerometers, they measure the sum of gravity and acceleration caused by movement. Separation of these vectors is a real-world engineering problem. The BNO055 motion sensor can provide the gravity vector caused by gravity and linear acceleration vector caused by movement separately in the sensor fusion mode.

The BNO055 motion sensor supports I²C, UART, and HID-I²C communication modes. We will use the I²C mode of the sensor to acquire inertial data. This operation will be explained in Chap. 4.

Pin layout of the BNO055 sensor board is tabulated in Table 2.2. Vin pin can be used to supply 5 V to the sensor board. Alternatively, the board can directly be powered from the 3Vo pin. The communication protocol for the board can be selected using PS0 and PS1 pins. The I²C mode is selected in case these pins are left floating. SDA and SCL pins belong to I²C communication. The 7-bit I²C address can be changed using the ADR pin. Finally, INT pin can be used for interrupt requests of the BNO055 sensor.

2.3.4 Relative Humidity and Temperature Sensor

We can use temperature and relative humidity data in machine learning applications. Therefore, we picked the STEVAL-MKI141V2 sensor board given in Fig. 2.5. This board has the HTS221 sensor to measure temperature and relative humidity. To be more specific, the HTS221 sensor can provide temperature and relative humidity data in a 16-bit resolution. It can operate in temperature ranges from −40° to +120°

Table 2.2 Pin layout of the
BNO055 sensor

Pin	Description
Vin	Voltage input
3Vo	3.3 V input/output
GND	Ground voltage
SDA	Serial data
SCL	Serial clock
RST	Reset input (active low)
PS0	Protocol select pin 0
PS1	Protocol select pin 1
INT	Interrupt output pin
ADR	I2C address select

Fig. 2.5
STEVAL-MKI141V2 sensor
board

Celsius. It can measure the temperature with an accuracy of $\pm0.5°$ Celsius from
15° to 40° Celsius. The sensor can measure relative humidity with an accuracy of
$\pm3.5\%$ rH from 20 to 80% rH. Sensor output data rate can be selected from 1 Hz to
12.5 Hz.

The HTS221 sensor supports I^2C and SPI communication modes. The communication mode can be selected by providing logic one or zero to the SPI enable pin
of the sensor. We will use I^2C communication for data acquisition throughout the
book.

Pin layout of the HTS221 sensor is tabulated in Table 2.3. 1.7 V to 3.6 V can
be supplied to the VDD pin of the sensor. SCL pin is used for providing serial
clock in both SPI and I^2C communication modes. HTS221 sensor can inform the
microcontroller when new measurement data is available by changing the logic level
on the DRDY pin. SDA pin is used to transfer serial data in I^2C mode and three-wire
SPI mode. The communication mode is selected as I^2C when SPI enable pin is logic
one and selected as SPI when SPI enable pin is logic zero.

Table 2.3 Pin layout of the HTS221 sensor

Pin	Description
VDD	Voltage input
SCL	I2C or SPI serial clock
DRDY	Data ready output
SDA	I2C serial data or three-wire SPI data input/output
GND	Ground
SPI enable	Logic one for I2C mode and logic zero for SPI mode

2.4   Summary of the Chapter

This book aims to blend theory with practice for machine learning applications. Therefore, we focused on the hardware to be used throughout the book in this chapter. To do so, we started with properties of the STM32 board and STM32 microcontroller. These will be the embedded hardware environments for implementation. Then, we explored sensors to be used in the book. We will be extensively using the hardware introduced in this chapter. Therefore, the reader can consult this chapter whenever needed.

Software to Be Used in the Book 3

3.1 Python on PC

Python is a prototyping language used in scientific and engineering applications. We will use it on PC throughout the book. We leave the Python installation and downloading steps to the reader. We also direct the reader to excellent books on Python programming language. Instead, we will introduce the NumPy and scikit-learn libraries for traditional machine learning algorithms in this section. We also formed a specific library to be used throughout the book. This library consists of support functions to embed the machine learning algorithm to the microcontroller. We will also briefly introduce it here. Besides traditional machine learning algorithms, we will also introduce deep learning and neural networks-based models in this book. To do so, we will benefit from the TensorFlow platform in Python. Since this platform is more advanced and requires specific functions to operate, we postpone its usage to Chap. 9.

3.1.1 The NumPy Library

We will extensively use the NumPy library while designing and testing our machine learning system on PC. This library will help us to represent the features as arrays and operations as matrices. Besides, it is also required for the scikit-learn library to be introduced next. Therefore, the reader should add the NumPy library to Python repository via executing `pip install NumPy` on the command window in the Windows operating system. The reader can import this library at the beginning of the Python code by adding the code line **import** `NumPy as np`.

© The Author(s), under exclusive license to Springer Nature Switzerland AG 2025 19
C. Ünsalan et al., *Embedded Machine Learning with Microcontrollers*,
https://doi.org/10.1007/978-3-031-70912-8_3

3.1.2   The scikit-learn Machine Learning Library

We will train and design our traditional machine learning systems on PC by the help of the scikit-learn library. Therefore, we should also add this library to our Python repository. To do so, the reader should execute `pip install -U scikit-learn` in the command window under the Windows operating system. The reader can import this library at the beginning of the Python code by adding the code line **from** sklearn **import** x by declaring which part of it will be used by x.

3.1.3   The Sklearn2c Library for Microcontrollers

We will form our machine learning systems in two parts: The first part, working on PC, will be for the training and/or designing step of the system. This part will be formed by Python codes extensively using scikit-learn and NumPy libraries. The second part, working on the target microcontroller, will be responsible for the inference operation of the trained/designed machine learning system. Since this part will be working on the microcontroller, we will form it in C language. We formed a library to create necessary C header and source files from the machine learning algorithm in Python. We call this library as sklearn2c, the machine learning library for embedded systems. The reader can install it by the command `pip install sklearn2c` from https://pypi.org/project/sklearn2c/. We will explain how to benefit from this library in detail in the following chapters.

3.2   The STM32CubeIDE Platform

C is the most effective language to program microcontrollers in terms of resource usage, speed of execution, and ease of programming. Therefore, it is taken as the defacto language for most microcontrollers. Here, we assume that the reader has sufficient knowledge on C programming techniques. If this is not the case, then we refer the reader to valuable books on this topic.

We will benefit from the STM32CubeIDE platform to program the STM32 microcontroller via C language. This platform has several advantages: First, it is free. Second, it is specifically developed for microcontrollers by STMicroelectronics. Third, it is possible to reach and modify all properties of the microcontroller including its peripheral units using STM32CubeIDE. Therefore, we pick it in this book. To note here, it is also possible to use C++ language to develop programs in STM32CubeIDE.

To get familiar with STM32CubeIDE, we will start with its installation. Afterward, we will explain how to create and manage a project. While doing so, we will not adjust peripheral units of the microcontroller. Instead, we will focus on fundamental properties of creating and executing a project in STM32CubeIDE. As we master this step, we will see how peripheral units can be modified in Sect. 3.2.5. To do so, we will benefit from STM32CubeMX which will be of great use in managing microcontroller hardware.

3.2.1   Downloading and Installing STM32CubeIDE

It may seem trivial; however, the first step in using the STM32CubeIDE platform is downloading and installing it. STM32CubeIDE can be downloaded from the website [41] after registering or logging in. When the download is complete, the reader should follow the steps given there for installation. When the installation is complete, the reader should set the workspace directory from the opening window. Afterward, STM32CubeIDE will launch.

3.2.2   Launching STM32CubeIDE

As the STM32CubeIDE launches, the welcome window will be as in Fig. 3.1. We will use this window for all our needs. Therefore, let's briefly summarize its basic sections.

As can be seen in Fig. 3.1, there exists the "Project Explorer" panel on the left side of the welcome window. The user can create a new project through it. As the new project is created, this panel summarizes all its properties. Code files will be opened as they are added to the project at the center of STM32CubeIDE.

Fig. 3.1 STM32CubeIDE welcome window

Hence, the reader will modify the code in this section. There will be other sections which summarize hardware properties of the microcontroller on the right side of the panel. The reader can choose available perspectives such as "C/C++" (default), "Debug," and "Device Configuration Tool" in STM32CubeIDE. There are also grouped buttons on top of the panel. These will be useful while debugging and modifying the code. There are tabs summarizing all build and debug operations at the bottom of the panel. We will explain all these properties in the following sections as we focus on parts of the project.

3.2.3   Creating a New Project

A project typically contains source and header and includes files. STM32CubeIDE generates an executable file from these to be embedded on the microcontroller. This section is about creating a C project. We have two options here: The first one does not deal with peripheral units and pin properties of the microcontroller. The second one deals with these. We will focus on the first option in this section to get familiar with the basics of STM32CubeIDE. We will focus on the second option in Sect. 3.2.5.

To create a new project, click "File," "New," and "STM32 Project." In the opening window, select the hardware platform to be used in the project. For our case, we should choose "STM32F746G-DISCO" from the "board selector" tab, and press "Next." Afterward, the "Project Setup" window opens as in Fig. 3.2.

Fig. 3.2 Project setup window

Fig. 3.3 Project explorer tab

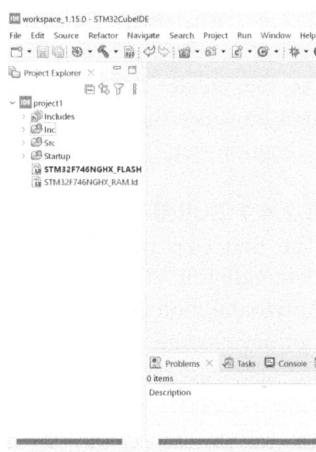

We should give a name to our project in this window. For our case, we set it as "project1." We should set the location of the project which may be done by the option "use default location." We should set the "Targeted Language" as C. We should set the "Targeted Binary Type" as "Executable." The final selection is titled "Targeted Project Type." Here, we have two options as "STM32Cube" and "Empty." The first option allows modifying peripheral units of the microcontroller. We will consider this option in Sect. 3.2.5. Hence, we select the "Empty" option to generate the project as for now. As we press the "Finish" button, a new project will be created which can be observed under the "Project Explorer" tab as in Fig. 3.3.

As can be seen in Fig. 3.3, STM32CubeIDE adds all necessary files to the project. The only step left is adding our C code to the "main.c" file generated under the "Src" folder. We will use the code given in Listing 3.1 as an example in this section. To do so, open the "main.c" file by double-clicking on it. Add your code to this file. Then, save it by clicking the "Save" button on the upper left corner of the menu.

Listing 3.1 The first C code for the STM32 microcontroller

```c
int a = 1;
int b = 2;
int c;

int main(void)
{
int d;

c = a + b;
d = c;

while (1);
}
```

3.2.4   Building, Debugging, and Executing the Project

As we create our project, the next step is its execution. To do so, we will
first consider the building and debugging steps. Afterward, we will focus on the
execution step.

3.2.4.1 Building and Debugging the Project

The first step in executing the code on an embedded platform is building and
debugging it. There is a button with hammer shape on the STM32CubeIDE
horizontal toolbar as "Build." As the user presses the right bottom arrow there, two
options emerge as "Debug" and "Release." The first option builds the project with
the necessary debug information. No optimization is done on the built output in this
default option. The second option is specific to deploying the built output to the
microcontroller. Hence, the main focus here is optimizing output of the project and
minimizing its size. Running steps in both options (such as warnings and errors) can
be observed in the "CDT Build Console" window. If there is an error in a code line,
then it is indicated by a red cross by it.

The second step in executing the code is debugging and loading it to the target
device. There is a button on the STM32CubeIDE horizontal toolbar as "Debug."
It can also be reached from the "Run" tab. As we press this button, a new
window appears asking for the "Edit Configuration" as in Fig. 3.4. We can make
all adjustments related to the debugging process within this window. The default
settings should be kept as they are in this window unless otherwise stated. As
we press "OK" in the "Edit Configuration" window, the code is debugged and
downloaded to the microcontroller. Afterward, the STM32CubeIDE interface turns
to the "Debug" perspective.

Fig. 3.4 The debug interface
menu

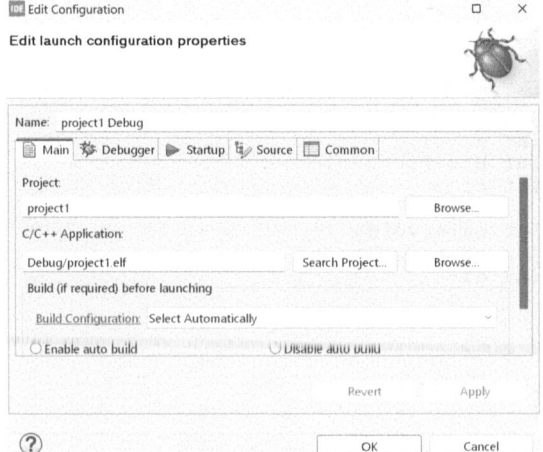

Fig. 3.5 The execution
session menu

3.2.4.2 Executing the Project

As the debugged C code is downloaded to the microcontroller, the next step is its execution. The code will be executed till the beginning of the main function at this stage. The microcontroller waits for further command. Buttons for program execution are located on the toolbar as in Fig. 3.5.

The name of each button can be observed by moving the cursor over it. These buttons and their functions are explained briefly in the list below:

- **Skip all breakpoints**: Disable all previously set breakpoints. We will talk about what a breakpoint is in the following paragraphs.
- **Terminate and relaunch**: If a modification is done in the program while in debug mode, then this button rebuilds and debugs it.
- **Resume**: Resumes execution of the code from the last executed location. Execution of the code continues until a breakpoint when this button is clicked.
- **Suspend**: Halts execution of the code. All windows used to observe software and hardware are updated with recent data.
- **Terminate**: Ends execution of the code. The debug session is also ended by this command.
- **Disconnect**: Disconnects the target device.
- **Step into**: Executes the next line of code. If this line calls a function, the compiler executes the next line in it. Then, it stops.
- **Step over**: Executes the next line of code. If this line calls a function, the compiler executes the function completely. Then, it stops.
- **Step return**: Completes execution of the function and exits the program.
- **Instruction stepping mode**: The C code is executed based on its generated assembly code within the step into, step over, and step return commands.

Observing variables, registers, or memory is important while developing the project. Code execution should halt in order to perform this operation. A breakpoint should be added to stop execution of the code at a specific code line. To do so, left-click on the desired code line, and select "Add Breakpoint..." from the pop-up window. A new window appears asking for the breakpoint type. Here, we can select the "Regular" type. As the breakpoint is added, a blue circle will appear by the code line to indicate that there is a breakpoint there. The inserted breakpoint can be deleted by double-clicking on it.

The "Expressions" window, given in Fig. 3.6, can be used to observe selected variables. Select the variable to be observed and right click on it to add a variable to this window. Select the "Add Watch Expression" option in the opening list. Then, click "OK." The reader can also double-click on the "Add new expression" button in the "Expressions" window and enter name of the variable to the opened box.

Fig. 3.6 The Expressions
window

Expression	Type	Value
(x)= a	int	1
(x)= b	int	2
(x)= c	int	0
(x)= d	int	0

We can define a variable either as local or as global in C language. As the name implies, the global variable is available to all code sections. The local variable is only available to the function it is defined in. Although local and global variables can be observed in the "Expressions" window, local variables can also be observed in the "Variables" window. Here, all local variables are automatically added to the mentioned window.

There is also "Live Expressions" window which can be enabled from the "Window/Show View" list besides other options. Variables are updated in real time while executing the code in this window. Hence, the variable can be observed without stopping the code.

3.2.5 Using STM32CubeMX to Modify Hardware of the Microcontroller

STM32CubeMX, available under STM32CubeIDE, can be used to control and modify hardware of the STM32 microcontroller. To note here, hardware properties can also be set via available libraries such as hardware abstraction layer (HAL). The advantage of STM32CubeMX is that operations are done visually on it. Moreover, it produces a template C code containing predefined functions to be used to control hardware properties. Therefore, we will use this option throughout the book.

3.2.5.1 Creating a New Project Using STM32CubeMX

To modify hardware of the microcontroller, we can create a new project as explained in Sect. 3.2.3 with one difference. Since we will modify hardware of the microcontroller, we will set the "Targeted Project Type" as "STM32Cube" in Fig. 3.2. Afterward, a new window appears asking for the "Firmware Library Package Setup." At this step, leave all default settings as they are and press "Finish." The next pop-up window asks whether to initialize all peripheral units with their default mode. Click "No" since we will modify necessary peripheral unit properties as we need them. Then, a window should open up as in Fig. 3.7.

We can set all system hardware properties through STM32CubeMX. Let's assume that we want to turn on the green LED on the STM32 board whenever the user button on the board is pressed. Let's create a simple project for this purpose. To do so, we should first modify the microcontroller pin properties. Hence, we should set the pin PI11 of the STM32 microcontroller as "GPIO_Input" and pin PI1 as "GPIO_Output." These modifications can be done in the STM32CubeMX interface easily. Within the "Pinout&Configuration" tab, a pin can be located by typing its

Fig. 3.7 STM32CubeMX
initial screen

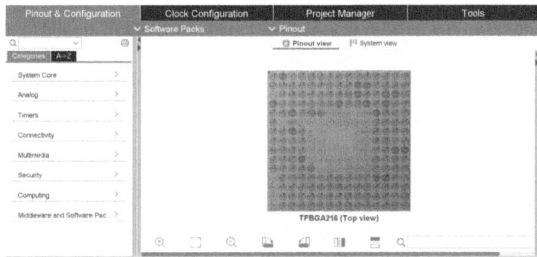

name in the search box at the bottom. Then, STM32CubeMX will show where the desired pin is. Locate the mentioned pins and left-click on them one by one. A pop-up window appears asking for which purpose each pin will be used for. Set the pin PI11 as "GPIO_Input" and pin PI1 as "GPIO_Output." Proceed to clock configuration by opening the "Clock Configuration" tab. We will benefit from the automatic clock setting property of STM32CubeMX here. By the way, this is the default setting, and we suggest using this option unless otherwise stated.

3.2.5.2 Generating and Modifying the Code

Now, we are ready to generate the code corresponding to the hardware setup. Before doing so, we can open the "Project Manager" tab. Here, we can set properties of our project from the "Project," "Code Generator," and "Advanced Settings" tabs. We will modify some of the options here in the following chapters. However, as for now, please keep all the settings as they are since they are generated automatically. The next step is generating the code related to the hardware setup. To do so, press the "Save" button. A pop-up window appears asking for generating the code. Press "OK" there. The second option to generate the code is to click "Generate Code" under the "Project" tab of the STM32CubeMX top menu. The third option is to use the keyboard shortcut by pressing ALT+K.

As the code is generated, we can open the "main.c" file under the "Src" folder. As this file is opened, the reader will observe that there are sections labeled "USER CODE BEGIN" and "USER CODE END" asking for the user to add the code snippets there. This is done to ensure that other parts of the code related to hardware setup are not modified by mistake. Finally, we will test the project. To do so, add the C code snippet in Listing 3.2 to the appropriate place in the opened "main.c" source file.

Listing 3.2 The C ode for turning on the LED when the button is pressed

```
/* Infinite loop */
/* USER CODE BEGIN WHILE */
while (1)
{
    if (HAL_GPIO_ReadPin(GPIOI, GPIO_PIN_11))
        HAL_GPIO_WritePin(GPIOI, GPIO_PIN_1, GPIO_PIN_SET);
    else
        HAL_GPIO_WritePin(GPIOI, GPIO_PIN_1, GPIO_PIN_RESET);
/* USER CODE END WHILE */
```

```
/* USER CODE BEGIN 3 */
}
/* USER CODE END 3 */
```

3.2.5.3 Executing the Project

As we debug the project and embed the generated code on the STM32 microcontroller, it will be ready to be executed. Here, we will follow the same steps
as in Sect. 3.2.4.2. The reader can check whether the code is running or not by
pressing and releasing the user button on the STM32 board. Every time the button
is pressed, the green LED on the board should turn on. This test ends our coverage
of STM32CubeMX.

3.2.6 Reaching the Microcontroller Hardware

The microcontroller hardware is configured via STM32CubeMX as we did in the
previous section. However, we only used the GPIO of the STM32 microcontroller
there. We will show how other peripheral units of the microcontroller can be
configured via the STM32CubeMX interface by developing two new projects in
this section. These projects will cover basic concepts of timer, GPIO, and interrupts,
which are frequently used in embedded systems. In the first project, we will show
how timer interrupts are enabled and used. In the second project, we will use the
user push button to handle an external GPIO interrupt.

We will toggle the green LED on the STM32 board using timer interrupts
in the first project. Therefore, the pin PI1 should be set as "GPIO Output" for
the LED as explained in Sect. 3.2.5.1. We can pick the TIM2 peripheral unit to
generate timer interrupts. Therefore, the unit should be configured at the desired
interrupt frequency. To do so, find the TIM2 peripheral unit in the "Pinout &
Configuration" section of STM32CubeMX. The TIM2 peripheral unit can also be
found by typing "TIM2" to the search box located in the upper left side of the
STM32CubeMX interface. After clicking the TIM2 peripheral unit, the "TIM2
Mode and Configuration" window appears. We will select the "Internal Clock" as
the clock source from the "Mode" section of the window as in Fig. 3.8.

After selecting the clock source, the configuration window for the TIM2
peripheral unit will appear. This window contains parameters of the peripheral unit.
We can configure the timer at the desired interrupt frequency by adjusting these
parameters. Therefore, we should understand what these parameters are. We can
find that the TIM2 peripheral unit is connected to the advance peripheral bus 1
(APB1) of the microcontroller from the datasheet of the STM32 microcontroller. All
clock frequencies of peripheral units can be adjusted from the clock configuration
window of the STM32CubeMX interface. We will leave the clock frequencies as
they are. However, we should know what the clock frequency of the APB1 Timer

Fig. 3.8 Mode section of the
TIM2 peripheral unit in
STM32CubeMX

TIM2 Mode and Configuration

Mode

Slave Mode   Disable

Trigger Source   Disable

Clock Source   Internal Clock

Channel1   Disable

Channel2   Disable

Channel3   Disable

Channel4   Disable

Combined Channels   Disable

☐ *Use ETR as Clearing Source*

☐ *XOR activation*

☐ One Pulse Mode

clock is since TIM2 will be clocked by it. As in the default case, APB1 Timer clock
frequency is 84 MHz. Then, we can return back to the TIM2 configuration.

To observe the green LED blinking, it is better to set the interrupt frequency as
1 Hz. In other words, LED will toggle every second. Prescaler parameter is used
to divide the internal clock frequency. We will set the prescaler to 83 to divide the
84 MHz internal clock by 84. The next step is setting the period parameter. Hence,
the interrupt frequency becomes 1 Hz. The period is the maximum value that the
counter can reach after it starts counting. When the value of the period is reached,
the timer will start counting again. As a result of prescaling the 84 MHz clock by
83, the counter register will be clocked with 1 MHz. When we set the period as
999999, it will take 1 second for the counter register to reach the period value.
The final configuration can be seen in Fig. 3.9. For more information on STM32
microcontroller timer modules, please see [35].

We configured the timer parameters up to this point. Since interrupts are handled
by the CPU, we should enable the corresponding interrupt line for the TIM2
peripheral unit there. The NVIC module in the CPU is dedicated for this purpose.
There is a tab as "NVIC Settings" in the STM32CubeMX interface of the TIM2
peripheral unit as shown in Fig. 3.10. We can activate the TIM2 interrupts in the
NVIC module by checking the checkbox for the "TIM2 global interrupt." We
can also set priority of the interrupt by giving any number from 0 to 15 to the
"preemption priority" from the configuration window of the NVIC module in the
STM32CubeMX interface. The priority level decreases as this number increases.
We will leave it as 0 for our case. Now, we are ready to generate the code.

The main code for toggling the LED on the STM32 board with timer interrupts is
given in Listing 3.3. In this code, we start the timer in interrupt mode by calling the
function HAL_TIM_Base_Start_IT(&htim2). The HAL library provides the function
void HAL_TIM_PeriodElapsedCallback(TIM_HandleTypeDef *htim). This function is

Fig. 3.9 Configuration of the
TIM2 peripheral unit in
STM32CubeMX

Fig. 3.10 Configuration of
the NVIC settings of the
TIM2 peripheral unit in
STM32CubeMX

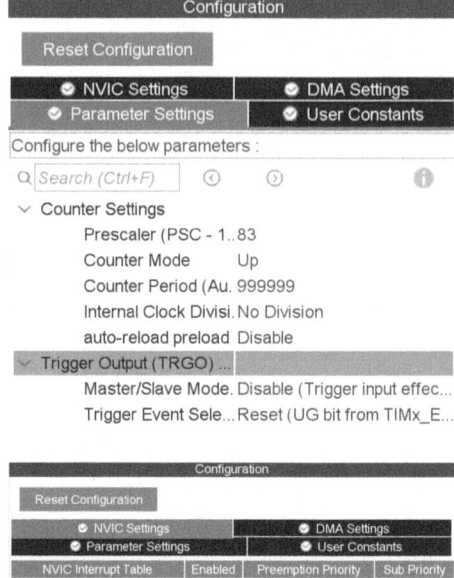

called when a timer interrupt occurs. HAL library also provides us which timer
peripheral unit caused the interrupt by giving the `htim` argument. We can understand
which timer caused the interrupt by checking this argument. Then, we can toggle
the LED if the interrupt is caused by the TIM2 timer.

Listing 3.3 The C code for toggling the LED with timer interrupts

```
/* USER CODE BEGIN 0 */
void HAL_TIM_PeriodElapsedCallback(TIM_HandleTypeDef *htim)
{
    if (htim->Instance == htim2.Instance)
    {
        HAL_GPIO_TogglePin(GPIOI, GPIO_PIN_1);
    }
}
/* USER CODE END 0 */

/* USER CODE BEGIN 2 */
HAL_TIM_Base_Start_IT(&htim2);
/* USER CODE END 2 */
```

Next, we will show how an external interrupt can be activated. To do so, we
can create a new STM32CubeIDE project. Here, we will use the external interrupt
to toggle the LED when the user push button on the STM32 board is pressed. We
know that the user push button is connected to pin PI11 and the LED is connected
to pin PI1. Therefore, we set the PI1 pin as "GPIO_Output" for the LED and PI11
pin as "GPIO_EXTI11" for the user push button.

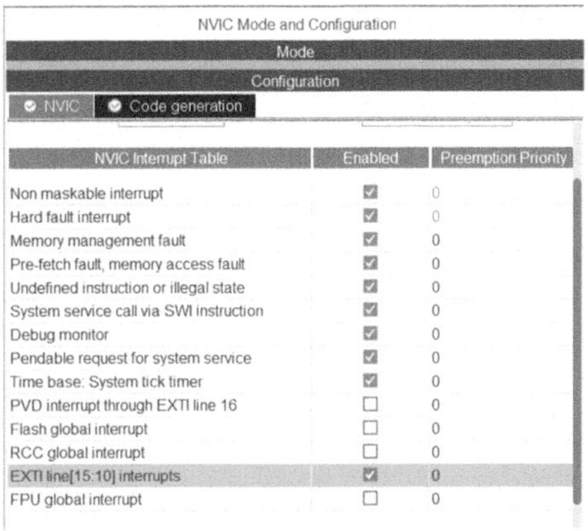

Fig. 3.11 Configuration of
the NVIC module for EXTI
in STM32CubeMX

NVIC Mode and Configuration		
Mode		
Configuration		
● NVIC ● Code generation		

NVIC Interrupt Table	Enabled	Preemption Priority
Non maskable interrupt	☑	0
Hard fault interrupt	☑	0
Memory management fault	☑	0
Pre-fetch fault, memory access fault	☑	0
Undefined instruction or illegal state	☑	0
System service call via SWI instruction	☑	0
Debug monitor	☑	0
Pendable request for system service	☑	0
Time base: System tick timer	☑	0
PVD interrupt through EXTI line 16	☐	0
Flash global interrupt	☐	0
RCC global interrupt	☐	0
EXTI line[15:10] interrupts	☑	0
FPU global interrupt	☐	0

Whenever an interrupt has to be activated, it should be enabled in the NVIC module. Therefore, we should select the "EXTI line[15:10] interrupts" checkbox under the "NVIC Mode and Configuration" window of the STM32CubeMX interface as in Fig. 3.11. By default, an external interrupt is generated when a rising edge is detected on pin PI11. As a side note, the button is connected in active high configuration on the STM32 board. In other words, the logic level remains at zero when the button is idle. When the button is pressed, logic level of the pin becomes logic one. Therefore, a rising edge occurs on the pin PI11 when the button is pressed. This can be changed from the "GPIO Mode and Configuration" window in STM32CubeMX. We left it as it is and generate the code.

The main code for toggling the LED with external interrupt is given in Listing 3.4. In this code, the CPU calls the HAL library function **void** HAL_GPIO_EXTI_Callback(uint16_t GPIO_Pin) when an external interrupt is requested by the GPIO. HAL library also provides us source pin of the interrupt via the function argument GPIO_Pin. Since the user push button is connected to the pin PI11, we can toggle the LED by checking it.

Listing 3.4 The C code for toggling the LED with external interrupts

```
/* USER CODE BEGIN 0 */
void HAL_GPIO_EXTI_Callback(uint16_t GPIO_Pin)
{
        if (GPIO_Pin == GPIO_PIN_11)
        {
                HAL_GPIO_TogglePin(GPIOI, GPIO_PIN_1);
        }
}
/* USER CODE END 0 */
```

Fig. 3.12 Mbed Studio
launch window

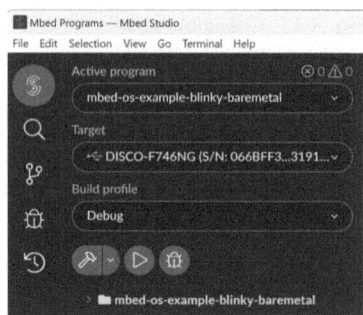

3.3 Mbed Studio on Desktop and Keil Studio in Cloud

Although STM32CubeIDE allows controlling all microcontroller properties, the
user may not need such a detailed setup in a project. Therefore, we introduce the
Mbed Studio and its web-based version Keil Studio Cloud in this section. The user
can form a project faster via these platforms. We handle these two platforms in one
section since they have the same interface. To be more specific, we will provide the
usage of Mbed Studio first. Then, if there are any differences on the Keil Studio
Cloud, we will highlight them.

3.3.1 Creating a New Project

The reader can download Mbed Studio from the website https://os.mbed.com/
studio/. The reader should follow the steps summarized there for installation when
the download is complete. The reader should login to the Mbed account when the
installation is complete. Afterward, the program will launch and the IDE will be as
in Fig. 3.12. Mbed Studio has a fairly compact interface as can be seen in this figure.
There exists a panel on the left side of the window. Through it, the user can create a
new project, select the target, and build the project for the selected target. The main
code file will be opened as it is added to the project at the center of the window.
There are tabs which will summarize build and debug operations at the bottom of
the panel.

We can create, build, debug, and execute a project under Mbed Studio. We
should connect our STM32 board to PC before creating the project. Mbed Studio
automatically detects the board and displays it under the "Target" section. Moreover,
the reader can check connection status of the board in the same section. If the board
connection is active, then the green USB sign should be visible in front of the target
board name.

We can create a new project under Mbed Studio by clicking on the "File" menu
and select the "New program" there. A new window opens up as in Fig. 3.13. Next,
we should select a template program. We can select "mbed-os-example-blinky-

Fig. 3.13 Creating the
project under Mbed Studio

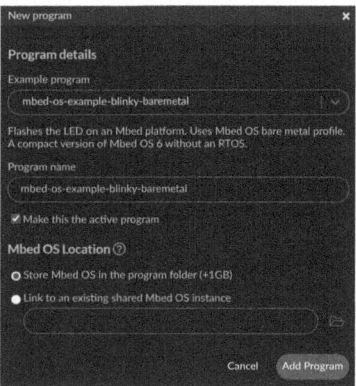

baremetal" for our case. We should give a name to our project in the "Program name" window. Afterward, we should press the "Add Program" button and let Mbed Studio import all necessary libraries.

The code to be executed can be found in the "main.cpp" file under our project. We will use the available code in our first project. We provide the content of the main file in Listing 3.5 for completeness.

Listing 3.5   The main file for our project in Mbed Studio

```
#include "mbed.h"

#define WAIT_TIME_MS 500
DigitalOut greenLED(LED1);

int main()
{
printf("This is the bare metal blinky example running on Mbed OS
    %d.%d.%d.\n", MBED_MAJOR_VERSION, MBED_MINOR_VERSION,
    MBED_PATCH_VERSION);

while (true)
{
greenLED = !greenLED;
thread_sleep_for(WAIT_TIME_MS);
}
}
```

3.3.2   Building, Debugging, and Executing the Project

There are three buttons on Mbed Studio left panel with the hammer, triangle, and bug shape to build, run, and debug the project, respectively. As the program in Listing 3.5 is built and executed following the mentioned steps, the green LED on the STM32 board should be blinking every 0.5 s.

3.3.3   Reaching the Microcontroller Hardware

We will use Mbed Studio to reach the microcontroller hardware in this section. We provide two examples showing how the timer interrupt and GPIO interrupt is set. We should create a new Mbed Studio project for the first example. Mbed provides us the Ticker class to set up a periodic interrupt in a given period. We should form a function to be executed periodically to use this class. We toggle the LED every 1 second in this function in our example.

The main code for toggling the LED using the Ticker class is given in Listing 3.6. In this code, we attach our **void** TICKER_Callback function to the Ticker to be called every second using the function attach. We formed the function **void** TICKER_Callback to toggle the LED. Run the code and observe the LED on the STM32 board.

Listing 3.6 The C++ code for toggling the LED using the Ticker class

```
#include "mbed.h"

DigitalOut led1(LED1);
Ticker ticker;

void TICKER_Callback(){
    led1 = !led1;
}

int main(){
    ticker.attach(&TICKER_Callback, 1);
    while (true){
        wait_us(1000);
    }
}
```

As the second example, we pick the GPIO interrupt usage under Mbed Studio. Mbed provides us the InterruptIn class to activate the GPIO interrupt on a given pin. A callback function has to be called when an interrupt occurs. In our example, we toggle the LED when the user push button on the STM32 board is pressed. Therefore, the content of our callback function will perform this operation.

The main code for toggling the LED is given in Listing 3.7. In this code, we first create an InterruptIn object for the user push button on the STM32 board which is predefined by BUTTON1. We then attach a callback function to be called when the rising edge occurs at the corresponding pin using the function rise. Hence, the function **void** BUTTON_Callback will be called when the interrupt occurs. Run the code and push the user button on the STM32 board to observe the LED.

Listing 3.7 The C++ code for toggling the LED using the InterruptIn class

```
#include "mbed.h"

DigitalOut led1(LED1);
InterruptIn button(BUTTON1);

void BUTTON_Callback(){
    led1 = !led1;
}

int main(){
    button.rise(&BUTTON_Callback);

    while (true){
        wait_us(1000);
    }
}
```

3.4 Application: Tools for Analyzing the Generated Code

There are tools for analyzing the generated C and C++ codes. We will consider them in this section. Our main focus will be measuring the execution time and memory usage of the code block of interest. Therefore, we will be able to measure the latency during inference step of a machine learning system.

3.4.1 Analyzing the C Code in STM32CubeIDE

We divide the code analysis in STM32CubeIDE into two parts: The first part is on measuring the execution time of a given code. The second part is on memory usage. We will cover both parts separately next.

3.4.1.1 Measuring the Execution Time

HAL library needs a timer to work properly. By default, STM32CubeMX uses the Systick timer as the "Timebase Source." Hence, the timer is initialized to generate 1-millisecond interrupts. These interrupts are counted by the HAL library from the beginning of the code. Thanks to this utility of the HAL library, we can keep track of the operations in terms of execution time. As a side note, the "Timebase Source" can be changed to any available timer on the microcontroller from the STM32CubeMX interface under the "SYS" tab.

The code for measuring the elapsed time for a given operation is as in Listing 3.8. In this code, we assign the value of the millisecond counter to our variable startTick before the operation. Then, we use floating point operations as a sample operation. We assign the value of the millisecond counter to another variable stopTick after the operation is complete. When we subtract startTick from stopTick, we get the elapsed time during our sample operation in milliseconds.

Listing 3.8 The C code for measuring the elapsed time for an operation

```
/* USER CODE BEGIN Includes */
#include <math.h>
/* USER CODE END Includes */

/* USER CODE BEGIN 0 */
float data[1000];
/* USER CODE END 0 */

/* USER CODE BEGIN 2 */
uint32_t startTick, stopTick, executionTime;
startTick = HAL_GetTick();
for (uint32_t i = 0; i < 1000; i++)
{
  data[i] = M_PI * M_E * i;
}
stopTick = HAL_GetTick();
executionTime = stopTick - startTick;
/* USER CODE END 2 */
```

3.4.1.2 Measuring Memory Usage

The reader can reach detailed memory usage of a code in STM32CubeIDE using its "Build Analyzer" window. Here, there are two sub-windows as "Memory Regions" and "Memory Details." In the first window, the reader can observe start and end addresses; free, used, and total memory size; and usage percentage of RAM, FLASH, and CCMRAM. In the second window, the reader can observe detailed usage of each memory block separately.

3.4.2 Analyzing the C++ Code in Mbed Studio

We can analyze the generated C++ code under Mbed Studio. Therefore, we will consider measuring the execution time and memory usage for a given code next.

3.4.2.1 Measuring the Execution Time

There is the Timer module to perform time measuring operations under Mbed Studio. This module has the functions start and stop to start and stop the timer, respectively. There is also the function elapsed_time to calculate time difference between starting and stopping points. We provide a usage example for these functions in Listing 3.9.

Listing 3.9 Measuring the execution time and memory usage in Mbed Studio

```
#include "mbed.h"

mbed_stats_heap_t heapInfo;
mbed_stats_stack_t stackInfo;
uint32_t i;
uint32_t data[1000];
```

```
Timer myTimer;

int main()
{
myTimer.start();
for (i = 0; i < 1000; i++)
{
data[i] = i;
}
myTimer.stop();
printf("Execution time in microseconds: %llu\n", myTimer.
    elapsed_time());

mbed_stats_heap_get(&heapInfo);
printf("Heap size: %ld\n", heapInfo.reserved_size);
printf("Used heap: %ld\n", heapInfo.current_size);

mbed_stats_stack_get(&stackInfo);
printf("Main stack size: %ld\n", stackInfo.reserved_size);
printf("Used main stack: %ld\n", stackInfo.max_size);

while (true);
}
```

3.4.2.2 Measuring Memory Usage

Mbed Studio provides a set of functions to monitor the stack and heap usage in run time. We should create an empty "mbed_app.json" file at the project root folder to use these functions. Then, we should add the below code block to this file:

```
{
"target_overrides": {
"*": {
"platform.all-stats-enabled": true
} } }
```

As we perform all the mentioned operations, we will be ready to measure the memory usage for a given code. We provide a sample code to show how these operations can be done in Listing 3.9. Please note that the same code also provides an example of measuring the execution time of a code block.

3.5   Summary of the Chapter

We introduced software platforms to be used throughout the book in this chapter. We covered this topic from two perspectives as Python on PC and C/C++ on the STM32 microcontroller. We benefit from Python and its libraries to design and test machine learning algorithms offline. We also introduced the embedded machine learning library to embed the developed machine learning code on PC to the microcontroller. As for programming the microcontroller, we covered the usage of the STM32CubeIDE, STM32CubeMX, Mbed Studio, and Keil Studio

Cloud platforms. We provided sample programs and detailed usage recipes for each platform. As the end-of-chapter application, we provided tools for analyzing the generated code. Hence, the reader can measure the memory usage and inference time of a given machine learning code. This will be of great help in the following chapters.

Data Acquisition from Sensors

4

4.1 Data Transfer Between the PC and STM32 Microcontroller

Although it is natural to process acquired data on the STM32 microcontroller, it is not easy to observe and analyze the data there. Therefore, we should transfer the acquired data from the STM32 microcontroller to PC. Likewise, we may need to process controlled data on the microcontroller. For such cases, we can prepare the data on PC and then transfer it to the STM32 microcontroller. We will handle both cases in this section.

4.1.1 Overview of Data Transfer via UART Communication

UART is generally preferred when fast data transfer is not required. Hence, it is used to form a communication link between the microcontroller and PC. Data transfer in UART can be done in frames. Each frame is composed of start bit, data, parity bit, and stop bit(s). Data can be composed of seven, eight, or nine bits. Start and stop bits are available in each frame. Parity bit can be used if needed. For more information on these topics, please see [61].

We provide the general setup for UART communication between two devices in Fig. 4.1. As can be seen in this figure, each device has two pins as Rx (receive) and Tx (transmit). They are connected such that the Rx pin of one device is connected to Tx pin of the other device.

The STM32 microcontroller has four UART modules called UART4, UART5, UART7, and UART8. Besides, the STM32 microcontroller has four USART modules called USART1, USART2, USART3, and USART6. These modules can be used in UART mode. Hence, there are a total of eight effective UART modules in the STM32 microcontroller. The USART1 module is connected to ST-Link. Therefore, it can communicate with PC through virtual com port.

Fig. 4.1 General setup for
UART communication
between two devices

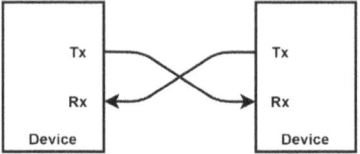

Fig. 4.2 USART1 mode and
configuration window

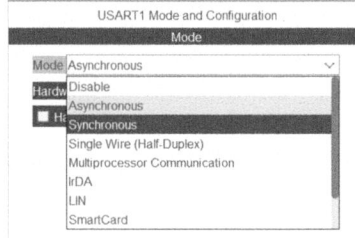

4.1.2   UART Communication with STM32CubeIDE

We will consider UART communication between the STM32 microcontroller and
PC via STM32CubeIDE in this section. Therefore, we will cover the setup and
usage. To make life easier, we formed a serial library to transfer data between the
microcontroller and PC. We will cover it in this section as well.

4.1.2.1 Setup in STM32CubeMX
We will show how UART module of the STM32 microcontroller is set up using
STM32CubeMX in this section. We assume that a new project has already been
created as explained in Sect. 3.2.3. As a reminder, peripheral units are listed under
the "Pinout & Configuration" section of the device configuration tool window of
STM32CubeMX. Units to enable communication are listed under the connectivity
category there. After clicking on USART1, "USART1 Mode and Configuration"
window will appear as in Fig. 4.2.

We will use asynchronous UART mode to transfer data between the STM32
microcontroller and PC. After selecting the asynchronous mode, the USART1
module is not only configured accordingly, but the corresponding pins are also set.
Hence, the pin PA9 is set to transmit data to PC, and the pin PB7 is set to receive data
from PC. Next, we should configure parameters of the USART1 module under the
"Parameters Settings" tab of the configuration menu as in Fig. 4.3. These parameters
are baud rate, word length, parity, stop bits, and others. Here, we will only set the
baud rate to 2,000,000 Bits/s (2 Mbps) and leave the rest as they are. This means
that the data length is selected as 8 bits, parity bit is set as none, and stop bit is set
as 1.

Fig. 4.3 Parameters settings tab

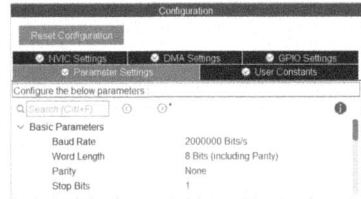

4.1.2.2 Serial Library for Data Transfer

We formed the lib_serial library to transfer data between the PC and STM32 microcontroller using HAL library UART functions. Our library consists of one source and header file as "lib_serial.c" and "lib_serial.h," respectively. These should be copied to the "Project/Core/Src" and "Project/Core/Inc" folders, respectively. Please note that the UART peripheral unit should be initialized to 2 Mbits/s baud rate from STM32CubeMX as explained in Sect. 4.1.2.1 when the lib_serial library is used.

Let's clarify how to handle data types in transfer before proceeding with library functions. We formed a C enumeration, called SERIAL_DataTypeDef, to separate data types. We provided this enumeration in Listing 4.1. The STM32 microcontroller will pick one data type from this enumeration to request data transmission or reception.

Listing 4.1 SERIAL_DataTypeDef enum

```
typedef enum\vspace{6pt}
{
        TYPE_U8         = 1,
        TYPE_S8         = 2,
        TYPE_U16        = 3,
        TYPE_S16        = 4,
        TYPE_U32        = 5,
        TYPE_S32        = 6,
        TYPE_F32        = 7,
} SERIAL_DataTypeDef;
```

We have two functions in our lib_serial library as in Listing 4.2. We can use the function LIB_SERIAL_Transmit to transfer data to a Python script running on PC. This function first transmits header bytes to inform the Python script that data will be sent to PC. Then, data type and length are transmitted. Finally, data is transmitted in bytes. We can use the function LIB_SERIAL_Receive to request data from PC. This function sends data request bytes to PC followed by data type and length. It then waits for data from PC. If no data is received, then it time outs in 10 seconds. Please note that the parameter length fed to these functions are the number of samples to be sent/requested.

Listing 4.2 lib_serial library functions.

```
int8_t LIB_SERIAL_Transmit(void *pData, uint32_t length,
    SERIAL_DataTypeDef type)
```

```
/*
Transmits data with its data type information in packets
pData: Pointer to data buffer of any type
length: Number of data in quantity (not bytes!)
type: Choose from SERIAL_DataTypeDef enum
*/

int8_t LIB_SERIAL_Receive(void *pData, uint32_t length,
    SERIAL_DataTypeDef type)
/*
Receives the data in packets
pData: Pointer to data buffer of any type
length: Number of data in quantity (not bytes!)
type: Choose from SERIAL_DataTypeDef enum
*/
```

4.1.3   Data Transfer Examples in STM32CubeIDE

We provide two examples on data transfer using STM32CubeIDE in this section.
The first example is on transmitting data from the STM32 microcontroller to PC.
The second example is on transmitting data from PC to the STM32 microcontroller.
The UART peripheral unit should be configured as explained in Sect. 4.1.2.1 in both
examples.

4.1.3.1 Data Transfer from the STM32 Microcontroller to PC

We provide the main code block to transfer data from the STM32 microcontroller
to PC in Listing 4.3. Here, the STM32 microcontroller transmits the numbers pi
and Euler number e to PC in a floating-point form every second. Please note that
we provide the address of our transmit array, length of data to be transmitted, and
data type to the function LIB_SERIAL_Transmit. We should provide these arguments
correctly since the PC side handles data according to them. The Python script in
Listing 4.7 should be running on PC to run this example. We will explain the
contents of this script in Sect. 4.1.6.

Listing 4.3 Serial data transfer example from the STM32 microcontroller to PC using
STM32CubeIDE

```
/* USER CODE BEGIN Includes */
#include <math.h>
#include "lib_serial.h"
/* USER CODE END Includes */

/* USER CODE BEGIN 0 */
const float txArray[2] = {M_PI, M_E};
/* USER CODE END 0 */

/* USER CODE BEGIN WHILE */
while (1)
{
/* USER CODE END WHILE */

/* USER CODE BEGIN 3 */
```

```
  LIB_SERIAL_Transmit((void*)txArray, 2, TYPE_F32);
  HAL_Delay(1000);
}
/* USER CODE END 3 */
```

4.1.3.2 Data Transfer from PC to the STM32 Microcontroller

We provide the main code block to transfer data from PC to the STM32 microcontroller in Listing 4.4. Here, the STM32 microcontroller sends a read request to PC every second by the function LIB_SERIAL_Receive. The STM32 microcontroller can receive any type and length of data from PC by providing the correct arguments to this function. To see the received bytes, run the project in debug mode and add a breakpoint to the line after LIB_SERIAL_Receive by double clicking its line number. Content of the rxArray can be seen by holding the mouse on it. The Python script in Listing 4.7 should be running on PC to receive data with the STM32 microcontroller. We will explain the contents of this script in Sect. 4.1.6.

Listing 4.4 Serial data transfer example from PC to the STM32 microcontroller using STM32CubeIDE

```
/* USER CODE BEGIN Includes */
#include "lib_serial.h"
/* USER CODE END Includes */

/* USER CODE BEGIN 0 */
uint8_t rxArray[12] = {0};
/* USER CODE END 0 */

/* USER CODE BEGIN WHILE */
while (1)
{
/* USER CODE END WHILE */
/* USER CODE BEGIN 3 */
  LIB_SERIAL_Receive(rxArray, 12, TYPE_U8);
  HAL_Delay(1000);
}
/* USER CODE END 3 */
```

4.1.4   UART Communication with Mbed Studio

We consider UART communication between the STM32 microcontroller and PC via Mbed Studio in this section. Therefore, we should first create an Mbed Studio project. We can use the lib_serial library introduced in Sect. 4.1.2.2 to make the data transfer process easier. We also have a specific library to be used only in Mbed Studio projects. This library is called lib_uart. It handles UART configuration and initialization operations explained in Sect. 4.1.2.1. Therefore, we should copy the files "lib_uart.h," "lib_uart.c," "lib_serial.h," and "lib_serial.c" under the root folder of our project. Please note that these four files should be copied under the project root folder when we transfer data between the STM32 microcontroller and PC under Mbed Studio in future examples. Then, we can use our lib_serial library functions listed in Listing 4.2 for data transfer.

4.1.5 Data Transfer Examples in Mbed Studio

We provide two examples on data transfer using Mbed Studio in this section. The first example is on transmitting data from the STM32 microcontroller to PC. The second example is on transmitting data from PC to the STM32 microcontroller. In both examples, we should include the required source and header files to the project as explained in Sect. 4.1.4.

4.1.5.1 Data Transfer from the STM32 Microcontroller to PC

As we did in Sect. 4.1.3.1, we transfer the number pi and Euler's number e to PC in floating-point form every second now using Mbed Studio. We provide the complete code prepared for this purpose in Listing 4.5. In this code, we initialize the UART peripheral unit with the function LIB_UART_Init. We then transmit the txArray to PC using the function LIB_SERIAL_Transmit. Please note that the Python script in Listing 4.7 should be running on PC to observe the numbers there. We will explain the contents of this script in Sect. 4.1.6.

Listing 4.5 Serial data transfer from the STM32 microcontroller to PC using Mbed Studio

```
#include "mbed.h"
#include "lib_uart.h"
#include "lib_serial.h"

const float txArray[2] = {3.14159265358979323846f,
    2.7182818284590452354f};

int main(void){
    LIB_UART_Init();
    while (1){
        LIB_SERIAL_Transmit((void*)txArray, 2, TYPE_F32);
        wait_us(1000000);
    }
}
```

4.1.5.2 Data Transfer from PC to the STM32 Microcontroller

We can also transfer data from PC to the STM32 microcontroller. We provide the complete code formed for this purpose in Listing 4.6. Here, the STM32 microcontroller will make a request to receive 12 bytes as in Sect. 4.1.3.2. If the received bytes equals the string "Hello World", then the LED on the STM32 board toggles. The Python script in Listing 4.7 should be running on PC to run this example. We will explain the contents of this script in Sect. 4.1.6.

Listing 4.6 Serial data transfer from PC to the STM32 microcontroller using Mbed Studio

```
#include "mbed.h"
#include "lib_uart.h"
#include "lib_serial.h"
#include <string.h>

uint8_t rxArray[12] = {0};
DigitalOut led1(LED1);

int main(void){
    LIB_UART_Init();

    while (1){
        LIB_SERIAL_Receive(rxArray, 12, TYPE_U8);
        if (std::strncmp((const char*)rxArray, "Hello World\n",
            12) == 0){
            led1 = !led1;
        }
        wait_us(1000000);
    }
}
```

4.1.6   Data Transfer at the PC Side

Data can be represented and processed easily in Python on PC. Besides, there are several data processing libraries in Python. Therefore, we pick Python as the data processing medium on PC. Since the STM32 microcontroller sends or receives data via UART communication, we should apply the corresponding settings in Python. Therefore, we should install the pySerial library on PC by executing the command `pip install pyserial` at the command window. We should also know the COM port on PC in which the STM32 board is connected to. Hence, we should check "Device Manager − > Ports (COM & LPT)" under the Windows operating system. There should be a device listed as "STMicroelectronics STLink Virtual COM Port ...." Please note that the same settings should be done for a PC with another operating system, such as Linux. We kindly ask the reader to check the operating system working principles for this purpose.

We formed the serial data transfer library py_serial in Python to simplify data transfer between the PC and STM32 microcontroller. This library consists of one Python script named "py_serial.py." Our library has four functions as follows: The function SERIAL_Init gets the COM port name as COMx. This function also creates the serial communication object. The function SERIAL_PollForRequest returns the request type (direction), data length, and type when data request comes. Data type is a number from the C enumeration given in Listing 4.1. The function SERIAL_PollForRequest also halts execution until a data transfer request arrives. The function SERIAL_Write sends data from PC to the STM32 microcontroller according to its size and data type. The function SERIAL_Read reads data from the STM32 microcontroller to PC according to its size and data type.

We formed a Python script, called "py_serialexample.py," to be used in data transfer examples in the following sections. Our script imports the serial data transfer library py_serial to use its functions. We provide this script, running on PC, in Listing 4.7. Here, the script waits until a write or read request comes from the STM32 microcontroller. The STM32 microcontroller makes a request to transmit data to PC in the write request. The STM32 microcontroller makes a request to receive data to PC in the read request. When the STM32 microcontroller requests to receive data, the Python script will transmit the string "Hello World" to the STM32 microcontroller. When the STM32 microcontroller requests to transmit data, the Python script reads and prints it on the terminal screen in PC.

Listing 4.7 Serial data transfer at the PC side

```
import py_serial
import numpy as np

py_serial.SERIAL_Init("COM4")

while 1:
    rqType, datalength, dataType = py_serial.
        SERIAL_PollForRequest()
    if rqType == py_serial.MCU_WRITES:
        data = py_serial.SERIAL_Read()

    elif rqType == py_serial.MCU_READS:
        sendArray = np.frombuffer(b'Hello World\n', dtype=np.
            uint8)
        py_serial.SERIAL_Write(sendArray)
```

We will modify the Python script "py_serialexample.py" in the following sections. Hence, it will be customized to acquire data from different sensors. Besides, printing part of the script will be also modified accordingly.

4.2   Acquiring Relative Humidity and Temperature Data

We introduced the STEVAL-MKI141V2 sensor in Sect. 2.3.4. Here, we cover its usage. Hence, we consider how to acquire the relative humidity and temperature data from this sensor. We use the STM32CubeIDE and Mbed Studio platforms for this purpose. We also handle transferring the acquired data to PC.

4.2.1   Hardware Setup

The STEVAL-MKI141V2 sensor supports SPI and I^2C digital communication modes. We pick I^2C to acquire data from the sensor in this book. There are VDD,

Fig. 4.4 Connections
between the STM32 board
and STEVAL-MKI141V2
sensor

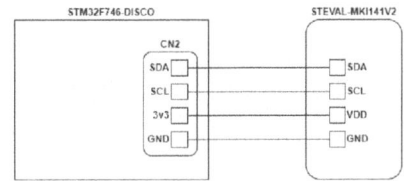

GND, INT, CS, SDA, and SCL pins on the sensor board. We use the SCL and SDA
pins for I^{2}C and VDD and GND pins for powering the sensor. The STM32 board
has an onboard connector, named CN2, to connect external I^{2}C devices. We use it to
connect the sensor to the board. Circuit diagram for this connection is as in Fig. 4.4.

4.2.2   Data Acquisition and Transfer with STM32CubeIDE

We formed the HTS221 library consisting of files "lib_hts221.h" and "lib_hts221.c"
to use the STEVAL-MKI141V2 sensor. Our library is built on the stm32-
hts221 library provided by STMicroelectronics in its GitHub repository. There
are four source and header files named "hts221.c," "hts221.h," "hts221_reg.c,"
and "hts221_reg.h" in this repository. We should download them. The header
files "hts221.h," "hts221_reg.h," and "lib_hts221.h" should be copied under the
"Core/Inc" folder of the project. The source files "hts221.c," "hts221_reg.c," and
"lib_hts221.c" should be copied under the "Core/Src" folder of the project.

The reader should enable the I2C1 peripheral unit to use our library. Furthermore,
its speed mode should be set as fast. Its speed frequency should be set to 400 KHz
from STM32CUBEMX. Then, STM32CubeIDE will include the I2C HAL library
to the project folder. It will initialize the I2C1 peripheral unit as configured in
STM32CUBEMX. Afterward, our HTS221 library will use the I2C HAL library
write and read functions.

We formed our HTS221 library to simplify the data acquisition process from the
sensor. In other words, we abstracted stm32-hts221 and I2C HAL libraries for the
user. There are three functions in our library. We will explain them next.

```
int8_t LIB_HTS221_Init(void);
/*
Initializes the HTS221 library, humidity and temperature sensors.
*/

int8_t LIB_HTS221_GetTemperature(float *temp);
/*
Reads the temperature value.
temp: Pointer to the temperature value.
*/

int8_t LIB_HTS221_GetHumidity(float *hum);
/*
Reads the humidity value.
temp: Pointer to the humidity value.
*/
```

The function int8_t LIB_HTS221_Init(**void**) initializes the stm32-hts221 library. It enables the temperature and humidity sensors of HTS221. If the function succeeds, then it returns zero. The function int8_t LIB_HTS221_GetTemperature(**float** *temp) reads the temperature data if it is ready to be read. The function expects a pointer to a float variable to update its value. It returns 0 if the temperature is successfully read or -1 if the data is not ready yet. The function int8_t LIB_HTS221_GetHumidity(**float** *hum) reads the humidity data and performs the same operations as the temperature function. We provide a usage example via the mentioned functions in Listing 4.8.

Listing 4.8 Acquiring the relative humidity and temperature data via STM32CubeIDE

```
/* USER CODE BEGIN Includes */
#include "lib_hts221.h"
/* USER CODE END Includes */

/* USER CODE BEGIN 0 */
float temperature;
float humidity;
/* USER CODE END 0 */

/* USER CODE BEGIN 2 */
LIB_HTS221_Init();
/* USER CODE END 2 */

/* USER CODE BEGIN WHILE */
while (1)
{
/* USER CODE END WHILE */

/* USER CODE BEGIN 3 */
  LIB_HTS221_GetHumidity(&humidity);
  LIB_HTS221_GetTemperature(&temperature);
  HAL_Delay(10);
}
/* USER CODE END 3 */
```

As we acquire data from the STEVAL-MKI141V2 sensor, the next step is its transfer to PC. Therefore, we will benefit from our lib_serial library. After creating a new STM32 project with USART1 and I2C1 peripheral units enabled, the next step is including library files under the appropriate project folders. In addition to the HTS221 library inclusion as explained in this section, we should also add the serial library files for data transfer as explained in Sect. 4.1.2.2. Next, we should call the library functions as in Listing 4.9. Please note that we provide the data type as TYPE_F32 and length as 2 to the function LIB_SERIAL_Transmit. In other words, we have two floating numbers to be sent to PC. We explain the corresponding Python script to run on PC in Sect. 4.2.4.

Listing 4.9 Transferring the acquired relative humidity and temperature data via STM32CubeIDE

```
/* USER CODE BEGIN Includes */
#include "lib_hts221.h"
```

```
#include "lib_serial.h"
/* USER CODE END Includes */

/* USER CODE BEGIN 0 */
float hts221[2] = {0};
/* USER CODE END 0 */

/* USER CODE BEGIN 2 */
LIB_HTS221_Init();
/* USER CODE END 2 */

/* USER CODE BEGIN WHILE */
while (1)
{
/* USER CODE END WHILE */

/* USER CODE BEGIN 3 */
LIB_HTS221_GetHumidity(&hts221[0]);
LIB_HTS221_GetTemperature(&hts221[1]);
LIB_SERIAL_Transmit(hts221, 2, TYPE_F32);
HAL_Delay(1000);
}
/* USER CODE END 3 */
```

4.2.3   Data Acquisition and Transfer with Mbed Studio

We can also use Mbed Studio to acquire and transfer data obtained from the STEVAL-MKI141V2 sensor. We will use the same functions introduced in Sect. 4.2.2 for this purpose. Therefore, we should copy the source and header files of our library to our project folder after creating an Mbed Studio project. The header files "hts221_reg.h," "hts221.h," "lib_hts221.h," and "lib_i2c.h" and source files "hts221_reg.c," "hts221.c," "lib_hts221.c," and "lib_i2c.c" should be copied under the project folder for data acquisition. The header files "lib_serial.h" and "lib_uart.h" and source files "lib_serial.c" and "lib_uart.c" should be copied under the same project folder for data transfer. We provide a usage example via the mentioned functions in Listing 4.10. We explain the corresponding Python script to run on PC in Sect. 4.2.4.

Listing 4.10 Acquiring and transferring the relative humidity and temperature data via Mbed Studio

```
#include "mbed.h"
#include "lib_hts221.h"
#include "lib_serial.h"
#include "lib_uart.h"
#include "lib_i2c.h"

float hts221[2] = {0};
int main()
{
    SCB_EnableDCache();
    SCB_EnableICache();
```

```
LIB_I2C1_Init();
LIB_UART_Init();
LIB_HTS221_Init();
while (true)
{
    LIB_HTS221_GetHumidity(&hts221[0]);
    LIB_HTS221_GetTemperature(&hts221[1]);
    LIB_SERIAL_Transmit(hts221, 2, TYPE_F32);
    HAL_Delay(1000);
}
}
```

4.2.4   Data Transfer at the PC Side

The Python script to read data sent by the STM32 microcontroller is given in
Listing 4.11. This script reads the humidity and temperature values from the
STM32 microcontroller and prints them on the PC terminal. We call this script
as "py_hts221.py." Please note that this script uses the functions of our py_serial
library for data transfer operations. Therefore, we recommend running this script in
the same directory as with the script "py_serial.py." The communication port should
also be checked in this script based on the one used in PC.

Listing 4.11 Acquiring the relative humidity and temperature data on the PC side

```
import py_serial
import numpy as np

py_serial.SERIAL_Init("COM4")

while 1:
    rqType, datalength, dataType = py_serial.
        SERIAL_PollForRequest()
    if rqType == 87:
        data = py_serial.SERIAL_Read()
        print("Humidity: " + str(data[0]))
        print("Temperature: " + str(data[1]))
```

4.3   Acquiring Accelerometer, Gyroscope, and Magnetometer Data

We introduced the BNO055 sensor in Sect. 2.3.3. Here, we cover its usage. Hence,
we consider how to acquire the accelerometer, gyroscope, and magnetometer data
from this sensor. We use the STM32CubeIDE and Mbed Studio platforms for this
purpose. We also handle transferring the acquired data to PC.

Fig. 4.5 Connections
between the STM32 board
and BNO055 sensor

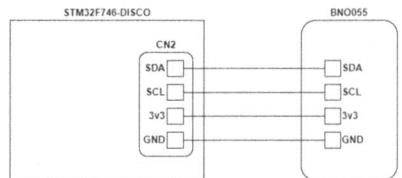

4.3.1 Hardware Setup

The BNO055 sensor supports I^{2}C and UART digital communication modes. We pick I^{2}C to acquire data from the sensor in this book. We use four pins of the BNO055 sensor to connect it to the STM32 board. These are 3V, GND, SCL, and SDA. The STM32 board has an onboard connector, named CN2, to connect external I^{2}C devices. We use it to connect the sensor to the board. Circuit diagram for this connection is as in Fig. 4.5.

4.3.2 Data Acquisition and Transfer with STM32CubeIDE

We formed the BNO055 library to speed up the data acquisition process from the BNO055 sensor. In this section, we explain the library functions and how to use them. Our library consists of files "lib_bno055.h" and "lib_bno055.h." Our library is built on the Bosch Sensortec BNO055 C library, provided in their GitHub repository named "BNO055_driver." Therefore, we should download this repository and include it to our project. The source file "bno055.c" should be copied under the "Core/Src" folder of our project. The header file "bno055.h" should be copied under the "Core/Inc" folder of our project.

The reader should enable the I2C1 peripheral unit to use our BNO055 library. Furthermore, its speed mode should be set as fast. Its speed frequency should be set to 400 KHz from the STM32CUBEMX window. Then, STM32CubeIDE will include the I2C HAL library to the project folder. It will initialize the I2C1 peripheral unit as configured from the STM32CUBEMX window. Afterward, our BNO055 library will use the I2C HAL library write and read functions.

We formed our library to simplify the data acquisition process from the sensor. In other words, we provide a higher-level code to make the data acquisition process easier. There are four functions in our library as listed in Listing 4.12. We explain them next.

Listing 4.12 The BNO055 library functions

```
int8_t LIB_BNO055_Init(void)
/*
Initializes the accelerometer, gyroscope, and magnetometer.
*/

int8_t LIB_BNO055_ReadAccelXYZ(float *x, float *y, float *z)
```

```
/*
Reads the accelerometer data.
x: Pointer to the accelerometer's x value.
y: Pointer to the accelerometer's y value.
z: Pointer to the accelerometer's z value.
*/

int8_t LIB_BNO055_ReadGyroXYZ(float *x, float *y, float *z)
/*
Reads the gyroscope data.
x: Pointer to the gyroscope's x value.
y: Pointer to the gyroscope's y value.
z: Pointer to the gyroscope's z value.
*/

int8_t LIB_BNO055_ReadMagXYZ(float *x, float *y, float *z)
/*
Reads the magnetometer data.
x: Pointer to the magnetometer's x value.
y: Pointer to the magnetometer's y value.
z: Pointer to the magnetometer's z value.
*/
```

The function `LIB_BNO055_Init` initializes the Bosch BNO055 library and enables the accelerometer, gyroscope, and magnetometer sensors. The functions `LIB_BNO055_ReadAccelXYZ`, `LIB_BNO055_ReadGyroXYZ`, and `LIB_BNO055_ReadMagXYZ` read the corresponding data from the sensor and write them to the addresses we provide. These functions require three addresses to store the x, y, and z values. We provide a data acquisition example via the mentioned functions in Listing 4.13.

Listing 4.13 Acquiring the accelerometer gyroscope and magnetometer data via STM32CubeIDE

```
/* USER CODE BEGIN Includes */
#include "lib_bno055.h"
/* USER CODE END Includes */

/* USER CODE BEGIN 0 */
float accel[3];
float gyro[3];
float mag[3];
/* USER CODE END 0 */

/* USER CODE BEGIN 2 */
LIB_BNO055_Init();
/* USER CODE END 2 */

/* USER CODE BEGIN WHILE */
while (1)
{
/* USER CODE END WHILE */

/* USER CODE BEGIN 3 */
  LIB_BNO055_ReadAccelXYZ(&accel[0], &accel[1], &accel[2]);
  LIB_BNO055_ReadGyroXYZ(&gyro[0], &gyro[1], &gyro[2]);
  LIB_BNO055_ReadMagXYZ(&mag[0], &mag[1], &mag[2]);
}
/* USER CODE END 3 */
```

In Listing 4.13, we formed a C structure to store and group the acquired BNO055 sensor data. This structure allocates memory space to store three floating-point values for the accelerometer, gyroscope, and magnetometer data separately. This data is also stored consecutively in memory to facilitate the data transmission process. The definition of this structure is given in Listing 4.14.

Listing 4.14   C structure for storing the BNO055 data

```
typedef struct
{
        float accel[3];
        float gyro[3];
        float mag[3];
}BNO055_F32DataTypeDef;
```

As we acquire data from the BNO055 sensor, the next step is its transfer to PC. Therefore, we will benefit from our lib_serial library at the STM32 microcontroller side. To do so, the UART peripheral unit should be enabled as explained in Sect. 4.1.2.1. The lib_serial library should be included to the project as explained in Sect. 4.1.2.2. The next step is creating an instance of the C structure in the main code. This structure is filled with the sensor data using the lib_bno055 library functions. Finally, we transmit data stored in this structure to PC. The main code is given in Listing 4.15. We explain the corresponding Python script to run on PC in Sect. 4.3.4.

Listing 4.15   Transferring the acquired accelerometer gyroscope and magnetometer data via STM32CubeIDE

```
/* USER CODE BEGIN Includes */
#include "lib_bno055.h"
#include "lib_serial.h"
/* USER CODE END Includes */

/* USER CODE BEGIN 0 */
BNO055_F32DataTypeDef bno055Data;
/* USER CODE END 0 */

/* USER CODE BEGIN 2 */
LIB_BNO055_Init();
/* USER CODE END 2 */

/* USER CODE BEGIN WHILE */
while (1)
{
/* USER CODE END WHILE */

/* USER CODE BEGIN 3 */
    LIB_BNO055_ReadAccelXYZ(&bno055Data.accel[0], &bno055Data.accel
        [1], &bno055Data.accel[2]);
    LIB_BNO055_ReadGyroXYZ(&bno055Data.gyro[0], &bno055Data.gyro
        [1], &bno055Data.gyro[2]);
    LIB_BNO055_ReadMagXYZ(&bno055Data.mag[0], &bno055Data.mag[1], &
        bno055Data.mag[2]);
    LIB_SERIAL_Transmit(&bno055Data, 9, TYPE_F32);
```

```
  HAL_Delay(1000);
}
/* USER CODE END 3 */
```

4.3.3   Data Acquisition and Transfer with Mbed Studio

We can also use Mbed Studio to acquire and transfer the BNO055 sensor data. Therefore, we should copy the required source and header files under the project folder. The "bno055.h" and "bno055.c" files should be downloaded as explained in Sect. 4.3.2 and copied under the project folder. Our library header files "lib_bno055.h," "lib_i2c.h," "lib_serial.h," and "lib_uart.h" and source files "lib_bno055.c," "lib_i2c.c," "lib_serial.c," and "lib_uart.c" should be copied under the project folder. We then use the same functions explained in Sect. 4.3.2. We provide a usage example via the mentioned functions in Listing 4.16. We explain the corresponding Python script to run on PC in Sect. 4.3.4.

Listing 4.16   Acquiring and transferring the accelerometer gyroscope and magnetometer data via Mbed Studio

```
#include "mbed.h"
#include "lib_bno055.h"
#include "lib_serial.h"
#include "lib_uart.h"
#include "lib_i2c.h"

BNO055_F32DataTypeDef bno055Data;

int main()
{
    SCB_EnableICache();
    SCB_EnableDCache();
    LIB_I2C1_Init();
    LIB_UART_Init();
    LIB_BNO055_Init();

    while (true)
    {
        LIB_BNO055_ReadAccelXYZ(&bno055Data.accel[0], &bno055Data
            .accel[1], &bno055Data.accel[2]);
        LIB_BNO055_ReadGyroXYZ(&bno055Data.gyro[0], &
            bno055Data.gyro[1], &bno055Data.gyro[2]);
        LIB_BNO055_ReadMagXYZ(&bno055Data.mag[0], &bno055Data
            .mag[1], &bno055Data.mag[2]);
        LIB_SERIAL_Transmit(&bno055Data, 9, TYPE_F32);
        HAL_Delay(1000);
    }
}
```

4.3.4 Data Transfer at the PC Side

The Python script to read data sent by the STM32 microcontroller is given in Listing 4.11. This script reads the accelerometer, gyroscope, and magnetometer values from the communication port and print them on the PC terminal. We call this script as "py_bno055.py." Please note that this script uses our py_serial library functions for data transfer operations. Therefore, we recommend running this script in the same directory as with the script "py_serial.py." The communication port should also be checked in this script based on the one used in PC.

Listing 4.17 Reading the accelerometer gyroscope and magnetometer data on PC side

```
import py_serial

py_serial.SERIAL_Init("COM4")

while 1:
    rqType, datalength, dataType = py_serial.
        SERIAL_PollForRequest()
    if rqType == 87:
        data = py_serial.SERIAL_Read()
        print("Acc X: " + str(data[0]) + " Acc Y: " + str(data
            [1]) + " Acc Z: " + str(data[2]))
        print("Gyro X: " + str(data[3]) + " Gyro Y: " + str(data
            [4]) + " Gyro Z: " + str(data[5]))
        print("Mag X: " + str(data[6]) + " Mag Y: " + str(data
            [7]) + " Mag Z: " + str(data[8]))
```

4.4 Acquiring Audio Signals

We handle audio signal acquisition and transfer from the STM32 microcontroller in this section. We perform the corresponding operation by using the STM32CubeIDE and Mbed Studio platforms as in previous sections. We also cover the procedure to transfer audio signals between the microcontroller and PC. Hence, we will be equipped with necessary tools to acquire and transfer audio signals at the end of this section.

4.4.1 Audio Signal Acquisition and Transfer with STM32CubeIDE

We explain how the audio signals can be acquired using STM32CubeIDE in this section. To do so, we should configure STM32CubeMX and include necessary libraries to our project. Therefore, we will first focus on how to set up the microcontroller in STM32CubeMX. Then, we will provide necessary libraries and code to acquire the audio signal from the STM32 onboard MEMS microphones. We will also transfer the acquired audio data to PC and save it as a wav file there.

4.4.1.1 Setup in STM32CubeMX

We provide a library to configure the STM32 hardware in Sect. 4.4.1.2. To include these dependencies to our project, we should enable SAI2, TIM1, and USART1 peripheral units from the STM32CubeMX interface. We should select the mode of SAI A as "Master" under the "SAI2 Mode and Configuration" window of STM32CubeMX. Hence, we can initialize the SAI2 peripheral unit. Then, we should select the "Clock Source" as "Internal Clock" under the "TIM1 Mode and Configuration" window of the STM32CubeMX interface. The USART1 module can be enabled as we did in Sect. 4.1.2.1.

Hardware initialization of the SAI2 peripheral unit will be done by code. The timer TIM1 is not used in operation. However, its drivers are required for the audio acquisition library. Therefore, we should configure STM32CubeMX such that it should not generate any code for the SAI2 and TIM1 peripheral units. As a result, STM32CubeIDE will only include HAL drivers for SAI and TIM. It will not generate any initialization code for them. To enable this feature, click on the "Project Manager" tab in STM32CubeMX. Then, click on the "Advanced Settings" to see code generation configurations. Uncheck the "Generate Code" boxes for SAI2 and TIM1 instances under the "Generated Function Calls" section. Afterward, we can generate the code.

4.4.1.2 Usage in STM32CubeIDE

Our audio acquisition library, called lib_audio, is built on the STM32 board audio BSP drivers. Therefore, we should include them to our project. Path of these BSP drivers in the computer file system can be found through STM32CubeIDE. To do so, click on "Window" on the top menu of STM32CubeIDE, and select "Preferences" from the dropdown list. Expand the "STM32Cube" tab and select the "Firmware Updater." The path in the text box labeled with "Firmware installation repository" is the path where drivers are located in the computer file system.

To include BSP drivers, right-click on the current project folder in "Project Explorer," and select "Import." Select the "File System" under "General" and click "Next." Then, click "Browse" to select the system directory. Go to the "Firmware" installation repository, and select the "STM32Cube_FW_F7_VX.XX.X" folder. Click on the "STM32746G-Discovery" folder under "STM32Cube_FW_F7_VX.XX.X/Drivers/BSP/," and select "stm32746g_discovery.c," "stm32746g _dis-covery.h," "stm32746g_discovery_audio.c," and "stm32746g_discovery_audio.h" files. Afterward, select the "wm8994" folder under the "BSP/Components" folder by checking the box near it. Finally, select the "audio.h" header file located under the path "BSP/Components/Common." Click on the "Finish" button to include BSP drivers to the project. We provide the file organization of the required BSP files under STM32CubeIDE in Fig. 4.6.

Our audio acquisition library, lib_audio, consists of two files "lib_audio.c" and "lib_audio.h." There are four functions in our library: The function LIB_AUDIO_Init initializes the SAI2 peripheral unit and onboard CODEC for 16 KHz audio signal acquisition. The function LIB_AUDIO_StartRecording starts audio recording. We should provide two arguments to this function as the pointer to data buffer

Fig. 4.6 File organization of
the required BSP files under
STM32CubeIDE

```
v  🖺 Drivers
   v  🗁 BSP
      v  🗁 Components
         v  🗁 Common
            >  🖹 audio.h
         v  🗁 wm8994
            >  🖹 wm8994.c
            >  🖹 wm8994.h
               ● Release_Notes.html
      v  🗁 STM32746G-Discovery
         >  🖹 stm32746g_discovery_audio.c
         >  🖹 stm32746g_discovery_audio.h
         >  🖹 stm32746g_discovery.c
         >  🖹 stm32746g_discovery.h
      >  🗁 CMSIS
      >  🗁 STM32F7xx_HAL_Driver
```

and size of recording. Audio recording stops after the audio data is acquired. The function LIB_AUDIO_PollForRecording waits for the audio recording process to complete. Audio recording can also be manually stopped with the function LIB_AUDIO_StopRecording. We provide detailed information on these functions in Listing 4.18.

Listing 4.18 Functions of our audio acquisition library

```
int8_t LIB_AUDIO_Init(void)
/*
Initializes SAI2 and CODEC for 16 KHz audio frequency
*/

void LIB_AUDIO_StartRecording(uint16_t *pData, uint32_t length)
/*
Starts audio recording
pData: Pointer to data buffer to be filled with audio data
length: Number of data in quantity (not bytes!)
*/

int8_t LIB_AUDIO_PollForRecording(uint16_t timeout)
/*
Starts audio recording
Note The timeout value must be large enough
timeout: Timeout in milliseconds
*/

void LIB_AUDIO_StopRecording(void)
/*
Stops audio recording
*/
```

The next step is to include our audio acquisition library to the project. To do so, copy the source file "lib_audio.c" under the folder "Core/Src" and "lib_audio.h" under the folder "Core/Inc" of the project. We can transfer the audio data to PC as we acquire it. To do so, we should include the lib_serial library to the project as explained in Sect. 4.1.2.2.

We formed a sample project to acquire and transfer audio data with the main code in Listing 4.19. Here, the audio signal is acquired with 16 KHz sampling frequency within our library. The acquired audio signal has two channels as right and left. This means that we should allocate a buffer with $16,000 \times 2$ 16-bit integer values (int16) for 1-second recording. In this example, we start the audio recording process for 3 seconds. Therefore, we allocate a buffer with $32,000 \times 3$ elements. After the audio signal is successfully acquired, we transfer it to PC. For this example to work, the Python script given in Sect. 4.4.3 should be running on PC.

Listing 4.19 Main code for audio data acquisition and transfer

```
/* USER CODE BEGIN Includes */
#include "lib_audio.h"
#include "lib_serial.h"
/* USER CODE END Includes */

/* USER CODE BEGIN 0 */
#define BUFFER_SIZE (32000 * 3)
uint16_t AudioBuffer[BUFFER_SIZE] = {0};
/* USER CODE END 0 */

/* USER CODE BEGIN 2 */
LIB_AUDIO_Init();
/* USER CODE END 2 */

/* USER CODE BEGIN WHILE */
while (1)
{
/* USER CODE END WHILE */

/* USER CODE BEGIN 3 */
  LIB_AUDIO_StartRecording(AudioBuffer, BUFFER_SIZE);
  if (LIB_AUDIO_PollForRecording(5000) == 0)
  {
      LIB_SERIAL_Transmit(AudioBuffer, BUFFER_SIZE, TYPE_S16);
  }
}
/* USER CODE END 3 */
```

4.4.2   Audio Signal Acquisition and Transfer with Mbed Studio

We explain acquiring audio signal from the STM32 board and transferring it to PC via Mbed Studio in this section. Here, we use our lib_audio audio acquisition library and its associated BSP drivers as in Sect. 4.4.1. As a reminder, these drivers can be found in STMicroelectronics STM32CubeF7 GitHub repository or in the folder of your STM32CubeIDE workspace located under the folder "Drivers/BSP."

We should create a new folder and name it as "BSP" after creating an Mbed Studio project. Then, we should copy the file "audio.h" to the folder "BSP/-

Fig. 4.7 Required BSP files

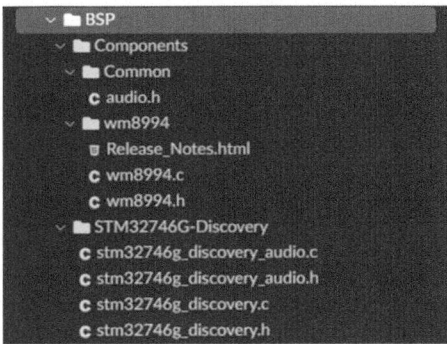

Components/Common." We should copy the files "wm8994.h" and "wm8994.c" to the folder "BSP/wm8994." We should copy the files "stm32746g_discovery.c," "stm32746g_discovery.h," "stm32746g_discovery_audio.c," and "stm32746g _discovery_audio.h" to the folder "BSP/STM32746G-Discovery." Paths of these files should be kept the same as in Fig. 4.7.

We will use our lib_audio library for audio data acquisition as explained in Sect. 4.4.1.2. The next step is copying our library files under the root folder of our project containing the "main.cpp" file. We copy the header files "lib_audio.h," "lib_serial.h," and "lib_uart.h" and source files "lib_audio.c," "lib_serial.c," and "lib_uart.c" to the project root folder. We provide the main code for data acquisition in Listing 4.20. This code acquires the audio signal for 3 seconds and transmits it to PC as in Listing 4.19. Please note that the Python script in Sect. 4.4.3 should be running on PC to run this example.

Listing 4.20 Audio data acquisition and transfer using Mbed Studio

```
#include "mbed.h"
#include "lib_serial.h"
#include "lib_uart.h"
#include "lib_audio.h"

#define BUFFER_SIZE (32000 * 3)

uint16_t AudioBuffer[BUFFER_SIZE] = {0};

int main(){
    SCB_EnableICache();
    SCB_EnableDCache();
    LIB_UART_Init();
    LIB_AUDIO_Init();

    while (true){
        LIB_AUDIO_StartRecording(AudioBuffer, BUFFER_SIZE);
        if (LIB_AUDIO_PollForRecording(5000) == 0){
            LIB_SERIAL_Transmit(AudioBuffer, BUFFER_SIZE,
                TYPE_S16);
        }
    }
}
```

4.4.3 Audio Signal Transfer at the PC Side

The Python script to read data sent by the STM32 microcontroller is given in Listing 4.21. This script reads the received data from the communication port as one-dimensional audio data and converts it to two-dimensional NumPy array. Then, it saves the audio data in wav audio format. We call this script as "py_audio.py." Please note that this script uses the py_serial library functions for data transfer operations. Therefore, we recommend running this script in the same directory as with the script "py_serial.py." The communication port should also be checked in this script based on the one used in PC.

Listing 4.21 Transferring the acquired audio signal at the PC side

```
import py_serial
import numpy as np
from scipy.io.wavfile import write

py_serial.SERIAL_Init("COM4")

sampleRate = 16000
while 1:
    rqType, datalength, dataType = py_serial.
        SERIAL_PollForRequest()
    if rqType == 87:
        data = py_serial.SERIAL_Read()
        channel1 = np.array(data[0::2], dtype=np.int16)
        channel2 = np.array(data[1::2], dtype=np.int16)
        data = np.transpose(np.array([channel1,channel2], dtype=
            np.int16))
        print(np.shape(data))
        write('test.wav', sampleRate, data.astype(np.int16))
```

4.5 Acquiring Digital Images

In this section, we will explain how an image is acquired and stored in the STM32 microcontroller. We pick the B-CAMS-OMV camera module, introduced in Sect. 2.3.2, as the embedded camera module to acquire the image in this book. We will explain how the STM32 microcontroller should be configured to capture the image from this module. Then, we will show how the acquired image is sent to PC. We will provide functions organized in libraries for all these operations.

4.5.1 Setting Up the B-CAMS-OMV Camera Module

We will use the B-CAMS-OMV camera module, consisting of an OV5640 camera, to acquire digital images. To do so, we should first explain how to set up the B-CAMS-OMV camera module in terms of hardware. Please note here: we should

Table 4.1 Image size standards for the camera module

Name	Size (pixels)	Constant Declaration
VGA width	640	IMAGE_RESOLUTION_VGA_WIDTH
VGA height	480	IMAGE_RESOLUTION_VGA_HEIGHT
QVGA width	320	IMAGE_RESOLUTION_QVGA_WIDTH
QVGA height	240	IMAGE_RESOLUTION_QVGA_HEIGHT
QQVGA width	160	IMAGE_RESOLUTION_QQVGA_WIDTH
QQVGA height	120	IMAGE_RESOLUTION_QQVGA_HEIGHT

have a 30-pin flexible flat cable to connect the camera module to the STM32 board. We should use the flat cable connector labeled as CN5 on the camera side. The connector ends are labeled as 1 at the left and 30 at the right side. We should set the flat cable connector labeled as P1 on the STM32 board side. After placing the STM32 board with the circuit side on top and LCD side on the bottom, the side labeled as P1 on the STM32 board should coincide with the side labeled as 1 on the camera module while connecting the flat cable. Hardware setup will be done for image acquisition after plugging in the flat cable to sockets.

4.5.2 Representing the Digital Image on the STM32 Microcontroller

Digital images have common attributes as height, width, and format. These should be taken into account to represent an image on the microcontroller. Therefore, we formed the lib_image library consisting of two files: "lib_image.c" and "lib_image.h." Our library functions depend on standard definitions. Memory size required to store an image depends on these attributes. Therefore, let's start with the image size.

There are image size standards for camera modules. The OV5640 camera supports the video graphics array (VGA) standard and its derivatives. To be more specific, we will use the VGA, quarter VGA (QVGA), and quarter QVGA (QQVGA) image size formats, tabulated in Table 4.1, throughout the book. We formed constant declarations for this purpose in the same table. We will use them in image- and camera-based operations throughout the book.

Camera modules have standard grayscale (monochrome) and color image formats. We will use the grayscale, RGB565, and RGB888 formats throughout the book. Bytes per pixel (bpp) for these formats are 1, 2, and 3, respectively. We formed constant declarations for the grayscale, RGB565, and RGB888 formats as IMAGE_FORMAT_GRAYSCALE, IMAGE_FORMAT_RGB565, and IMAGE_FORMAT_RGB888, respectively. As an example, the constant IMAGE_FORMAT_RGB565 indicates that we are using a color image represented by red, green, and blue components. Moreover, the number of bits assigned to these color bands are five, six, and five, respectively. The same format also applies to the other mentioned color constants.

We can represent an image in a structured form to hold all the necessary information about the image. We provide such a structure to be used in this book in Listing 4.22. As can be seen here, we keep the image width, height, format, and size in the structure. Moreover, the image itself should to be stored in RAM. Therefore, we should declare a C/C++ array having sufficient size to store it. Memory address of the declared array should also be initialized in our structure given in Listing 4.22.

Listing 4.22 Structure to hold image information

```
typedef struct
{
        uint8_t *pData;
        uint16_t width;
        uint16_t height;
        IMAGE_Format format;
        uint32_t size;
}IMAGE_HandleTypeDef;
```

The structure in Listing 4.22 will be used to transfer the image with all its information. Therefore, we should initialize it by code. We provide a C function to initialize the structure with the required information in Listing 4.23.

Listing 4.23 C function to initialize the image structure

```
int8_t LIB_IMAGE_InitStruct(IMAGE_HandleTypeDef * img, uint8_t *
    pImg, uint16_t height, uint16_t width, IMAGE_Format format)
/*
Initialize the image structure with required information
img: Pointer to image structure
pImg: Pointer to image buffer
height: height of the image
width: width of the image
format: Choose IMAGE_FORMAT_GRAYSCALE, IMAGE_FORMAT_RGB565, or
    IMAGE_FORMAT_RGB888
*/
```

We provide a usage example for our image structure and its initialization function for a VGA color image with RGB565 format next. Here, we first define an image via the struct IMAGE_HandleTypeDef. Then, we set image properties with the function LIB_IMAGE_InitStruct:

```
/* USER CODE BEGIN PV */
IMAGE_HandleTypeDef img;
/* USER CODE END PV */

/* USER CODE BEGIN 2 */
LIB_IMAGE_InitStruct(&img, (uint8_t*)pImage,
    IMAGE_RESOLUTION_VGA_HEIGHT, IMAGE_RESOLUTION_VGA_WIDTH,
    IMAGE_FORMAT_RGB565);
/* USER CODE END 2 */
```

4.5.3 Image Acquisition Library

We formed the lib_ov5640 image acquisition library to make the image acquisition operation easier. This library has two files as "lib_ov5640.h" and "lib_ov5640.c." These files consist of all the required definitions and functions to initialize and start image acquisition from the OV5640 camera. We will use this library in both STM32CubeIDE and Mbed Studio projects.

Our image acquisition library depends on other support libraries. Therefore, let's start with explaining them. Our first library is stm32-ov5640. We should download it from the STMicroelectronics GitHub page with the same repository name (stm32-ov5640). There are two source files, "ov5640.c" and "ov5640_reg.c," and two header files, "ov5640.h" and "ov5640_reg.h," in this repository. Files in this repository are used to handle I²C operations and initialize the camera according to the given pixel format and image resolution. Our second library is lib_mpu having two files. They consist of all the required definitions and functions to set up the memory protection unit while reaching the external memory unit. Therefore, we explain the usage of this library in detail in Sect. 4.6. Our third library is BSP_SDRAM. Functions in this library are used to initialize the SDRAM. Therefore, we explain the usage of this library in detail in Sect. 4.6. Our fourth library is lib_image. We explained this library in detail in Sect. 4.5.2.

Now, we are in a position to explain the setup and working principles of our image acquisition library. First, we should download the "stm32-ov5640" repository from the STMicroelectronics GitHub page. We should add the source and header files to appropriate folders in our project. Next, we should include the image representation files to the project as explained in Sect. 4.5.2. Once these steps are complete, dependencies are successfully included to the project. Next, we should include our image acquisition library, lib_ov5640, to the project.

There are two important enumeration definitions in the lib_ov5640 library. The first enumeration LIB_OV5640_Resolution initializes the camera with the desired image resolution. The second enumeration LIB_OV5640_Format initializes the camera with the desired image format. We will use the definitions OV5640_RESOLUTION_R320x240, OV5640_RESOLUTION_R480x272, and OV5640_RESOLUTION_R640x480 for initializing the image resolution. We will use the definitions OV5640_FORMAT_RGB565 and OV5640_FORMAT_RGB888 for initializing the image format. The reader should not confuse the definitions in the lib_image library summarized in Sect. 4.5.2 and the ones here. We had to form two different definitions due to predefined values in the OV5640 camera.

There are five functions defined in the lib_ov5640 library for camera operations: The function LIB_OV5640_Init initializes the camera with desired format and resolution. The function LIB_OV5640_StartContinuos starts continuous acquisition of frames. The function LIB_OV5640_CaptureSnapshot captures a snapshot. The function LIB_OV5640_Stop stops image acquisition. The function LIB_OV5640_GetFrameCount returns the number of frames acquired from the camera. These functions are provided in detail in Listing 4.24.

Listing 4.24 Image acquisition library functions

```
int8_t LIB_OV5640_Init(LIB_OV5640_Resolution resolution,
    LIB_OV5640_Format format)
/*
Initializes the OV5640 library and the camera.
resolution: from LIB_OV5640_Resolution enum
format: from LIB_OV5640_Format enum
*/

int8_t LIB_OV5640_StartContinuous(IMAGE_HandleTypeDef * img)
/*
Starts the DCMI module for continuous capture.
img: pointer to image object
*/

int8_t LIB_OV5640_CaptureSnapshot(IMAGE_HandleTypeDef * img,
    uint32_t timeout)
/*
Starts the DCMI module for only one shot.
img: pointer to image object
timeout: max time allowed in ms to capture one shot
*/

int8_t LIB_OV5640_Stop(void)
/*
Stops the DCMI module.
*/

uint32_t LIB_OV5640_GetFrameCount(void)
/*
Returns the total number of captured frames.
*/
```

4.5.4   Image Acquisition and Transfer with STM32CubeIDE

We will explain the required steps for acquiring and transferring digital images through STM32CubeIDE in this section. We perform these operations in four steps: First, we set up the STM32CubeMX interface. Next, we set up the external SDRAM memory, explained in detail in Sect. 4.6, for image acquisition. Then, we introduce the image transfer library. Finally, we provide the main code to acquire and transfer the digital image.

4.5.4.1 Setup in STM32CubeMX and STM32CubeIDE

We should activate the DCMI and FMC peripheral units for image acquisition. To activate the DCMI, its mode should be selected as "Slave 8 bits External Synchro" under the "DCMI Mode" tab in STM32CubeMX. To activate the FMC, expand the "SDRAM 1" option under the "FMC Mode" tab, and choose "SDCKE0+SDNE0" for the parameter "Clock and chip enable." DCMI and FMC peripheral units will be configured by code. We also need the UART peripheral unit for image transfer. Therefore, it should be configured as explained in Sect. 4.1.2.1. Then, we can run code generation in STM32CubeMX. The acquired image will have large size. Therefore, we need the external SDRAM on the STM32 board for image acquisition and storage. We will explain the usage of this module in detail in Sect. 4.6.

4.5.4.2 Image Transfer Library

We formed the lib_serialimage library to simplify image transfer between the PC and STM32 microcontroller. In this section, we will explain this library. To note here, this library will be used in both STM32CubeIDE- and Mbed Studio-based projects.

The lib_serialimage image transfer library consists of one source and header file as "lib_serialimage.c" and "lib_serialimage.h," respectively. The library has two functions as LIB_SERIAL_IMG_Transmit and LIB_SERIAL_IMG_Receive. Both functions take a pointer to the structure IMAGE_HandleTypeDef as argument. The function LIB_SERIAL_IMG_Transmit transmits the image pixel data with its resolution and color format information stored in the structure. This function will be useful to form an image dataset to be stored in PC. The function LIB_SERIAL_IMG_Receive requests an image from the PC with specified image resolution and color format in the structure. This function can be used to test the machine learning method with the dataset available on PC. We provide these functions in Listing 4.25.

Listing 4.25 Image transfer library functions

```
int8_t LIB_SERIAL_IMG_Transmit(IMAGE_HandleTypeDef * img)
/*
Transmits an image with its required information
img: Pointer to image structure
*/

int8_t LIB_SERIAL_IMG_Receive(IMAGE_HandleTypeDef * img)
/*
Receives an image with its required information
img: Pointer to image structure
*/
```

4.5.4.3 Examples on Image Acquisition and Transfer

We provide two examples on image acquisition and transfer in this section. The first example shows how an image is acquired with the OV5640 camera and transferred to PC. The second example shows how an image is transferred to the STM32 microcontroller from PC. We introduced several libraries with their header and source files till here. To run the following examples, we summarized these files in Table 4.2 with the paths to be copied in the project.

We provide the main code to transmit an image acquired from the OV5640 camera to PC in Listing 4.26. Here, we first allocate a buffer from external SDRAM

Table 4.2 Files and their corresponding paths

/Core/Src	/Core/Inc	/Drivers/BSP/STM32746G-Discovery
lib_image.c	lib_image.h	stm32746g_discovery_sdram.c
lib_mpu.c	lib_mpu.h	stm32746g_discovery_sdram.h
lib_ov5640.c	lib_ov5640.h	
lib_serialimage.c	lib_serialimage.h	
ov5640_reg.c	ov5640_reg.h	
ov5640.c	ov5640.h	

with size sufficient to store a VGA image with resolution 640 × 480 in RGB565 format. Then, we create an `IMAGE_HandleTypeDef` structure to store the image information. Finally, we capture the image from the camera module and transfer it to PC through UART.

Listing 4.26 Main code to acquire and transfer an image using STM32CubeIDE

```
/* USER CODE BEGIN Includes */
#include "../../Drivers/BSP/STM32746G-Discovery/
    stm32746g_discovery_sdram.h"
#include "lib_image.h"
#include "lib_ov5640.h"
#include "lib_serialimage.h"
#include "lib_mpu.h"
/* USER CODE END Includes */

/* USER CODE BEGIN 0 */
__attribute__((section(".sdram_data"))) volatile uint8_t pImage
    [640*480*2];
IMAGE_HandleTypeDef img;
/* USER CODE END 0 */

/* USER CODE BEGIN 2 */
LIB_MPU_Init();
LIB_IMAGE_InitStruct(&img, (uint8_t*)pImage,
    IMAGE_RESOLUTION_VGA_HEIGHT, IMAGE_RESOLUTION_VGA_WIDTH,
    IMAGE_FORMAT_RGB565);
BSP_SDRAM_Init();
LIB_OV5640_Init(OV5640_RESOLUTION_R640x480, OV5640_FORMAT_RGB565)
    ;
/* USER CODE END 2 */

/* USER CODE BEGIN WHILE */
while (1)
{
/* USER CODE END WHILE */

/* USER CODE BEGIN 3 */
    if (!LIB_OV5640_CaptureSnapshot(&img, 5000))
    {
        LIB_SERIAL_IMG_Transmit(&img);
    }
}
/* USER CODE END 3 */
```

We can also transfer an image from PC to the STM32 microcontroller. We provide one such example in Listing 4.27. This code first requests an image from PC. If the image is successfully received, then the microcontroller transfers the same image back to PC. Please note that the Python script in Listing 4.31 should be running on PC for the overall operation to work.

Listing 4.27 Main code to receive image using STM32CubeIDE

```
/* USER CODE BEGIN Includes */
#include "../../Drivers/BSP/STM32746G-Discovery/
    stm32746g_discovery_sdram.h"
#include "lib_image.h"
#include "lib_serialimage.h"
#include "lib_mpu.h"
/* USER CODE END Includes */

/* USER CODE BEGIN 0 */
__attribute__((section(".sdram_data"))) volatile uint8_t pImage
    [640*480*2];
IMAGE_HandleTypeDef img;
/* USER CODE END 0 */

/* USER CODE BEGIN 2 */
LIB_MPU_Init();
LIB_IMAGE_InitStruct(&img, (uint8_t*)pImage, 480, 640,
    IMAGE_FORMAT_RGB565);
BSP_SDRAM_Init();
/* USER CODE END 2 */

/* USER CODE BEGIN WHILE */
while (1)
{
/* USER CODE END WHILE */

/* USER CODE BEGIN 3 */
    if (LIB_SERIAL_IMG_Receive(&img) == SERIAL_OK)
    {
        LIB_SERIAL_IMG_Transmit(&img);
    }
}
/* USER CODE END 3 */
```

4.5.5 Image Acquisition and Transfer with Mbed Studio

We will explain how to acquire and transfer images acquired from the OV5640 camera using Mbed Studio in this section. To do so, we should first add the external SDRAM BSP driver to our project. Therefore, we should create a folder named BSP. Then, we should copy the files "stm32746g_discovery_sdram.c" and "stm32746g_discovery_sdram.h" available under the STM32CubeF7 firmware package to the folder "BSP/STM32746G-Discovery." The folder layout will be as in Fig. 4.8.

Fig. 4.8 Folder layout for the BSP driver usage under Mbed Studio

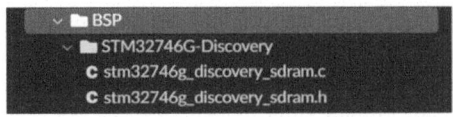

Table 4.3 Required source and header files

Source File	Header File
lib_image.c	lib_image.h
lib_mpu.c	lib_mpu.h
lib_ov5640.c	lib_ov5640.h
lib_serialimage.c	lib_serialimage.h
lib_uart.c	lib_uart.h
ov5640_reg.c	ov5640_reg.h
ov5640.c	ov5640.h

We should copy our source and header files, responsible for initializing and starting peripheral units used, after including the BSP driver for external SDRAM. We provide the list of files in Table 4.3. These files must be copied under the Mbed project root folder.

We provide the main code for image acquisition and transfer with Mbed Studio in Listing 4.28. Here, we form a pointer keeping the start address of the external SDRAM. We use the memory region starting from that address to store the image. We benefit from our image acquisition library, introduced in Sect. 4.5.3, and serial image transfer library, introduced in Sect. 4.5.4.2, to acquire and transfer images. Please note that the Python script in Listing 4.31 should already be running on PC to run this example.

Listing 4.28 Main code to acquire and transfer an image using Mbed Studio

```
#include "mbed.h"
#include "BSP/STM32746G-Discovery/stm32746g_discovery_sdram.h"
#include "lib_mpu.h"
#include "lib_uart.h"
#include "lib_ov5640.h"
#include "lib_image.h"
#include "lib_serialimage.h"

const uint8_t * imageBuffer = (uint8_t *)0xC0000000;
IMAGE_HandleTypeDef img;

int main(){
    SCB_EnableICache();
    SCB_EnableDCache();
    LIB_UART_Init();
    LIB_MPU_Init();
    BSP_SDRAM_Init();
    LIB_IMAGE_InitStruct(&img, (uint8_t*)imageBuffer,
        IMAGE_RESOLUTION_VGA_HEIGHT, IMAGE_RESOLUTION_VGA_WIDTH,
        IMAGE_FORMAT_RGB565);
    LIB_OV5640_Init(OV5640_RESOLUTION_R640x480,
        OV5640_FORMAT_RGB565);
```

```
while (true){
    if (!LIB_OV5640_CaptureSnapshot(&img, 5000)){
        LIB_SERIAL_IMG_Transmit(&img);
    }
}
}
```

We can also transfer an image from PC to the STM32 microcontroller. We provide the main code for receiving an image at the STM32 microcontroller in Listing 4.29. Here, the microcontroller requests an image from PC according to the information initialized in the image structure. If the image is successfully received, then it is transferred back to PC. As in previous examples, the Python script given in Listing 4.31 should be running on PC to run this example.

Listing 4.29 Main code to receive an image using Mbed Studio

```
#include "mbed.h"
#include "BSP/STM32746G-Discovery/stm32746g_discovery_sdram.h"
#include "lib_mpu.h"
#include "lib_uart.h"
#include "lib_image.h"
#include "lib_serialimage.h"

const uint8_t * imageBuffer = (uint8_t *)0xC0000000;
IMAGE_HandleTypeDef img;

int main(){
    SCB_EnableICache();
    SCB_EnableDCache();
    LIB_UART_Init();
    LIB_MPU_Init();
    BSP_SDRAM_Init();
    LIB_IMAGE_InitStruct(&img, (uint8_t*)imageBuffer,
        IMAGE_RESOLUTION_VGA_HEIGHT, IMAGE_RESOLUTION_VGA_WIDTH,
        IMAGE_FORMAT_RGB565);

    while (true){
        if(LIB_SERIAL_IMG_Receive(&img) == SERIAL_OK){
                LIB_SERIAL_IMG_Transmit(&img);
        }
    }
}
```

4.5.6 Image Transfer at the PC Side

We should know how to read and save an image with Python on the PC side before the image transfer operations. Therefore, we will use OpenCV available in Python which allows us to read images and display them on PC. The reader can install OpenCV by the command pip install opencv-python in the command line on PC.

As a side note, OpenCV is also available in C++, Java, and JavaScript languages. The reader can benefit from them if needed.

We next provide the Python script to read an image and display it on PC in Listing 4.30. Here, we import OpenCV first. Then, we read the color image, display it on PC, wait for a key press, and close the window. Please make sure that the image to be read is in the same folder as with the executed Python script or the exact address should be given for the image file to be read by OpenCV.

Listing 4.30 Reading and displaying an image via OpenCV in Python

```
import cv2

#read the color image
img = cv2.imread('mandrill.tif')

#display the color image
cv2.imshow('Color Image',img)
cv2.waitKey(0)

cv2.destroyAllWindows()

#write the image to a file
#cv2.imwrite('mandrill_store.tif',img)
```

We can use the function imwrite in OpenCV to save the image. We should uncomment the last line of the script in Listing 4.30 for this purpose. Hence, the reader can save the image to the working directory. This will be helpful when we receive an image from the STM32 microcontroller.

The next step is receiving images captured from the STM32 microcontroller. We formed the Python script, called "py_serialimg.py," to read image data from serial terminal and convert it to a NumPy array to make the image transfer process simple. Once we get this array, we can use OpenCV functions for other operations. Our Python script "py_serialimg.py" consists of four functions. These are SERIAL_Init, SERIAL_IMG_PollForRequest, SERIAL_IMG_Read, and SERIAL_IMG_Write. The function SERIAL_Init initializes the serial port. The function SERIAL_IMG_PollForRequest waits for a microcontroller request. The microcontroller can either request to receive an image from PC or transmit an image to PC. The function SERIAL_IMG_Write transfers the image, with its path given as an argument, to the microcontroller. The function SERIAL_IMG_Read receives the image from microcontroller and displays it on PC. We provide the Python script in Listing 4.31 which uses the "py_serialimg.py" script. This code handles both image reception and transmission at the PC side.

Listing 4.31 Python script for image transfer at the PC side

```
import py_serialimg
import numpy as np

py_serialimg.SERIAL_Init("COM4")

mandrill = "mandrill.tif"
```

```
while 1:
    rqType, height, width, format = py_serialimg.
        SERIAL_IMG_PollForRequest()
    if rqType == py_serialimg.MCU_WRITES:
        img = py_serialimg.SERIAL_IMG_Read()
    elif rqType == py_serialimg.MCU_READS:
        img = py_serialimg.SERIAL_IMG_Write(mandrill)
```

4.6 Application: Accessing the External SDRAM on the STM32 Board

Storing and processing images require significant amount of memory in the microcontroller. For example, we need $640 \times 480 \times 2$ bytes, approximately 614 Kbytes, of memory space to store a VGA image in RGB565 pixel format. Such memory space cannot be allocated from the internal RAM of the STM32 microcontroller with 320 Kbytes RAM. Hence, we should store the captured images in the external SDRAM available on the STM32 board. To do so, we should first overwrite the default linker file with the one containing definitions of external SDRAM memory. As a side note, linker files are used to specify memory addresses and size for different memory sections. We provide the required linker file in the accompanying book website as "STM32F746NGHX_FLASH.ld." We will also use the memory protection unit (MPU) of the STM32 microcontroller to configure the external memory access. Therefore, we formed two files as "lib_mpu.c" and "lib_mpu.h." We have only one function defined in these files, called LIB_MPU_Init. This function configures the MPU for external SDRAM memory usage.

We use the flexible memory controller (FMC) peripheral unit of the STM32 microcontroller to access the external SDRAM on the STM32 board. We will initialize the FMC using BSP library. To import the required BSP files, right-click the project folder and select import. Select the "File System" under "General" and click "Next." Then, click on "Browse" to select a system directory. Go to the "Firmware" installation repository and select the folder "STM32Cube_FW_F7_VX.XX.X." Click on the "STM32746G-Discovery" folder under "STM32Cube_FW_F7_VX.XX.X/Drivers/BSP/." Then, select files "stm32746g_discovery_sdram.h" and "stm32746g_discovery_sdram.c." Click on the "Finish" button to include the BSP drivers to the project.

4.7 Summary of the Chapter

We handled data acquisition from sensors in this chapter. Since machine learning methods depend on data, this is the first step for their operation. To do so, we divided the data acquisition operation into five parts: First, we handled data transfer between the PC and microcontroller. Second, we evaluated data acquisition from the relative humidity and temperature sensor. Third, we covered data acquisition from the accelerometer, gyroscope, and magnetometer sensor. Fourth, we introduced audio signal acquisition. Fifth, we focused on the B-CAMS-OMV camera module and explore acquiring digital images from it. In all these operations, we followed the same strategy as hardware setup, data acquisition and transfer at the microcontroller side, and data transfer at the PC side. This structure allows the reader to form a complete setup to acquire and transfer data between the microcontroller and PC. As a final note, we will extensively use the acquired data in the following chapters. Therefore, the topics explained in this section have utmost importance.

Introduction to Machine Learning

5

5.1 Random Number Generation as Pseudosensor Data

Data acquisition from sensors may not be possible during the design or test phase of a machine learning system. Moreover, we may need controlled experiments in a fast and reliable manner while designing or testing the system. Therefore, we should generate our own data. Random number generation comes into play here. It allows us to generate controlled data. In this section, we will first introduce background on random number generation in digital systems. Then, we will show how to generate random numbers on PC and STM32 microcontroller via Python and C languages, respectively. Finally, we will obtain sample probability density function (pdf) of generated random data via histogram formation.

5.1.1 Random Number Generation Methods

Digital systems can generate random numbers using algorithms known as pseudo-random number generators (PRNG). The basic idea behind PRNG is to use a deterministic algorithm to generate sequence of numbers having statistical properties of random numbers. To achieve this, PRNGs use mathematical functions that transform a seed value into a new number, which is then used as the seed in the next iteration of the algorithm. The output sequence is determined by the seed value and algorithm used. PRNGs can provide sufficiently random values for many practical applications where true randomness is not necessary.

Random numbers are used in several applications in digital systems. Therefore, methods have been developed to generate them. Linear congruential generators (LCG) are one of the simplest and most widely used PRNG algorithms. LCG uses a simple formula that involves multiplying the previous number in the sequence by a constant, adding another constant to the result, and then taking modulo of the result with a number. The modulo operation ensures that output is within a specific range. The formula for LCG is $x[n + 1] = (a * x[n] + c) \mod m$ where $x[n]$ is the current number in the sequence. $x[n + 1]$ is the next number in the sequence. a, c, and m are constants chosen to produce good statistical properties in the output sequence. There is also a more complex algorithm for PRNG called Mersenne Twister. This algorithm generates a longer sequence of numbers than LCG. It is the default random number generator used in most programming languages including C, C++, and Python.

We can take random numbers as samples (or observations) obtained from a random variable from a mathematical perspective. We introduce what the random variable is in Sect. A.1. The reader may consult the mentioned section for a summary or reference books on probability theory for a more detailed explanation on these concepts. We should emphasize one important property here. Each random variable has an associated probability density function (pdf) with it. The pdf summarizes characteristics of the random variable, hence the random sample. We provide two well-known pdfs as Gaussian and uniform in Sect. A.1 for the sake of completeness.

5.1.2 Random Number Generation on PC

We can benefit from the NumPy library in Python to generate random numbers on PC. To do so, we can use the functions `np.random.normal` and `np.random.uniform` to generate random numbers from the Gaussian and uniform pdf, respectively. There are also other pdfs available under the NumPy library besides these two pdfs. The reader can check them from [22].

We can start by generating random data from Gaussian pdf. As the first step, let's generate one-dimensional data from two different pdfs. The first Gaussian pdf has mean and standard deviation values as 2 and 1, respectively. The second Gaussian pdf has mean and standard deviation values as 5 and 0.5, respectively. We provide the corresponding Python script in Listing 5.1. Here, we set the sample size to 100 for each pdf.

Listing 5.1 Random number generation from one-dimensional Gaussian pdfs

```
import numpy as np

np.random.seed(42)   # For reproducibility

data_size=100

x1 = np.random.normal(2, 1, data_size)
x2 = np.random.normal(5, 5, data_size)
```

We can generate two-dimensional Gaussian data via providing the mean vector and covariance matrix. We formed the function `Gaussian2D(mean, E, theta, len)` to generate such data in Listing 5.2. This function takes inputs mean value and covariance matrix. The parameter `theta` rotates the covariance matrix clockwise.

The parameter `data_size` specifies the number of samples to be generated. In the given example, the code will generate 1000 samples with their mean centered at $[-2, -2]$ and covariance matrix rotated 45 degrees around the xy-axis. Similarly, the next 1000 samples will be generated with the mean value $[2, 2]$ and same covariance matrix but rotated -45 degrees around xy-axis. We provide the generated samples in Fig. 5.1. We will use this data in classification and clustering applications in the following chapters. Therefore, the generated data has been saved to "npy" files within the code.

Listing 5.2 Random number generation from two-dimensional Gaussian pdfs

```
import os.path as osp
import numpy as np
from sklearn.model_selection import train_test_split

np.random.seed(42)   # For reproducibility

DATA_SIZE = 1000
CLASSIFICATION_DATA_DIR = "classification_data"

def Gaussian2D(mean, L, theta, data_size):
    c, s = np.cos(theta), np.sin(theta)
    R = np.array([[c, -s], [s, c]])
    cov = R @ L @ R.T
    return np.random.multivariate_normal(mean, cov, data_size)

mean = [-2, -2]
E = np.diag([1, 10])
theta = np.radians(45)
class1_samples = Gaussian2D(mean, E, theta, DATA_SIZE)

mean = [2, 2]
theta = np.radians(-45)
class2_samples = Gaussian2D(mean, E, theta, DATA_SIZE)

samples = np.concatenate([class1_samples, class2_samples])
labels = [0] * 1000 + [1] * 1000

train_samples, test_samples, train_labels, test_labels =
    train_test_split(
    samples, labels, test_size=0.2
)
# Save training data to file
np.save(osp.join(CLASSIFICATION_DATA_DIR, "cls_train_samples.npy"
    ), train_samples)
np.save(osp.join(CLASSIFICATION_DATA_DIR, "cls_test_samples.npy")
    , test_samples)
np.save(osp.join(CLASSIFICATION_DATA_DIR, "cls_train_labels.npy")
    , train_labels)
np.save(osp.join(CLASSIFICATION_DATA_DIR, "cls_test_labels.npy"),
    test_labels)
```

We can also use the generated random data as noise affecting a function. This setup will be extensively used in regression applications in the following chapters. We pick two examples to explain this concept better. Our first example is the line

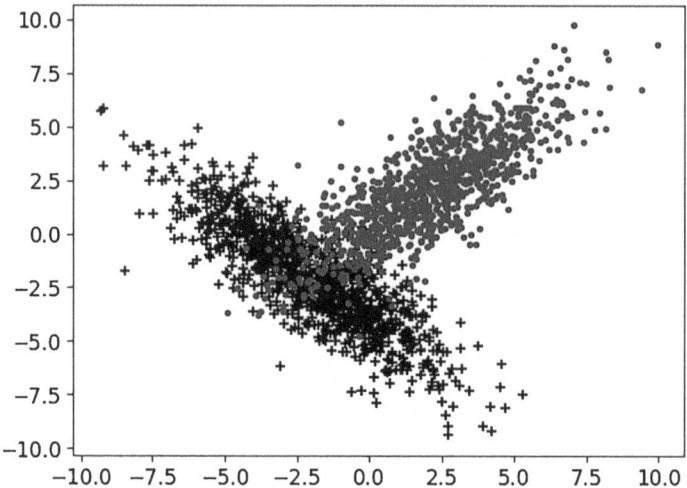

Fig. 5.1 Two-dimensional random data generated from Gaussian pdfs

data $y = 3x + 4$. We can add Gaussian random noise to it. Hence, it becomes
$y = 3x + 4 + \varepsilon$ where ε represents the added Gaussian noise term. Assume that the
Gaussian noise has parameters mean and standard deviation as 0 and 1, respectively.
Our second example is the sine data $y = sin(\pi/2x + \varepsilon)$. Assume the same noise
parameters in this example.

We provide the Python script to generate noisy data in Listing 5.3. In this code,
random noise for line data is generated by the function np.random.normal with zero
mean and 0.5 standard deviation. The noise term in the sine data is again Gaussian
with zero mean and 0.2 standard deviation. These noise terms are added to the
functions. We provide the generated samples this way in Fig. 5.2. We will use these
data in regression applications in the following chapters. Therefore, the generated
data has been saved to "npy" files within the code.

Listing 5.3 Random number generation for regression operations

```
import os.path as osp
import numpy as np

np.random.seed(42)   # For reproducibility

DATA_SIZE = 200
REGRESSION_DATA_DIR = "regression_data"

# Generate noisy line data
samples = np.array(range(DATA_SIZE)) / DATA_SIZE * 10
samples = np.expand_dims(samples, axis = 1) #Alternatively, np.
    reshape(samples, (DATA_SIZE, 1))
noise = np.random.normal(0, 0.5, (DATA_SIZE, 1))
line_values = 3 * samples + 4 + noise

# Generate noisy sine data
```

```
noise = np.random.normal(0, 0.2, (DATA_SIZE, 1))
sine_values = np.sin(np.pi / 2 * samples) + noise

np.save(osp.join(REGRESSION_DATA_DIR, "reg_samples.npy"), samples
    )
np.save(osp.join(REGRESSION_DATA_DIR,"reg_line_values.npy"),
    line_values)
np.save(osp.join(REGRESSION_DATA_DIR, "reg_sine_values.npy"),
    sine_values)
```

5.1.3 Random Number Generation on the STM32 Microcontroller

We can also generate random numbers on the STM32 microcontroller. The dedicated peripheral unit available on the microcontroller is called the random number generator (RNG). This unit samples an analog noise source and postprocesses it for random number generation. As a result, it produces 32-bit random numbers. We will use this peripheral unit to generate random numbers on the STM32 microcontroller with STM32CubeIDE and Mbed Studio development environments.

5.1.3.1 Random Number Generation via STM32CubeIDE

We should first activate the RNG peripheral unit to generate random numbers. To do so, please select the checkbox "Activated" under "Mode" window of RNG unit in STM32CubeMX. A clock issue may occur in the clock configuration window after activation. Click on the "Clock Configuration" tab of STM32CubeMX to solve this issue. Select the "PLLQ" clock signal for the "CLK48 Clock Mux." Then, value of the clock divider "Q" in PLL block can be increased until the issue is solved. Afterward, code generation can be executed.

We formed the random number generation library lib_rng consisting of two files "lib_rng.h" and "lib_rng.c." These files should be copied under the folders "/Core/Inc" and "/Core/Src" of the project, respectively. Our random number generation library consists of two functions. The function LIB_RNG_Init initializes the RNG peripheral unit on the STM32 microcontroller. If it has been initialized in

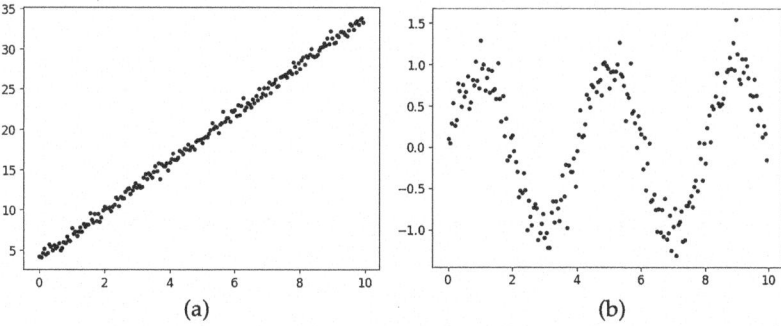

Fig. 5.2 Generated random numbers for regression applications. (**a**) Line data with noise. (**b**) Sine data with noise

STM32CubeMX beforehand, then we should not use this function. We will use our random number generation library under Mbed Studio as well. There, we have to use the function `LIB_RNG_Init`. The function `LIB_RNG_GetRandomNumber` returns the 32-bit random number generated by RNG.

We provide a sample code to generate a random number every second using STM32CubeIDE in Listing 5.4. The generated random number is also transferred to PC in the code. Therefore, we should activate the UART peripheral unit and include the serial library to our project as explained in Sect. 4.1.2.

Listing 5.4 Random number generation example on STM32CubeIDE

```
/* USER CODE BEGIN Includes */
#include "lib_rng.h"
#include "lib_serial.h"
/* USER CODE END Includes */

/* USER CODE BEGIN WHILE */
while (1)
{
    uint32_t randomNumber = LIB_RNG_GetRandomNumber();
    LIB_SERIAL_Transmit(&randomNumber, 1, TYPE_U32);
    HAL_Delay(1000);
    /* USER CODE END WHILE */

    /* USER CODE BEGIN 3 */
}
/* USER CODE END 3 */
```

We provide the Python script to run on PC along with the code in Listing 5.5. We should include the data transfer library to our Python script as explained in Sect. 4.1.6.

Listing 5.5 Python script to print random numbers generated by the STM32 microcontroller

```
import py_serial

py_serial.SERIAL_Init("COM4")

while 1:
    rqType, datalength, dataType = py_serial.
        SERIAL_PollForRequest()
    if rqType == py_serial.MCU_WRITES:
        data = py_serial.SERIAL_Read()
```

5.1.3.2 Random Number Generation via Mbed Studio

We will use the lib_rng library to generate random numbers in Mbed Studio as in STM32CubeIDE. Therefore, we should copy the files "lib_rng.h" and "lib_rng.c" under the project root folder. We should also include the data transfer library to our project as explained in Sect. 4.1.4.

We formed a sample code to generate random numbers using the RNG peripheral unit under Mbed Studio in Listing 5.6. Data transfer to PC is also handled in the same code. Therefore, the Python script in Listing 5.5 should be running on PC along with our sample code.

Listing 5.6   Random number generation example on Mbed Studio

```
#include "mbed.h"
#include "lib_rng.h"
#include "lib_serial.h"
#include "lib_uart.h"

int main(){
    LIB_RNG_Init();
    LIB_UART_Init();

    while (true){
        uint32_t randomNumber = LIB_RNG_GetRandomNumber();
        LIB_SERIAL_Transmit(&randomNumber, 1, TYPE_U32);
        wait_us(1000000);
    }
}
```

5.1.4   Histogram Formation from Generated Random Numbers

We can form sample pdf from generated random data or any data acquired by a sensor. This allows us to observe characteristics of the data. One way of forming this sample pdf is by constructing a histogram. To do so, we should divide range of sensor data into bins. The histogram is formed by counting the number of samples falling into each bin. As we divide the number of samples in bins by the total number of samples at hand, we will have a normalized histogram. This representation satisfies all the constraints to be a pdf. Hence, this normalized histogram is generally called as the sample pdf.

We can represent the histogram formation operation in algorithmic way as follows. Assume that we applied quantization to data at hand such that we have integer values within range 0 and $K - 1$. Assume further that we have N such samples. We can use the pseudocode in Algorithm 1 to form the histogram, denoted as $h[k]$, for $k = 0, \ldots K - 1$. In Algorithm 1, x_i refers to the quantized ith sample from our dataset. Here, we get the value and increment the corresponding histogram bin by one. Consequently, we obtain the total number of samples for each bin in the histogram.

We did not consider how the quantization operation is done while forming the histogram. We perform quantization by the ADC module of the microcontroller if we use a sensor providing analog data. If the sensor provides digital data, then the ADC operation is done within it. For both cases, we will start with digital data.

One important issue in histogram formation is selecting an appropriate bin size for the data at hand. We form three histograms with different bin sizes for the

---

Algorithm 1 Pseudocode to generate histogram from given samples

---

for $i = 0 \rightarrow N$ **do**
$\quad k = x_i$
$\quad h[k] \leftarrow h[k] + 1$
end for

Gaussian random data generated in Sect. 5.1.2. We first pick a small bin size and provide the formed histogram in Fig. 5.3a. Since the bin size is small, the formed histogram becomes sparse. We next pick the bin size to be large. We provide the formed histogram in Fig. 5.3c. As can be seen in this figure, the obtained histogram is crude. Finally, we set an appropriate bin size as in Fig. 5.3b. Hence, we can observe the distribution of data better.

5.2   From Sensor Data to Feature Extraction

We considered how different sensors can be used to acquire data in Chap. 4. We can use this data directly in machine learning applications. Therefore, we will first introduce the definition of feature based on raw data. The raw sensor data may not

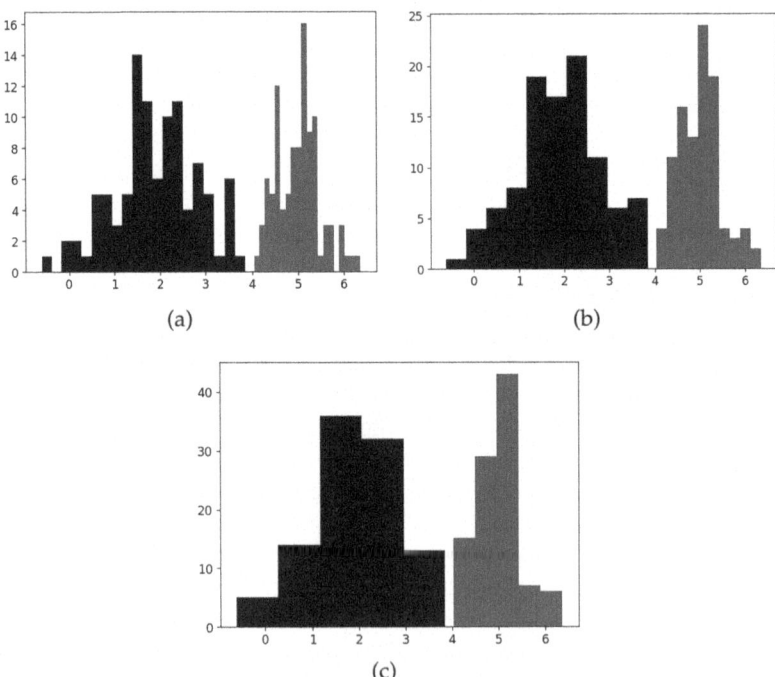

Fig. 5.3 Histogram formation with different bin sizes. (**a**) Histogram with 20 bins. (**b**) Histogram with 10 bins. (**c**) Histogram with 5 bins

be used directly in machine learning applications for some cases. Hence, it should be processed. This operation is called feature extraction. We will consider it in this section. Related to this, we will also introduce the feature space concept.

5.2.1 Definition of Feature

We can call raw or processed sensor data fed to a machine learning system as feature. The machine learning system depends on such features to operate. In other words, the system knows the outside world via features fed to it.

We can explain the feature by a real-life example. Assume that we would like to classify temperature of the CPU on a PC as "hot" or "normal." Therefore, we will need a temperature sensor placed on the CPU. In this scenario, the measured value via the sensor will be the feature. Our machine learning system, classifier to be introduced in Chap. 6, will process feature values to decide on whether the temperature of the CPU is "hot" or "normal." Hence, it can start the fan if the temperature is classified as hot.

We can consider yet another real-life example on the definition of feature. We can use a sensor to measure ambient temperature of a room. The measured value here will be the feature. We can form a machine learning system, regressor to be introduced in Chap. 7, to predict the temperature value next hour. Here, the system will use the feature values in time to predict the next value.

We can use random number generation, introduced in Sect. 5.1, to form features in line with pseudosensor data. Hence, we can associate the feature definition by a random variable given in appendix. However, we will not pursue this path in this book since we are more focused on practice than theory. Therefore, we will directly use features as they are in our embedded machine learning system.

5.2.2 Feature Extraction via Processing Raw Sensor Data

Raw data acquired by a sensor may not be directly used in machine learning applications as a feature. There are two main reasons for this. First and most important of all, raw data may be excessively large. Second, one sample of the raw data may not be informative on its own. We can summarize the raw sensor data by passing it through a function or transformation such that it becomes condensed and more informative. Hence, it can be useful in a machine learning system. The general name for this operation is feature extraction. As a result, we transform the acquired raw data from the sensor such that it can be used in machine learning applications. Let's consider three such examples.

Our first example on feature extraction is based on accelerometer data. Assume that we acquire data from the accelerometer in time as an array. A single sample from the array may not be useful in a machine learning system. Therefore, we may need to handle part of the array and generate a representative value from it. We can use this value as feature. Hence, we use an array of sensor readings to extract the feature value. We will consider feature extraction methods this way in our real-life example in Sect. 5.4.

Our second example on feature extraction is based on audio signals. We can acquire an audio signal from the microphone as an array. A single sample from the array may not be useful in a machine learning system. Therefore, we can extract feature values from the array and use them in the machine learning system. We will consider feature extraction methods this way in our real-life example in Sect. 5.5.

Our third example on feature extraction is based on digital images. We can acquire a digital image from a camera. Individual pixels of the acquired image may not be useful as feature values in a machine learning system. Hence, we should apply a transformation to extract features from pixel values. We will consider feature extraction methods this way in our real-life example in Sect. 5.6.

Unfortunately, there is no unique or golden feature extraction method in literature for the sensor data considered in previous three examples. Instead, there are several methods having different approaches proposed for this purpose. We will introduce a representative set of methods in Sects. 5.4, 5.5, and 5.6 for the accelerometer data, audio signals, and digital images, respectively. In general, there is no unique feature extraction method for the data obtained from different sensors. Therefore, it is not possible to give a single recipe. The reader should check for an appropriate feature extraction method in literature for the given problem and data source.

We will see that feature extraction in traditional and neural network-based methods may differ. As an example, convolutional neural network (CNN)-based systems, to be introduced in Chap. 13, will have their own feature extraction block within themselves. We postpone this usage till that chapter for the sake of clarity.

5.2.3   Feature Space Formation

A single feature may not be sufficient alone to solve a machine learning problem. Therefore, we may need more than one feature for operation. When we represent these features as coordinates, we form a feature space. There is no limitation on the number of features to be used at the same time in machine learning applications. However, we are bound by the three-dimensional space for visualization. Therefore, we will provide at most three-dimensional feature spaces in the following sections.

We can take the two-dimensional Gaussian number generation example given in Listing 5.2. As explained in Sect. 5.1.2, generated data is two-dimensional. Hence, we can assume that each dimension corresponds to a feature. Therefore, the feature space is formed as in Fig. 5.1. The first dimension corresponds to the first feature. The second dimension corresponds to the second feature. We can assume these two dimensions to be perpendicular and form the feature space as such.

5.3 Normalizing Feature Values

Feature values obtained directly from sensors or extracted from raw data may need preprocessing before being used in machine learning systems. This preprocessing operation has several aims such as eliminating noise terms or scaling feature values. Noise terms may degrade performance of the machine learning system. Non-scaled features may cause problems when they are used together since the feature with higher range may dominate the other. Therefore, we will use two methods as normalizing min-max values and z-score normalization in this book. Let's consider them next.

5.3.1 Normalizing Min-Max Values

Min-max normalization is the first method to perform range normalization. It scales the original feature values as $f' = \frac{f - \min(f)}{\max(f) - \min(f)} + \min(f)$ where f' is the normalized feature value and f is the original feature value. The feature range after min-max normalization is $[0, 1]$. This range is suitable for most machine learning methods in literature. However, the min-max scaling method is vulnerable to outliers in the feature set, since it is affected by the minimum or maximum outliers. Therefore, the reader should check and discard outliers before applying this method.

We can explain the min-max normalization method on the generated random data from Listing 5.2. As can be seen in Fig. 5.1, samples are spread. We can normalize the feature space by the min-max scaling method as in Listing 5.7. Here, we benefit from the scikit-learn library.

Listing 5.7 Normalizing the two-dimensional random numbers by the min-max method

```
import numpy as np
from sklearn.preprocessing import MinMaxScaler

np.random.seed(42)   # For reproducibility

def Gaussian2D(mean,E,theta,data_size):
    c, s = np.cos(theta), np.sin(theta)
    R = np.array([[c, -s], [s, c]])
    cov=R@L@R.T
    return np.random.multivariate_normal(mean, cov, data_size)

DATA_SIZE=1000
mean = [-2, -2]
E = np.diag([1,10])
theta=np.radians(45)
class1_samples = Gaussian2D(mean,E,theta,DATA_SIZE)

mean = [2, 2]
theta=np.radians(-45)
class2_samples = Gaussian2D(mean,E,theta,DATA_SIZE)

F=np.concatenate((class1_samples,class2_samples))
```

```
scaler = MinMaxScaler()
scaler = scaler.fit(F)
G=scaler.transform(F)
```

As we apply the min-max normalization method to the two-dimensional Gaussian data, we obtain the result as in Fig. 5.4. Please note that this data is transferred in the range [0, 1] for both features. Therefore, normalization prevents any dominance in the feature space.

5.3.2   Z-Score Normalization

Z-score normalization is the second method to perform feature range normalization. This approach uses standard deviation and mean of the feature such that it has zero mean and unit standard deviation after normalization. Hence, z-score normalization is applied as

$$f' = \frac{f - \mu_f}{\sigma_f} \tag{5.1}$$

where f' is the normalized feature value, f is the original feature value, μ_f is the mean for feature f, and σ_f is the standard deviation for f. Majority of feature values will be within range $[-1, 1]$ after normalization. This can overcome shortcoming of the min-max scaling originating from outliers.

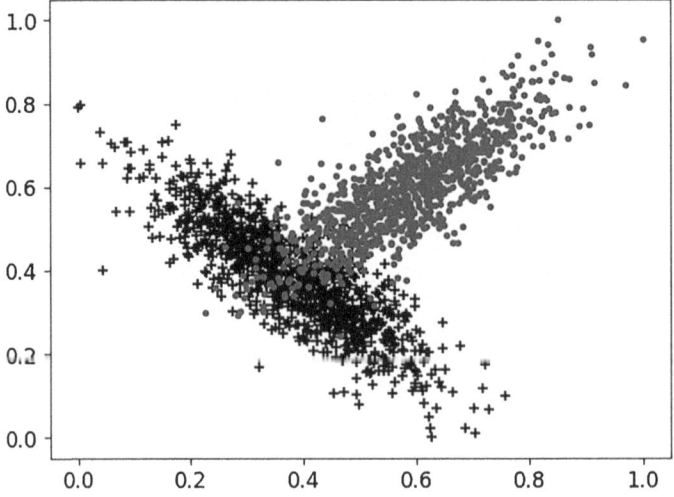

Fig. 5.4 Min-max normalization applied to the two-dimensional Gaussian data

There is one critical issue while using z-score normalization. This method assumes that feature values are normally distributed. If this condition is not satisfied, then the method may distort feature values. The reader should take this into account while applying the method.

We can explain the z-score normalization method on the generated random data from Listing 5.2. As can be seen in the feature space of the generated data in Fig. 5.1, the dimensions are not scaled. We can normalize the feature space by the z-score normalization method as in Listing 5.8. Here, we benefit from the scikit-learn library.

Listing 5.8 Normalizing the two-dimensional random numbers by the z-score normalization method

```
import numpy as np
from sklearn.preprocessing import StandardScaler

np.random.seed(42)  # For reproducibility

def Gaussian2D(mean,E,theta,data_size):
    c, s = np.cos(theta), np.sin(theta)
    R = np.array([[c, -s], [s, c]])
    cov=R@L@R.T
    return np.random.multivariate_normal(mean, cov, data_size)

DATA_SIZE=1000
mean = [-2, -2]
E = np.diag([1,10])
theta=np.radians(45)
class1_samples = Gaussian2D(mean,E,theta,DATA_SIZE)

mean = [2, 2]
theta=np.radians(-45)
class2_samples = Gaussian2D(mean,E,theta,DATA_SIZE)

F=np.concatenate((class1_samples, class2_samples))
scaler = StandardScaler()
scaler = scaler.fit(F)
G=scaler.transform(F)
```

As can be seen in Fig. 5.1, one axis has narrow range compared to the other. As we apply the z-score normalization method to the two-dimensional Gaussian data, we obtain the result as in Fig. 5.5. Please note that the data is centered around 0 and standard deviation of the data is scaled to 1.

5.4   Application: Feature Extraction from Accelerometer Data

Feature extraction from accelerometer data is a required step in various applications, including human activity recognition, health monitoring, and motion analysis. Here, the goal is to transform raw sensor data to informative features that capture relevant patterns or characteristics. As an example, human activity recognition aims to understand and interpret movement patterns. We will provide a classification

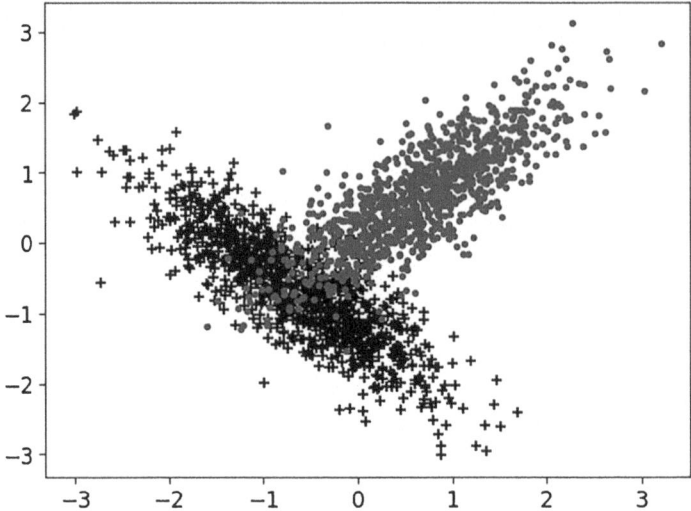

Fig. 5.5 Z-score normalization applied to the two-dimensional Gaussian data

example for this application in Sect. 6.6. The first step is feature extraction in this application. The second step is classification. Therefore, we will cover the basics of feature extraction from accelerometer data in this section. To do so, we will first download an available dataset. Then, we will extract features from samples in Python on PC. Afterward, we will implement the feature extraction step in C language. Hence, the application can be realized on the microcontroller.

5.4.1   The WISDM Dataset

We will benefit from the WISDM dataset in this application. This dataset consists of 1,098,207 samples collected from smart phone accelerometers. Collected samples belong to six classes, with their percentage in the dataset, as walking (38.6%), jogging (31.2%), upstairs (11.2%), downstairs (9.1%), sitting (5.5%), and standing (4.4%). The reader can download the dataset from [18]. Once the files are extracted from the downloaded archive file, the user will see the dataset files. Here, the file "WISDM_ar_v1.1_raw_about.txt" keeps information about data such as class distribution, sampling rate, contributed users, and data format. The file "WISDM_ar_v1.1_raw.txt" contains single sample per row, and attributes are separated by comma as   [user],[activity],[timestamp],[x-acceleration],[y-acceleration], and [z-acceleration].

We formed the function read_data for reading and formatting the raw data. We provide the function in Listing 5.9. Here, the built-in Pandas library function read_csv is used to automatically read comma-separated files. This function also allows assigning a name to each attribute. We define these as `column_names` in the code. The object `DataFrame` is created with named columns after calling the function `read_csv` with the parameter `names`. The `read_csv` function only handles commas as delimiters by default. However, there is a semicolon at the end of each line in the raw data file. Therefore, we added the code line `df["z-axis"] = df["z-axis"].str.replace(";", "").astype(float)` to remove the semicolon from `DataFrame` and convert it to float type at the same time.

Listing 5.9   The Python function for reading accelerometer data from the WISDM dataset

```
import pandas as pd

def read_data(file_path):
    column_names = ["user", "activity", "timestamp", "x-accel", "
        y-accel", "z-accel"]
    df = pd.read_csv(file_path, header=None, names=column_names)
    df["z-axis"] = df["z-axis"].str.replace(";", "").astype(float
        )
    df.dropna(inplace=True)
    print(f"Number of columns in the dataframe: {df.shape[1]}")
    print(f"Number of rows in the dataframe: {df.shape[0]}")
    df.head()
    return df
```

The dataset may have imperfections with NA (not available) or NaN (not a number) values. We should eliminate them for reliable operation. The Pandas library has a built-in function `dropna` to eliminate such samples. Therefore, we used the argument `inplace = True` of this function in Listing 5.9. Next, the number of columns (attributes) and samples (rows) are printed within the function `read_data`. Finally, `df.head` prints the first five elements from the dataset in the function.

5.4.2   Segmentation and Feature Extraction

As we format the samples in the dataset, we can extract features from them. Feature extraction can be done on a sequence of samples in segmented form. Moreover, samples in one segment should belong to one class only. Therefore, a specific time interval should be specified, and features should be extracted from these segments to achieve this goal. To do so, we formed the function `create_features` in Listing 5.10. This function first segments sequential samples. Then, it creates features based on these segments.

Listing 5.10   The Python function for extracting features from the accelerometer data

```
import numpy as np
import pandas as pd

def create_features(df, time_steps, step_size):
    x_segments = []
```

```python
y_segments = []
z_segments = []
labels = []
for i in range(0, len(df) - time_steps, step_size):
    xs = df["x-axis"].values[i : i + time_steps]
    ys = df["y-axis"].values[i : i + time_steps]
    zs = df["z-axis"].values[i : i + time_steps]

    count_per_label = df["activity"][i : i + time_steps].
        value_counts()
    label_count = count_per_label.iloc[0]
    if label_count == time_steps:
        label = count_per_label.index[0]
        x_segments.append(xs)
        y_segments.append(ys)
        z_segments.append(zs)
        labels.append(label)

# Bring the segments into a better shape
segments_df = pd.DataFrame({"x_segments": x_segments,
                            "y_segments": y_segments,
                            "z_segments": z_segments}
)
feature_df = pd.DataFrame()
# mean
feature_df['x_mean'] = segments_df["x_segments"].apply(lambda
    x: x.mean())
feature_df['y_mean'] = segments_df["y_segments"].apply(lambda
    x: x.mean())
feature_df['z_mean'] = segments_df["z_segments"].apply(lambda
    x: x.mean())

# positive count
feature_df['x_pos_count'] = segments_df["x_segments"].apply(
    lambda x: np.sum(x > 0))
feature_df['y_pos_count'] = segments_df["y_segments"].apply(
    lambda x: np.sum(x > 0))
feature_df['z_pos_count'] = segments_df["z_segments"].apply(
    lambda x: np.sum(x > 0))

# converting the signals from time domain to frequency domain
    using FFT
FFT_SIZE = time_steps // 2 + 1
x_fft_series = segments_df["x_segments"].apply(lambda x: np.
    abs(np.fft.fft(x))[1:FFT_SIZE])
y_fft_series = segments_df["y_segments"].apply(lambda x: np.
    abs(np.fft.fft(x))[1:FFT_SIZE])
z_fft_series = segments_df["z_segments"].apply(lambda x: np.
    abs(np.fft.fft(x))[1:FFT_SIZE])

# FFT std dev
feature_df['x_std_fft'] = x_fft_series.apply(lambda x: x.std
    ())
feature_df['y_std_fft'] = y_fft_series.apply(lambda x: x.std
    ())
feature_df['z_std_fft'] = z_fft_series.apply(lambda x: x.std
    ())

# FFT Signal magnitude area
feature_df['sma_fft'] = x_fft_series.apply(lambda x: np.sum(
    abs(x)/50)) + y_fft_series.apply(lambda x: np.sum(abs(x)
    /50)) \
```

```
                        + z_fft_series.apply(lambda x: np.sum(abs
                          (x)/50))

  labels = np.asarray(labels)

  return feature_df, labels
```

Before explaining the feature extraction part in the function `create_features`, let's first focus on its three inputs. The first input `df` determines the `DataFrame` store location. The second input `time_steps` determines how many data samples should be in each segment. The third input `step_size` determines the difference between the beginning time of consecutive segments. Assume that we have a dataset with 160 samples. We want to retrieve segments including 80 samples with step size 40. Then, we obtain segments from the dataset with element indices [0,80], [40,120], and [80,160].

Four variables are created under the function `create_features` as `x_segments`, `y_segments`, `z_segments`, and `labels`. They capture overall segment lists from input data. Segments for each axis are saved into corresponding variables `xs`, `ys`, and `zs`. Then, each segment is checked whether it contains a sample from another class by ensuring `label_count == time_steps`, as there is no separation in the dataset. Here, `label_count` represents the number of samples belonging to one class. If all samples belong to one class, then the corresponding segment is appended to the segments list. Next, these segments are converted to a `DataFrame` object for simpler calculation using Pandas library. We also create another `DataFrame` object called `feature_df` to store features.

We extract four features per axis in the function `create_features`. These features are average acceleration, count of positive values, standard deviation in frequency domain, and signal magnitude area in frequency domain. The first three features are calculated for each axis. The last feature, signal magnitude area in frequency domain, is calculated using summation of elements in the frequency domain. In the code, we benefit from the fast Fourier transform (FFT) to obtain frequency information. For more information on the extracted features, please see [17].

In Listing 5.11, we merge the functions introduced in the previous paragraphs to form the training and test dataset for the human activity recognition application. Here, we first import functions for reading and transforming data into features. Therefore, we set `TIME_PERIODS=80` and `STEP_DISTANCE=40`. Raw data is read and the user IDs greater than 28 are separated for test dataset using the WISDM data path. Thus, two `DataFrame` objects and their labels are created, one for training and the other for testing.

Listing 5.11 The main Python script to form the training and test dataset for the human activity recognition application

```
import os.path as osp
from data_utils import read_data
from feature_utils import create_features

DATA_PATH = osp.join("WISDM_ar_v1.1", "WISDM_ar_v1.1_raw.txt")
```

```
TIME_PERIODS = 80
STEP_DISTANCE = 40
data_df = read_data(DATA_PATH)
df_train = data_df[data_df["user"] <= 28]
df_test = data_df[data_df["user"] > 28]

train_segments_df, train_labels = create_features(df_train,
    TIME_PERIODS, STEP_DISTANCE)
test_segments_df, test_labels = create_features(df_test,
    TIME_PERIODS, STEP_DISTANCE)

print("Train samples shape: ", len(train_segments_df))
print("Train labels shape: ", train_labels.shape)
print("Test samples shape: ", len(test_segments_df))
print("Test labels shape: ", test_labels.shape)
```

5.4.3 Implementation on the Microcontroller

We should apply the same feature extraction steps on the microcontroller to form the human activity recognition application there. To do so, we will provide complete STM32CubeIDE and Mbed Studio projects in separate folders in the book repository. Here, we will only focus on the feature extraction part of the code.

While implementing the feature extraction part of the code, we benefit from the CMSIS-DSP library since it has built-in FFT functions. Therefore, we should include it to our code by #include "arm_math.h". The CMSIS-DSP library is also useful while extracting features based on mean and standard deviation.

We provide part of the feature extraction code to run on the microcontroller in Listing 5.12. Here, a fixed size array is formed to store accelerometer data. Then, mean and positive counts for each axis are calculated. For FFT, arm_rfft_fast_instance_f32 is formed to keep the FFT size information. After initialization of this object, FFT is calculated by the function arm_rfft_fast_f32. The functions arm_std_f32 and arm_abs_f32 are used to obtain standard deviation and absolute value of the signal in frequency domain. Finally, the signal magnitude area is calculated by aggregating the absolute value of each axis.

Listing 5.12 Feature extraction from the accelerometer data in C language

```
/* USER CODE BEGIN 2 */
        int arr_size = 64; // Must be 32,64,128,..., 4096
        float **acc_data = get_data(arr_size);
        float fft_output[3][arr_size];
        float fft_ssd[3];
        float fft_abs[3][arr_size];
        float sma_x = 0, sma_y = 0, sma_z = 0, sma;
        float x_mean = 0, y_mean, z_mean;
        int x_pos = 0, y_pos = 0, z_pos = 0;
        for(int i = 0; i < arr_size; i++){
                x_mean += acc_data[0][i];
                y_mean += acc_data[1][i];
                z_mean += acc_data[2][i];
```

```
                    x_pos += acc_data[0][i] > 0;
                    y_pos += acc_data[1][i] > 0;
                    z_pos += acc_data[2][i] > 0;

        }
        arm_rfft_fast_instance_f32 fft;
        arm_status res = arm_rfft_fast_init_f32(&fft, arr_size);
        if (res != 0){
    printf("FFT failed. Exiting...\n");
                exit(1);
        }

        for(int i = 0; i < 3; i++){
                float *fft_input = acc_data[i];
                arm_rfft_fast_f32(&fft, fft_input, fft_output[i],
                    0);
                arm_std_f32(fft_output[i], arr_size / 2, &fft_ssd
                    [i]);
                arm_abs_f32(fft_output[i],fft_abs[i], arr_size);
        }

        for(int i = 0; i < arr_size; i++){
                sma_x += fft_abs[0][i];
                sma_y += fft_abs[1][i];
                sma_z += fft_abs[2][i];
        }

        sma = (sma_x + sma_y + sma_z) / arr_size;

        free(acc_data);
/* USER CODE END 2 */
```

5.5   Application: Feature Extraction from Audio Signals

Human auditory system can process sound waves between frequencies 50 and
20,000 Hz. Audio signals are electronic representation of sound waves. In order to
process these signals on a microcontroller, we should convert them to digital form
as explained in Sect. 4.4. Then, we can extract features from these digital signals.
Afterward, we can form various applications from the extracted features. We will
provide one such application as keyword spotting from audio signals in Sect. 6.7. In
this section, we will provide ways of using an available dataset for this application
and extracting features from it.

5.5.1   The Free Spoken Digit Dataset

We will benefit from the free spoken digit dataset (FSDD) in this application.
This dataset consists of spoken digits recordings as wav files with 8 kHz sampling rate. The recordings are trimmed to have near minimal silence at the
beginning and end sections. Files in the dataset are named in the format "digit-
Label_speakerName_index.wav." The FSDD consists of 10 digits spoken by six

people. Each digit has 50 recordings per person. Therefore, each digit has 300 recordings. The dataset can be downloaded from [16].

5.5.2 Mel-Frequency Cepstral Coefficients as Features

Feature extraction is a crucial step in the keyword spotting application. The goal here is to extract useful features from the audio signal those can be fed to a machine learning model to build intelligent systems. Mel-frequency cepstral coefficient (MFCC) is one of the most used method to extract features from audio signals. MFCC coefficients contain information about rate changes in different spectrum bands. If a coefficient has positive value, then majority of the spectral energy is concentrated in low-frequency regions. If a coefficient has negative value, then it indicates that most of the spectral energy is concentrated in high-frequency regions. These coefficients are used to represent spectral characteristics of audio signals in a way that is well-suited for various machine learning tasks, such as speech recognition and music analysis.

MFCC transforms raw audio signals into a compact representation that captures important frequency and temporal information. They emphasize audio signal properties important for human audio perception while discarding less relevant information. MFCC feature extraction consists of different levels. First, we should form a spectrogram from audio signal samples. The spectrogram is the time-varying frequency response of the audio signal. Therefore, the signal should be framed by a window. Then, FFT is applied on each frame to calculate its frequency spectrum. Hence, we can observe frequency values of the signal in time. This operation is also called short-time Fourier transform (STFT). Typically, size of the frame and frequency spectrum components are selected as the order of two, such as 256 and 512. After the STFT operation, Mel spaced filter banks are applied. The Mel spaced filter bank is a set of triangular filters where varying frequency ranges are mapped to logarithmic Mel scale. Then, discrete cosine transform (DCT) is applied to the filtering results to obtain high- and low-frequency changes in the audio signal.

CMSIS-DSP library has the required functions to create MFCC features from the audio signal directly. `arm_mfcc_f32` provides MFCC features for the first frame. To do so, we should first install the "cmsisdsp package" by the pip command `pip install cmsisdsp` under Python. Hence, we will be able to use CMSIS-DSP library functions under Python. This way, we can guarantee that the same CMSIS-DSP library functions will be executed for both Python and C codes.

We provide the Python script to calculate MFCC coefficients, hence features, in Listing 5.13. Here, we first define the required parameters as FFT size, number of Mel filters, number of DCT outputs, and framing window. Afterward, we form Mel filters by the function `melFilterMatrix` from CMSIS-DSP library. Then, we use the function `dctMatrix` in the CMSIS-DSP library to compute DCT filter coefficients. Next, we read recorded audio signals by the function `scipy.io.wavfile`.

Listing 5.13 Loading the FSDD dataset and extracting MFCC features from it

```
import os
import scipy.signal as sig
from mfcc_func import create_mfcc_features

RECORDINGS_DIR = "recordings"
recordings_list = [os.path.join(RECORDINGS_DIR, recording_path)
    for recording_path in os.listdir(RECORDINGS_DIR)]

FFTSize = 1024
sample_rate = 8000
numOfMelFilters = 20
numOfDctOutputs = 13
window = sig.get_window("hamming", FFTSize)
create_mfcc_features(recordings_list, FFTSize, sample_rate,
    numOfMelFilters, numOfDctOutputs, window)
```

5.5.3 Implementation on the Microcontroller

We should apply the same feature extraction steps on the microcontroller to form the keyword spotting application there. To do so, we will provide complete STM32CubeIDE and Mbed Studio projects in separate folders in the book repository. Here, we will only focus on the feature extraction part of the code.

While implementing the feature extraction part of the code, we benefit from the CMSIS-DSP library since it has built-in FFT functions. Therefore, we should include it to our code by `#include "arm_math.h"`. The CMSIS-DSP library is also useful while extracting MFCC features. We provide the function declarations to calculate precomputed MFCC filter banks and DCT filters in the header file given in Listing 5.14. These functions are the ones used in the Python script in Listing 5.13. Moreover, we use the function `mfcc_compute` to benefit from these predefined filters and calculate MFCC of a one-dimensional signal.

Listing 5.14 The header file to extract MFCC features in C language

```
#ifndef INC_MFCC_H_
#define INC_MFCC_H_

#include <math.h>
#include <stdio.h>
#include <stdlib.h>
#include <string.h>
#include <stdbool.h>
#include "stm32f746xx.h"
#include "arm_math.h"

#define SAMP_FREQ 8000
#define NUM_FBANK_BINS 40
#define MEL_LOW_FREQ 20
#define MEL_HIGH_FREQ 4000

#define M_2PI 6.283185307179586476925286766559005
```

```
struct MFCCInstance{
  const float *dct_matrix;
  float *windowCoefs;
  uint32_t *filterPos;
  uint32_t *filterLengths;
  uint32_t fftLen;
  uint32_t nbMelFilters;
  uint32_t nbDCtOutputs;
  arm_rfft_fast_instance_f32 *rfft;
};

typedef struct MFCCInstance mfcc_instance;

void init_mfcc_instance(mfcc_instance *S, uint32_t fftLen,
    uint32_t nbMelFilters, uint32_t nbDctOutputs);
float frequencyToMelSpace(float freq);
float melSpaceToFrequency(float mels);
float* create_dct_matrix(int numOfDctOutputs, int numOfMelFilters
    );
int create_mel_fbank(int FFTSize, int n_mels, uint32_t **filtPos,
    uint32_t **filtLen, float **packedFilters);
void mfcc_compute(const mfcc_instance S, const int16_t *
    audio_data, int frame_len, float *mfcc_out);

#endif /* INC_MFCC_H_ */
```

The function declarations done in the header file are defined in Listing 5.15. Here, the function mfcc_compute takes the variables mfcc_instance, audio_data, frame_len, and mfcc_out as input. The variable mfcc_instance keeps the information of Mel filter banks, DCT matrix, and FFT parameters used to compute MFCC. The variable audio_data keeps the input data that we will use. The variable frame_len keeps the number of samples in the audio signal. The variable mfcc_out keeps the output MFCC array. When the function is fed with the mfcc_output array, it fills the array with results.

Listing 5.15 The source file to extract MFCC features in C language

```
#include "mfcc.h"

void init_mfcc_instance(mfcc_instance *S, uint32_t fftLen,
            uint32_t nbMelFilters, uint32_t nbDctOutputs) {

    //create window function
    float *window_func = malloc(sizeof(float) * fftLen);
    for (int i = 0; i < fftLen; i++)
            window_func[i] = 0.5 - 0.5 * cos(M_2PI * ((float)
                i) / (fftLen));

    //create mel filterbank
    S->filterPos = malloc(nbMelFilters * sizeof(int));
    S->filterLengths = malloc(nbMelFilters * sizeof(int));
    S->nbMelFilters = nbMelFilters;
    S->nbDctOutputs = nbDctOutputs;
    S->windowCoefs = window_func;
```

```c
        create_mel_fbank(fftLen, nbMelFilters, &S->filterPos, &S
            ->filterLengths, &S->windowCoefs);

        //create DCT matrix
        S->dct_matrix = create_dct_matrix(nbMelFilters,
            nbDctOutputs);

        //initialize FFT
//        S->rfft = arm_rfft_fast_instance_f32;
        arm_rfft_fast_init_f32(S->rfft, fftLen);
}

float frequencyToMelSpace(float freq) {
        return 1127.0 * logf(1.0 + freq / 700.0);
}

float melSpaceToFrequency(float mel) {
        return 700.0 * (expf(mel / 1127.0) - 1.0);
}

float* create_dct_matrix(const int numOfDctOutputs, const int
    numOfMelFilters) {
        float *dct_matrix = malloc(
                        sizeof(float) * numOfDctOutputs *
                            numOfMelFilters);
        float norm_mels = sqrt(2.0 / numOfMelFilters);
        for (int mel_idx = 0; mel_idx < numOfMelFilters; mel_idx
            ++) {
                for (int dct_idx = 0; dct_idx < numOfDctOutputs;
                    dct_idx++) {
                        float s = (mel_idx + 0.5) /
                            numOfMelFilters;
                        dct_matrix[dct_idx * numOfMelFilters +
                            mel_idx] = (cosf(
                                    dct_idx * M_PI * s) *
                                        norm_mels);
                }
        }

        return dct_matrix;
}

int create_mel_fbank(int FFTSize, int n_mels, uint32_t **filtPos,
                uint32_t **filtLen, float **packedFilters) {

        int half_fft_size = FFTSize / 2;
        float filters[n_mels][half_fft_size + 1];
        float spectrogram_mel[half_fft_size];

        float fmin_mel = frequencyToMelSpace(MEL_LOW_FREQ);
        float fmax_mel = frequencyToMelSpace(MEL_HIGH_FREQ);
        float freq_step = SAMP_FREQ / FFTSize;

        for (int freq_idx = 1; freq_idx < half_fft_size + 1;
            freq_idx++) {
                float linear_freq = freq_idx * freq_step;
                spectrogram_mel[freq_idx - 1] =
                    frequencyToMelSpace(linear_freq);
        }

        float mel_step = (fmax_mel - fmin_mel) / (n_mels + 1);
        int totalLen = 0;
        for (int mel_idx = 0; mel_idx < n_mels; mel_idx++) {
```

```c
        float mel = mel_step * mel_idx;
        bool startFound = false;
        int startPos = 0, endPos = 0, curLen = 0;
        for (int freq_idx = 0; freq_idx < half_fft_size;
            freq_idx++) {
                float upper = (spectrogram_mel[freq_idx]
                    - mel) / mel_step;
                float lower = (mel - spectrogram_mel[
                    freq_idx]) / mel_step + 2;
                float filter_val = fmaxf(0.0, fminf(upper
                    , lower));
                filters[mel_idx][freq_idx + 1] =
                    filter_val;
                if (!startFound & (filter_val != 0.0)) {
                        startFound = true;
                        startPos = freq_idx + 1;
                }

                else if (startFound & (filter_val == 0.0)
                    ) {
                        endPos = freq_idx;
                        break;
                }
        }
        curLen = endPos - startPos + 1;
        *filtLen[mel_idx] = (endPos - startPos + 1);
        *filtPos[mel_idx] = startPos;
        *packedFilters = realloc(*packedFilters,
                        (totalLen + curLen) * sizeof(
                            float));
        if (*packedFilters == NULL) {
                printf("Memory allocation failed\n");
                return 1;
        }

        memcpy(*packedFilters + totalLen, &filters[
            mel_idx][startPos],
                        curLen * sizeof(float));
        totalLen += curLen;
    }

    return 0;
}

void mfcc_compute(const mfcc_instance S, const int16_t *
    audio_data, int frame_len, float *mfcc_out) {

    int32_t i, j, bin;
    float frame[frame_len];
    float buffer[frame_len];
    float mel_energies[S.nbMelFilters];

    for (i = 0; i < frame_len; i++) {
            frame[i] = (float) audio_data[i] / (1 << 15);
    }
    //Fill up remaining with zeros

    for (i = 0; i < frame_len; i++) {
            frame[i] *= S->windowCoefs[i];
    }

    //Compute FFT
    arm_rfft_fast_f32(S.rfft, frame, buffer, 0);
```

```
//Convert to power spectrum
//frame is stored as [real0, realN/2-1, real1, im1, real2
    , im2, ...]
int32_t half_dim = frame_len / 2;
float first_energy = buffer[0] * buffer[0], last_energy =
    buffer[1]
                 * buffer[1];  // handle this special case
for (i = 1; i < half_dim; i++) {
        float real = buffer[i * 2], im = buffer[i * 2 +
            1];
        buffer[i] = real * real + im * im;
}
buffer[0] = first_energy;
buffer[half_dim] = last_energy;

float sqrt_data;
//Apply mel filterbanks
for (bin = 0; bin < S.nbMelFilters; bin++) {
        j = 0;
        float mel_energy = 0;
        int32_t first_index = S.filterPos[bin];
        int32_t length = S.filterLengths[bin];
        for (i = first_index; i <= first_index + length;
            i++) {
                arm_sqrt_f32(buffer[i], &sqrt_data);
                mel_energy += (sqrt_data) * S.windowCoefs
                    [bin * first_index + j];
                j++;
        }
        mel_energies[bin] = mel_energy;

        //avoid log of zero
        if (mel_energy == 0.0)
                mel_energies[bin] = 1e-7f;
}

//Take log
for (bin = 0; bin < S.nbMelFilters; bin++)
        mel_energies[bin] = logf(mel_energies[bin]);

//Take DCT. Uses matrix mul.
for (i = 0; i < S.nbDCtOutputs; i++) {
        float sum = 0.0;
        for (j = 0; j < S.nbMelFilters; j++) {
                sum += S.dct_matrix[i * S.nbMelFilters +
                    j] * mel_energies[j];
        }
        mfcc_out[i] = sum;
}

}
```

We provide the main file to call MFCC calculation functions in Listing 5.16. In the code, we should include the header file "mfcc.h" and create mfcc_instance and mfcc_out array under the main function. mfcc_instance can be formed with arbitrary parameters. Here, we define frame length as 1024, number of Mel filters as 20, and number of DCT outputs as 13. As the function mfcc_compute is called, the results will be saved to the array mfcc_out.

Listing 5.16 Feature extraction from audio signals in C language

```
/* USER CODE BEGIN Includes */
#include "mfcc.h"
#include "data.h"
/* USER CODE END Includes */

/* USER CODE BEGIN 2 */
mfcc_instance S;
float **mfcc_out;
int frame_len = 1024;
init_mfcc_instance(S, frame_len, 20, 13);
mfcc_compute(S, audio_data, frame_len, &mfcc_out);
/* USER CODE END 2 */
```

5.6   Application: Feature Extraction from Digital Images

Visual information is very important for relevant machine learning systems. One such application is the handwritten digit recognition. We will explain this application in detail in Sect. 6.8. There, we will recognize the handwritten digit written on a paper or another medium. To do so, we should first acquire the digital image using a camera as explained in Sect. 4.5. Usage of the raw image may not be suitable for the application. Therefore, we can extract features from the image at hand. In this section, we will provide ways of using an available dataset for this application and extracting features from it.

5.6.1   The MNIST Dataset

We will benefit from the MNIST dataset in this application. This dataset consists of 70,000 grayscale images of handwritten digits from 0 to 9. Images in the dataset are 28×28 pixels in size. They are normalized and centered in a fixed-size box. The MNIST dataset is considered as a classic benchmark in computer vision.

The reader can download the dataset from [19]. There are four files on the website as training images, training labels, test images, and test labels. These files are in ubyte binary format and need to be read accordingly. Further information can be found under "FILE FORMATS FOR THE MNIST DATABASE" section of the website. We formed the following code block in Listing 5.17 to read these files. The code reads the files and extracts metadata from them. Please note that the MNIST dataset is available under TensorFlow. We postpone this usage till Chap. 9.

Listing 5.17 Reading MNIST ubyte files

```
import numpy as np

def load_images(path):
    with open(path, "rb") as f:
```

```
        buffer = f.read()[16:]
        images = np.frombuffer(buffer, dtype=np.uint8).reshape
            (-1, 28, 28)
    return images

def load_labels(path):
    with open(path, "rb") as f:
        buffer = f.read()[8:]
        labels = np.frombuffer(buffer, dtype=np.uint8)
    return labels
```

5.6.2 Hu Moments as Features

Hu moments are widely used in computer vision for shape matching and object recognition [14]. They are robust to common image transformations. Hence, they can be used to recognize an object when viewed from different angles, distances, or orientations. Hu moments are calculated from central moments which depend on raw moments of an image. Therefore, let's first review them.

Raw moments of an image can be calculated as

$$M_{pq} = \sum_x \sum_y x^p y^q I(x, y) \tag{5.2}$$

where $I(x, y)$ is the image with coordinates (x, y).

Centroid coordinates $(\bar{x}, \bar{y})$ can be calculated using raw moments as

$$\bar{x} = \frac{M_{10}}{M_{00}} \tag{5.3}$$

$$\bar{y} = \frac{M_{01}}{M_{00}} \tag{5.4}$$

where M_{00} is the total intensity (or area if it is a binary image), M_{10} is the sum over x-axis, and M_{01} is the sum over y-axis.

Central moments are constructed from raw moments. They are translation invariant as they are calculated about the mean as

$$\mu_{pq} = \sum_x \sum_y (x - \bar{x})^p (y - \bar{y})^q f(x, y) \tag{5.5}$$

To achieve scale invariance, central moments are normalized by a scale factor, which is usually the total area of the image. Achieving rotation invariance is more complex. It involves building algebraic combinations of complex moments, so that the rotational component disappears. Hu moment invariants are a set of seven numbers derived from central moments that are invariant under translation (shifting the object left/right or up/down), scale (enlarging or shrinking the object), and rotation.

OpenCV has built-in functions to calculate moment invariants in Python. We provide the usage of these functions to calculate image, central, and Hu moments in Listing 5.18. Here, we first define data paths and load the MNIST dataset. Please note that we use the function load_images for image files and load_labels file for the label files. The function np.empty is only used to create placeholder arrays for huMoments. The function cv2.moments takes the image and binary flag as input. Then, it creates image moments (M_{pq}), central moments (μ_{pq}), and scale-invariant moments. These calculated moments are kept as a dictionary and saved as train_moments and test_moments separately. For each image, Hu moments are calculated using the previously computed moments and assigned to the Hu moments matrix.

Listing 5.18 Hu moments calculation under OpenCV

```
import os
from mnist import load_images, load_labels
import numpy as np
import cv2

train_img_path = os.path.join("MNIST-dataset", "train-images.idx3
    -ubyte")
train_label_path = os.path.join("MNIST-dataset", "train-labels.
    idx1-ubyte")
test_img_path = os.path.join("MNIST-dataset", "t10k-images.idx3-
    ubyte")
test_label_path = os.path.join("MNIST-dataset", "t10k-labels.idx1
    -ubyte")

train_images = load_images(train_img_path)
train_labels = load_labels(train_label_path)
test_images = load_images(test_img_path)
test_labels = load_labels(test_label_path)

train_huMoments = np.empty((len(train_images),7))
test_huMoments = np.empty((len(test_images),7))

for train_idx, train_img in enumerate(train_images):
    train_moments = cv2.moments(train_img, True)
    train_huMoments[train_idx] = cv2.HuMoments(train_moments).
        reshape(7)

for test_idx, test_img in enumerate(test_images):
    test_moments = cv2.moments(test_img, True)
    test_huMoments[test_idx] = cv2.HuMoments(test_moments).
        reshape(7)
```

5.6.3   Implementation on the Microcontroller

We should apply the same feature extraction steps on the microcontroller to form the handwritten digit recognition there. To do so, we will provide complete

STM32CubeIDE and Mbed Studio projects in separate folders in the book repository. Here, we will only focus on the feature extraction part of the code.

While implementing the feature extraction part of the code, we should include
required libraries in corresponding regions in the main file. We should include
the #include <math.h> header file to calculate Hu moments. Then, we can define
moment arrays as in Listing 5.19.

Listing 5.19 Arrays to be defined for moment calculation

```
/* USER CODE BEGIN PV */
float moments[4][4];
float nu[4][4];
float mu[4][4];
float hu_moments[7];
int numCols = 100;
int numRows = 100;
/* USER CODE END PV */
```

After defining the arrays, moments can be calculated using the formula in Eq. 5.2,
and central moments can be calculated using Eq. 5.5. We provide the C code to
calculate moments and Hu moments in Listing 5.20. The results are saved to arrays
inside the functions calculate_moments and calculate_hu_moments, respectively.
Here, the moments array represents raw moments. The mu array represents central
moments. The eta array represents scale and translation-invariant moments.

Listing 5.20 Raw and Hu moments calculation in C language

```
/* USER CODE BEGIN PFP */
void calculate_moments(float img[]){
    for(int c = 0; c < numCols; c++) {
        for (int r = 0; r < numRows; r++){
            for(int i = 0; i < 3; i ++){
                for(int j = 0; j < 3 - i; j++){
                    moments[i][j] += pow(c, i) * pow(r,j) * img[c
                        * numRows + r];
                }
            }
        }
    }

    float centroid_x = moments[1][0] / moments[0][0];
    float centroid_y = moments[0][1] / moments[0][0];
    mu[1][1] = fmax(moments[1][1] - centroid_x * moments[0][1],0)
        ;
    mu[2][0] = fmax(moments[2][0] - centroid_x * moments[1][0],0)
        ;
    mu[0][2] = fmax(moments[0][2] - centroid_y * moments[0][1],0)
        ;
    mu[3][0] = fmax(moments[3][0] - 3 * centroid_x * moments
        [2][0] + 2 * pow(centroid_x, 2) * moments[1][0], 0);
    mu[2][1] = fmax(moments[2][1] - 2 * centroid_x * moments
        [1][1] - centroid_y * moments[2][0] + 2 * pow(centroid_x,
        2) * moments[0][1],0);
```

```
    mu[1][2] = fmax(moments[1][2] - 2 * centroid_y * moments
        [1][1] - centroid_x * moments[0][2] + 2 * pow(centroid_y,
        2) * moments[1][0],0);
    mu[0][3] = fmax(moments[0][3] - 3 * centroid_y * moments
        [0][2] + 2 * pow(centroid_y, 2) * moments[0][1], 0);
    nu[2][0] = mu[2][0] / pow(moments[0][0], 2);
    nu[1][1] = mu[1][1] / pow(moments[0][0],2);
    nu[0][2] = mu[0][2] / pow(moments[0][0], 2);
    nu[3][0] = mu[3][0] / pow(moments[0][0], 2.5);
    nu[2][1] = mu[2][1] / pow(moments[0][0], 2.5);
    nu[1][2] = mu[1][2] / pow(moments[0][0], 2.5);
    nu[0][3] = mu[0][3] / pow(moments[0][0], 2.5);
}

void calculate_hu_moments(){
hu_moments[0] = nu[2][0] + nu[0][2];
hu_moments[1] = pow(nu[2][0] - nu[0][2], 2) + 4 * pow(nu[1][1],
    2);
hu_moments[2] = pow(nu[3][0] -3 * nu[1][2], 2) + pow(3 * nu[2][1]
    -nu[0][3], 2);
hu_moments[3] = pow(nu[3][0] + nu[1][2], 2) + pow(nu[2][1] + nu
    [0][3], 2);
hu_moments[4] = (nu[3][0] - 3 * nu[1][2])* (nu[3][0] + nu[1][2])*
    (pow(nu[3][0] + nu[1][2], 2) - 3 * pow(nu[2][1] + nu[0][3],
    2)) + (3 * nu[2][1] - nu[0][3])* (nu[2][1] + nu[0][3])* (3 *
    pow(nu[3][0] + nu[1][2], 2) - pow(nu[2][1] + nu[0][3],2));
hu_moments[5] = (nu[2][0] -nu[0][2])* (pow(nu[3][0]+nu[1][2],2) -
    pow(nu[2][1] + nu[0][3],2)) + 4 * nu[1][1] * (nu[3][0] + nu
    [1][2]) * (nu[2][1]+nu[0][3]);
hu_moments[6] = (3 * nu[2][1] - nu[0][3])* (nu[3][0] + nu[1][2])*
    (pow(nu[3][0] + nu[1][2], 2) - 3 * pow(nu[2][1] + nu[0][3],
    2))- (nu[3][0]-3 * nu[1][2]) * (nu[2][1]+nu[0][3]) * (3 * pow
    (nu[3][0]+nu[1][2],2)- pow(nu[2][1]+nu[0][3], 2));
}

/* USER CODE END PFP */
```

After calculating the moments, these function are called inside the main function adding the code block to corresponding part shown in Listing 5.21. Ultimately, raw moments, translation- and scale-invariant moments, and Hu moments are calculated for given image in this scenario.

Listing 5.21 Feature extraction from digital images in C language

```
/* USER CODE BEGIN 1 */
float img[numRows * numCols];
calculate_moments(img);
calculate_hu_moments();
/* USER CODE END 1 */
```

5.7 Summary of the Chapter

We considered three key concepts used in machine learning methods in this chapter. First, we covered random number generation as pseudosensor data. This will help us in testing machine learning methods when there is no active sensor around. Second, we considered feature extraction as a way of summarizing the acquired sensor data. This will act as a bridge between Chap. 4 on data acquisition and the following chapters. Third, we evaluated sensor data normalization. As end of chapter applications, we considered feature extraction from accelerometer data, audio signals, and digital images. We will extensively use these extracted features in the applications of the following chapters.

Classification

6

6.1   What Is a Classifier?

Classification aims to infer class label of an unknown observation summarized by its features. Hence, we have the feature space. There are limited number of classes to be inferred from. The classifier, as the machine learning system performing the classification operation, receives feature values as input. It generates the class label as output. Hence, the unknown observation is classified. The classifier should have been trained beforehand for this purpose. This is done by training data samples. Therefore, an expert has checked each sample from the training data and assigned a class label to it.

Let's consider the classification operation on a real-life example. Assume that we would like to develop an embedded machine learning system to decide on the activity of an object. The object can either stand steady or there are movements that indicate fall. Our aim is to form a fall detection system to infer whether the object has fallen based on measurements obtained from it. We can get the accelerometer data for this purpose as explained in Chap. 5. We will extract feature values from them to be fed to the classifier. There are two output conditions of the object as "standing still" and "falling." These are the two-class labels to be fed to output via the classifier.

During training, we should group feature values representing the standing still and fall classes. An expert should have labeled these features accordingly. Then, the classifier should be trained via these labeled features. After the training step, the classifier starts working by receiving unknown feature values. Each feature value, formed from the accelerometer data, is fed to the classifier. It then decides whether the object is standing still or falling.

We can represent our fall detection example in mathematical terms as follows. Assume that there are two features $x = [x_1, x_2]$ to be used in the classification operation. There are two classes labeled as C_1, standing still, and C_2, falling. Each class has its own training feature set labeled by an expert. This training set can be

© The Author(s), under exclusive license to Springer Nature Switzerland AG 2025   105
C. Ünsalan et al., *Embedded Machine Learning with Microcontrollers*,
https://doi.org/10.1007/978-3-031-70912-8_6

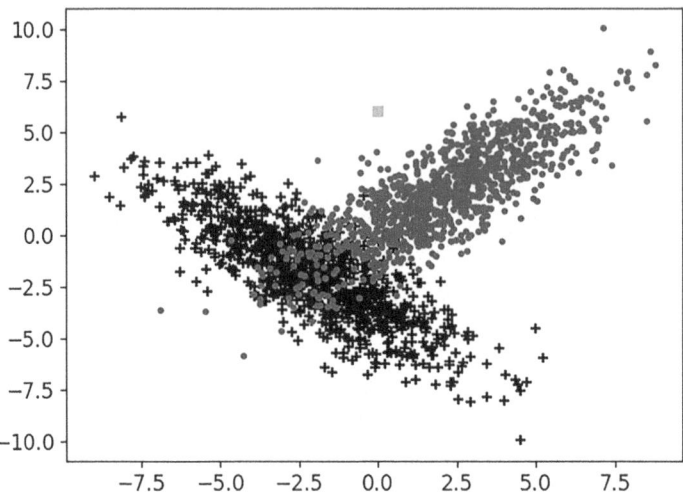

Fig. 6.1 Two-dimensional feature space with unknown sample to be classified

used to form a classifier which divides the feature space $x_1 - x_2$ into two regions by
a decision boundary. Hence, we can say that output of the training operation is the
formed decision boundary. As the training ends, the classifier will be ready to make
inference. Hence, when an unknown feature value comes, the classifier will decide
on its class whether being C_1 or C_2.

We provide the two-dimensional random data generated in Sect. 5.1.2 to explain
the classification operation better. We provide the modified dataset in Fig. 6.1. We
can explain the classification operation on this figure as follows. There are two
classes, with dataset plotted in light and dark gray, in the feature space. There is an
unknown sample or test data, plotted as gray square, in this feature space. Based on
these definitions, we expect the classifier to decide on the class label of the unknown
sample as either C_1 or C_2.

Although we focused on the two-class case in previous paragraphs, the classifier
can be formed for M classes as well. There, the feature space will be partitioned into
M regions. The decision boundary will also be formed accordingly. We will provide
one such example in Sect. 6.8 for classifying handwritten digit images acquired from
camera.

We will introduce four different classifiers throughout the chapter as Bayes, kNN,
SVM, and decision tree. Each classifier has its own methodology to form the decision boundary to partition the feature space for classification. We will summarize
advantages and disadvantages of each classifier in the following sections. Hence,
the reader can pick the suitable one for the application at hand.

6.1.1 Classifier Types

There are four classifiers to be considered in this chapter. We can group them into two categories as parametric and nonparametric. Next, we will explain general characteristics of both groups. Hence, the reader can understand how they can be implemented on an embedded system.

A parametric classifier has a decision boundary which can be constructed as a function with explicit parameters. Even if the parameter calculation step may change among different classifiers, the final decision boundary will be a parametric function of features. As an example, the decision boundary formed by the naive Bayes classifier will be a straight line (in two-dimensional feature space) as we will see in Sect. 6.2. In Sect. 6.4, we will see that the decision boundary formed by the SVM classifier will also be a straight line (in two-dimensional feature space) when the kernel trick is not applied.

Parametric classifiers have global representation ability since they have a function as the decision boundary. In other words, a parametric classifier can be used to partition the overall feature space. However, the classifier has limited modelling capability since it is bound by a predefined model while forming the decision boundary. Implementing parametric classifiers on an embedded system is another topic. Therefore, we will consider each parametric classifier implementation separately in the related section.

A nonparametric classifier does not have any parametric function to represent the decision boundary between classes. Instead, the classifier directly depends on training data. The kNN and decision tree classifiers, to be introduced in Sects. 6.3 and 6.5, fall into the nonparametric classifier category.

Nonparametric classifiers depend on local information. The advantage of this property is that nonparametric classifiers have strong modeling capability since they are not bound by any predefined model while forming the decision boundary. However, nonparametric classifiers do not have global representation ability. Therefore, they cannot be reliably used to partition the overall feature space in the same way as in parametric classifiers. Implementing nonparametric classifiers requires more resources in general since we do not have a parametric function at hand. We will consider each nonparametric classifier implementation separately in the related section.

6.1.2 Classifier Performance Measures

We should consider how to measure performance of a given classifier before focusing on different classifier types. Performance measurement can be done after training, during the inference phase. We also call this as testing phase of the classifier.

We should have labeled feature values at hand to test a given classifier and measure its performance. These feature values should be different from the ones used in the training phase. Since we have labeled feature values, we know the expected output from them. We feed feature values to the classifier and let it decide on the class label. We perform this operation for all test samples and calculate four values as true positive (TP), false positive (FP), true negative (TN), and false negative (FN).

Let's assume that we need to detect the activity class labels "standing still" and "falling" by using the accelerometer data. Thus, there are two labels to consider in this case. Let's explain TP, FP, TN, and FN based on our fall detection example which is a binary classification problem. Hence, there are only two possible outcomes for each prediction as positive or negative.

A TP occurs when the classifier predicts a label and the sample actually belongs to that label. In other words, a true positive occurs when the classifier correctly predicts the positive class. In our case, the classifier is trained to infer the "falling" class based on accelerometer data as an example. It correctly identifies the fall movement as "falling" class for the test sample. Hence, this is a TP. A TN occurs when the classifier correctly predicts the negative class. In our case, the classifier is trained to infer the "falling" class based on accelerometer data. It correctly infers the standing still condition as the "standing still" class. Hence, this is a TN. A FP occurs when the classifier incorrectly predicts the positive class. In our case, the classifier incorrectly infers the standing still movement as falling. FP is also known as "Type I error" or "false alarm." A FN occurs when the classifier incorrectly predicts the negative class. In our case, the classifier incorrectly infers the falling movement as standing still. FN is also known as "Type II error" or "miss."

There are measures based on TP, TN, FP, and FN in the literature. Precision, recall, and F1 score are metrics commonly used to evaluate performance of the classifier. We briefly summarize them next.

Precision is the ratio of true positives to the total number of positive predictions made by the model. It measures the proportion of positive predictions that are actually true positive. A high precision score indicates that the classifier is making accurate positive predictions. The formula for precision is

$$Precision = \frac{TP}{TP + FP} \tag{6.1}$$

Precision is useful when the cost of false positives is high. For example, a false-positive result could cause unnecessary treatment or further testing in a medical diagnosis scenario.

Recall is the ratio of true positives to the total number of actual positive cases in the dataset. It measures the proportion of actual positive cases that the classifier correctly identified as positive. A high recall score indicates that the classifier is able to identify most of the positive cases. The formula for recall is

$$Recall = \frac{TP}{TP + FN} \tag{6.2}$$

Recall is useful when the cost of false negatives is high. For example, a false negative could result in important emails being classified as spam and not delivered to the recipient in a spam detection scenario. In general, it is important to consider both precision and recall when evaluating the performance of a classifier, as they provide complementary information about the classifier's ability to accurately classify positive and negative cases.

F1 score is the harmonic mean of precision and recall. It is a measure of balance between precision and recall. A high F1 score indicates that the classifier is able to achieve both high precision and recall. The formula for the F1 score is

$$F1 = 2 \times \frac{Precision \times Recall}{Precision + Recall} \qquad (6.3)$$

F1 score is useful when there is an imbalance between the number of positive and negative cases in the dataset. It is a more balanced metric that takes the precision and recall into account. All the measures introduced in this section help the reader to point what is the main weakness of a given classifier. This may lead to modifying the classifier.

6.2 Bayes Classifier

Bayes classifier is fairly easy to implement and train. Hence, we pick it as our first classifier. We will start with explaining the theoretical background for the Bayes classifier in this section. Then, we will show how to train the classifier in Python on PC. Afterward, we will introduce ways of deploying the trained classifier to the STM32 microcontroller. Finally, we will measure performance of the deployed classifier on the microcontroller.

6.2.1 Theoretical Background

We can benefit from probability theory, specifically conditional probability and Bayes theorem, to form theoretical background for the Bayes classifier. Therefore, we can represent the classification operation as a random experiment with M outputs $C_1, \cdots, C_M$ based on the definitions in appendix. These outputs correspond to class labels to be assigned to an unknown input sample. Each output has prior probability assigned to it as $P(C_1), \cdots, P(C_M)$. We know that $\sum_{m=1}^{M} P(C_m) = 1$ from probability theory. In other words, the Bayes classifier should assign the unknown sample to one of the classes. We can assign the unknown sample to the class with highest prior probability. We also have feature values, as observations, to help us during classification. Therefore, we would like to assign the unknown sample to one of the classes based on these observations.

To explain the operations better, let's pick the classification problem with two classes, C_1, C_2, and two features, x_1, x_2. We can represent the classification problem as a random experiment with the sample space $C = \{C_1, C_2\}$ with features represented as $\mathbf{x} = [x_1, x_2]^T$. We can assign the unknown sample $\mathbf{x}T$ to the class with highest posterior probability. We can represent this class assignment rule by the help of conditional probability as follows. Assign $\mathbf{x}$ to the class C_1 if $P(C_1|\mathbf{x}) > P(C_2|\mathbf{x})$. Otherwise, assign $\mathbf{x}$ to the class C_2.

We can use the Bayes theorem to represent the posterior probability values as

$$P(C_1|\mathbf{x}) = \frac{P(\mathbf{x}|C_1)}{P(\mathbf{x})} P(C_1) \tag{6.4}$$

$$P(C_2|\mathbf{x}) = \frac{P(\mathbf{x}|C_2)}{P(\mathbf{x})} P(C_2) \tag{6.5}$$

Based on these definitions, we can represent the class assignment rule as

$$\text{Assign } \mathbf{x} \text{ to } \begin{cases} C_1; \text{ if } \frac{P(\mathbf{x}|C_1)}{P(\mathbf{x})} P(C_1) > \frac{P(\mathbf{x}|C_2)}{P(\mathbf{x})} P(C_2) \\ C_2; \text{ otherwise} \end{cases} \tag{6.6}$$

Since $P(\mathbf{x})$ is positive, we can discard it from both sides of the inequality. Moreover, we can take natural logarithm of both sides. Hence, we will have

$$\text{Assign } \mathbf{x} \text{ to } \begin{cases} C_1; \text{ if } \log(P(\mathbf{x}|C_1)) + \log(P(C_1)) > \log(P(\mathbf{x}|C_2)) + \log(P(C_2)) \\ C_2; \text{ otherwise} \end{cases}$$

$$\tag{6.7}$$

Equation 6.7 is the most general form of the Bayes classifier for two classes. Unfortunately, it is not useful in this form. Therefore, we should represent the likelihood terms $P(\mathbf{x}|C_1)$ and $P(\mathbf{x}|C_2)$ with known pdfs. The consensus in the literature is using Gaussian pdf for the likelihood values. Hence, we will have

$$P(\mathbf{x}|C_1) = \frac{1}{2\pi |\mathbf{\Sigma}_1|^{1/2}} \frac{-1}{2} \exp\left((\mathbf{x} - \boldsymbol{\mu}_1)^T \mathbf{\Sigma}_1^{-1} (\mathbf{x} - \boldsymbol{\mu}_1) \right) \tag{6.8}$$

$$P(\mathbf{x}|C_2) = \frac{1}{2\pi |\mathbf{\Sigma}_2|^{1/2}} \frac{-1}{2} \exp\left((\mathbf{x} - \boldsymbol{\mu}_2)^T \mathbf{\Sigma}_2^{-1} (\mathbf{x} - \boldsymbol{\mu}_2) \right) \tag{6.9}$$

where $\mathbf{\Sigma}_1$ and $\mathbf{\Sigma}_2$ are the covariance matrices for classes C_1 and C_2, respectively. $\boldsymbol{\mu}_1$ and $\boldsymbol{\mu}_2$ are the mean vector values for the classes C_1 and C_2, respectively. These values are obtained by the parameter estimation method given in appendix.

Mathematical derivations up to now may be confusing. However, we will have simpler implementation formulas from this point on. Therefore, the first step is to have three different assumptions for the covariance matrices used in operation as

$$\mathbf{\Sigma}_1 = \mathbf{\Sigma}_2 = \sigma^2 \mathbf{I} \tag{6.10}$$

$$\mathbf{\Sigma}_1 = \mathbf{\Sigma}_2 = \mathbf{\Sigma} \tag{6.11}$$

$$\mathbf{\Sigma}_1 \text{ and } \mathbf{\Sigma}_2 \text{ arbitrary} \tag{6.12}$$

Based on these three assumptions, we will have three different Bayes classifiers.

Let's start with the first case given in Eq. 6.10. This case is also called naive Bayes classifier since it assumes that features are independent (e.g., non-diagonal elements are zero). We can use this assumption in Eqs. 6.8 and 6.9 to make necessary simplifications. Moreover, we can assume that $P(C_1) = P(C_2)$. Hence, prior probability values do not favor one class over another. As a result, we can rewrite the class assignment rule in Eq. 6.7 as

$$\text{Assign } \mathbf{x} \text{ to } \begin{cases} C_1; \text{if } (\mathbf{x} - \boldsymbol{\mu_1})^T (\mathbf{x} - \boldsymbol{\mu_1}) < (\mathbf{x} - \boldsymbol{\mu_2})^T (\mathbf{x} - \boldsymbol{\mu_2}) \\ C_2; \text{otherwise} \end{cases} \tag{6.13}$$

Let's explain Eq. 6.13, hence the naive Bayes classifier working principles, in simpler terms. We should first calculate the sample mean value for each class from its training samples. This is the training phase of the classifier. During inference, we should assign the unknown sample to the class with the closest sample mean value in feature space. The closeness is measured by the Euclidean distance $(\mathbf{x} - \boldsymbol{\mu_i})^T (\mathbf{x} - \boldsymbol{\mu_i})$ for $i = 1, 2$. In other words, we represent each class by its mean vector in the feature space. Mean values are calculated by parameter estimation methods. Hence, we have a parametric classifier. The unknown sample is assigned to the class with the closest sample mean vector.

Let's obtain the decision boundary for the naive Bayes classifier. The decision boundary will be equidistant points from $\boldsymbol{\mu_1}$ and $\boldsymbol{\mu_2}$ in the feature space. We can pick the feature space in Fig. 6.1 and form the decision boundary by the formula in Eq. 6.13 using mean value of features. We do not need covariance matrix for this case since the decision is made based on the distance to the class mean values. We provide the formed decision boundary in Fig. 6.2. We will cover the training step for this example in the next section.

We can expand the naive Bayes classifier to M classes in a straightforward manner. To assign the unknown sample to one of the M classes, we should calculate its Euclidean distance to all class sample mean values. Then, we should assign it to the class with the closest sample mean value in the feature space. We can also expand the naive Bayes classifier to N-dimensional feature space. Here, the only modification will be calculating the N-dimensional mean vector for each class. Besides, the remaining steps will be the same.

The assumption in the naive Bayes classifier may be restrictive. Therefore, we can take the second assumption in Eq. 6.11 for the covariance matrix. For the two-class case, Eq. 6.7 becomes

$$\text{Assign } \mathbf{x} \text{ to } \begin{cases} C_1; \text{if } (\mathbf{x} - \boldsymbol{\mu_1})^T \boldsymbol{\Sigma}^{-1} (\mathbf{x} - \boldsymbol{\mu_1}) < (\mathbf{x} - \boldsymbol{\mu_2})^T \boldsymbol{\Sigma}^{-1}(\mathbf{x} - \boldsymbol{\mu_2}) \\ C_2; \text{otherwise} \end{cases}$$
$$\tag{6.14}$$

In Eq. 6.14, $(\mathbf{x} - \boldsymbol{\mu_i})^T \boldsymbol{\Sigma}^{-1}(\mathbf{x} - \boldsymbol{\mu_i})$ for $i = 1, 2$ corresponds to Mahalanobis distance. Therefore, all explanations in the naive Bayes classifier will hold here. The only difference will be how the distance from class centers are calculated. In

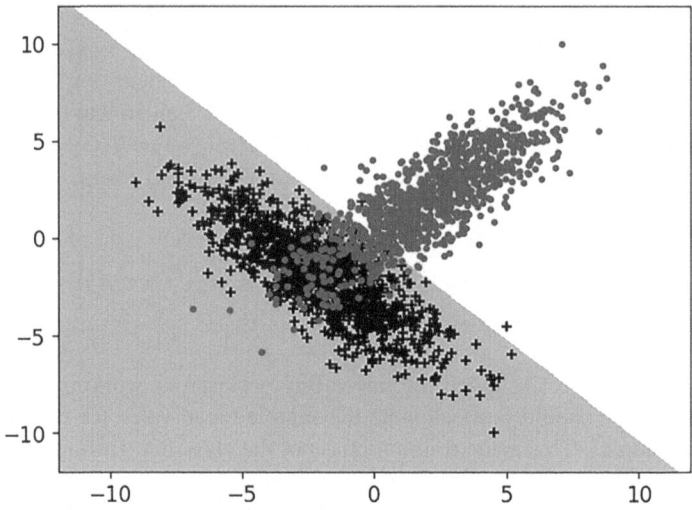

Fig. 6.2 Decision boundary formed by the Bayes classifier, where $\Sigma_1 = \Sigma_2 = \sigma^2 I$

Mahalanobis distance, the effect of each dimension is normalized by the inverse of its variance.

Let's obtain the decision boundary for the Bayes classifier with the assumption in Eq. 6.11. The decision boundary will be equidistant points, in terms of the Mahalanobis distance, from $\boldsymbol{\mu}_1$ and $\boldsymbol{\mu}_2$ in the feature space. We can pick the feature space in Fig. 6.1 and form the decision boundary by the formula in Eq. 6.14 using the mean and covariance matrix value of features. We provide the formed decision boundary in Fig. 6.3. We will cover the training step for this example in the next section.

We can expand the Bayes classifier to M classes with the assumption in Eq. 6.11 in a straightforward manner. To assign the unknown sample to one of the M classes, we should calculate its Mahalanobis distance to all class sample mean values. Then, we should assign it to the class with the closest sample mean value in the feature space. We can also expand the naive Bayes classifier to N-dimensional feature space. Here, the only modification will be calculating the N-dimensional mean vector for each class and $N \times N$ covariance matrix. Besides, the remaining steps will be the same.

Finally, we can take the third assumption in Eq. 6.12 and form the Bayes classifier accordingly. As in the second case, we will calculate the Mahalanobis distance between class centers and unknown sample point. Here, each class will have its own covariance matrix in calculations. Based on this case, Eq. 6.7 becomes

$$\text{Assign } \mathbf{x} \text{ to } \begin{cases} C_1; \text{ if } (\mathbf{x} - \boldsymbol{\mu}_1)^T \boldsymbol{\Sigma}_1^{-1} (\mathbf{x} - \boldsymbol{\mu}_1) < (\mathbf{x} - \boldsymbol{\mu}_2)^T \boldsymbol{\Sigma}_2^{-1} (\mathbf{x} - \boldsymbol{\mu}_2) \\ C_2; \text{ otherwise} \end{cases}$$

$$(6.15)$$

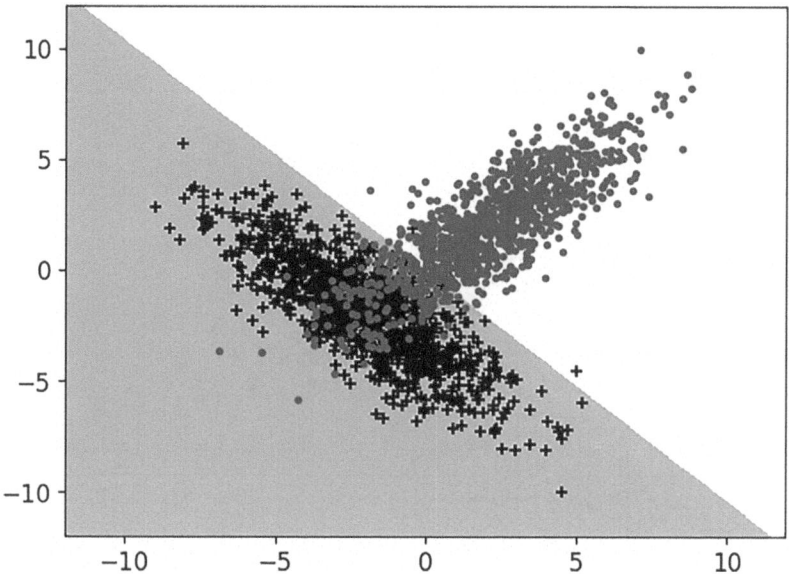

Fig. 6.3 Decision boundary formed by the Bayes classifier, where $\Sigma_1 = \Sigma_2 = \Sigma$

In Eq. 6.15, the Mahalanobis distance is also used. However, each class has its own covariance matrix in calculation. Therefore, all explanations in the naive Bayes classifier will hold here. The only difference will be how the distance from class centers are calculated.

Let's obtain the decision boundary for the Bayes classifier with the assumption in Eq. 6.12. The decision boundary will be equidistant points, in terms of Mahalanobis distance, from μ_1 and μ_2 in the feature space. We can pick the feature space in Fig. 6.1 and form the decision boundary by the formula in Eq. 6.15 using the mean and covariance matrix value of the features. We provide the formed decision boundary in Fig. 6.4. We will cover the training step for this example in the next section.

We can expand the Bayes classifier to M classes with the assumption in Eq. 6.12 in a straightforward manner. To assign the unknown sample to one of the M classes, we should calculate its Mahalanobis distance to all class sample mean values. Then, we should assign it to the class with the closest sample mean value in the feature space. We can also expand the naive Bayes classifier to N-dimensional feature space. The only modification will be calculating the N-dimensional mean vector for each class and $N \times N$ covariance matrix. Besides, the remaining steps will be the same.

6.2.2 Training the Classifier in Python

As explained in the previous section, we should know the sample mean and covariance matrix values for each class based on the Bayes classifier type. Therefore,

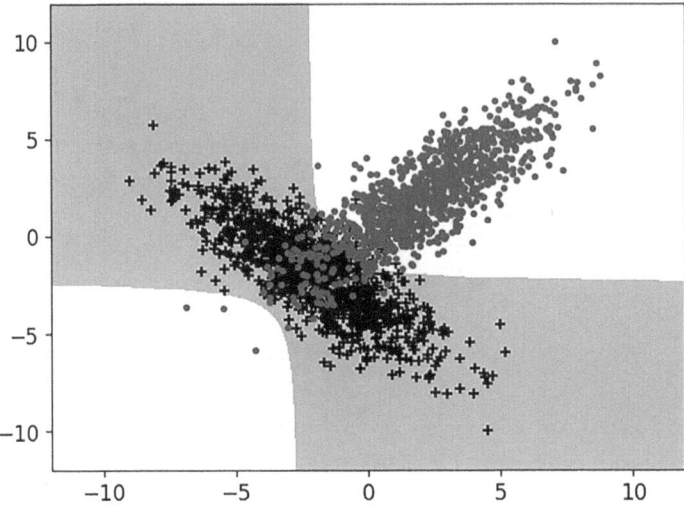

Fig. 6.4 Decision boundary formed by the Bayes classifier, where Σ_1 and Σ_2 are arbitrary

training the classifier means calculating these parameters from the labeled training set. Let's provide an example on the training operation. In order to explain the topic better, let's obtain training samples for two classes as introduced in Fig. 6.1. To note here, this data is generated by the Python script in Listing 5.2. We can summarize both classes with parameters

$$\mu_1 = \begin{bmatrix} -2 \\ -2 \end{bmatrix} \qquad \mu_2 = \begin{bmatrix} 2 \\ 2 \end{bmatrix} \tag{6.16}$$

and

$$\Sigma_1 = \begin{bmatrix} 5.5 & -4.5 \\ -4.5 & 5.5 \end{bmatrix} \qquad \Sigma_2 = \begin{bmatrix} 5.5 & 4.5 \\ 4.5 & 5.5 \end{bmatrix} \tag{6.17}$$

We can form the Bayes classifier from the labeled training samples at hand using the sklearn2c library. We provide such an example in Listing 6.1. Here, we first load the training data and corresponding labels by the NumPy function load. Afterward, we form the Bayes classifier using the class BayesClassifier. This class has the parameter case to decide on the classifier type based on Eqs. 6.10, 6.11, and 6.12. In our example, we set it as case = 3. The formed classifier can be trained by the function train. This function also takes the optional parameter save_path to save the trained classifier to a file. Therefore, we feed the train_samples data with the label set train_labels and save_path parameter in our example. As the operation ends, the trained classifier will be saved as the "bayes_classifier.joblib" file under the folder "classification_models." We will use the saved model for inference next.

Listing 6.1 Training the Bayes classifier

```
import os.path as osp
import numpy as np
from sklearn2c import BayesClassifier

train_samples = np.load(osp.join("classification_data", "
    cls_train_samples.npy"))
train_labels = np.load(osp.join("classification_data", "
    cls_train_labels.npy"))

bayesian = BayesClassifier(case=3)
model_save_path = osp.join("classification_models", "
    bayes_classifier.joblib")
bayesian.train(train_samples, train_labels, save_path=
    model_save_path)
```

In order to apply inference on the saved classifier in Listing 6.1, we should load the test data and then proceed with inference. We provide such an example in Listing 6.2. Here, the `test_samples` data and `test_labels` are loaded. Then, the classifier is loaded by the function `load` using the saved model path as parameter. Afterward, the Bayes classifier predicts the class of `test_samples` using the function `predict`.

Listing 6.2 Inference with the Bayes classifier in Python

```
import os.path as osp
import numpy as np
from sklearn2c import BayesClassifier

test_samples = np.load(osp.join("classification_data","
    cls_test_samples.npy"))
test_labels = np.load(osp.join("classification_data","
    cls_test_labels.npy"))

bayes_classifier = BayesClassifier.load(osp.join("
    classification_models", "bayes_classifier.joblib"))
likelihood = bayes_classifier.predict(test_samples)
print(likelihood)
```

6.2.3 Deploying the Trained Classifier to the Microcontroller

We trained the Bayes classifier by given data in Python on PC and saved the trained classifier in the previous section. The next phase is deploying it to the STM32 microcontroller such that we can use the trained classifier in an actual classification scenario. Therefore, we should generate the C header and source files of the trained classifier on PC. To do so, we should first load the trained model using the function `load`. We should generate the C header and source files automatically by the `BayesExporter` class function `export`. In other words, this function automatically generates the C header file that declares configurations of our Bayes classifier. It also generates the C source file which stores the information needed to run inference on the microcontroller. We provide a sample Python script to export the classifier parameters in Listing 6.3.

Listing 6.3 Exporting the Bayes classifier

```
import os.path as osp
from sklearn2c import BayesClassifier

model_path = osp.join("classification_models", "bayes_classifier.
    joblib")
export_path = osp.join("exported_models", "classification", "
    bayes_cls_config")

bayesian = BayesClassifier.load(model_path)
bayesian.export(export_path)
```

In Listing 6.3, the function export generates the C header, "bayes_cls_config.h," and source file, "bayes_cls_config.c." It then saves them to the given export path. Hence, the classifier can be deployed to the microcontroller. The generated C header file for the example considered in this chapter is given in Listing 6.4. This file only keeps variable names and their size. It will be used by other files to include the corresponding source file given in Listing 6.5. All variables declared in Listing 6.4 are assigned here.

Listing 6.4 Header file for the Bayes classifier configuration

```
#ifndef BAYES_CLS_CONFIG_H_INCLUDED
#define BAYES_CLS_CONFIG_H_INCLUDED
#define NUM_CLASSES 2
#define NUM_FEATURES 2
#define CASE 3
extern float MEANS[NUM_CLASSES][NUM_FEATURES];
extern const float CLASS_PRIORS[NUM_CLASSES];
extern const float INV_COVS[NUM_CLASSES][NUM_FEATURES][
    NUM_FEATURES];
extern const float DETS[NUM_CLASSES];
#endif
```

Listing 6.5 Source file for the Bayes classifier configuration

```
#include "bayes_cls_config.h"
float MEANS[NUM_CLASSES][NUM_FEATURES] =
    {{-2.04824109,-1.90133108}, { 2.00999565, 1.99583551}};
const float CLASS_PRIORS[NUM_CLASSES] = {0.50375,0.49625};
const float INV_COVS[NUM_CLASSES][NUM_FEATURES][NUM_FEATURES] =
    {{{ 0.54067245, 0.42838192},{ 0.42838192, 0.54200413}},...};
const float DETS[NUM_CLASSES] = { 9.12944908,10.42477877};
```

6.2.3.1 STM32CubeIDE Project Settings for Deployment

In order to deploy the trained classifier to the microcontroller, we should copy the generated header and source files to the "Core/Inc" and "Core/Src" folders of the STM32CubeIDE project, respectively. Then, the project explorer window should be as in Fig. 6.5.

Besides the generated "bayes_cls_config.h" and "bayes_cls_config.c" files, we should also use the "bayes_cls_inference.h" and "bayes_cls_inference.c" files in our project. In these files, the function bayes_cls_predict is defined to implement the Bayes classifier using the generated model parameters. This function makes use of the CMSIS DSP library for matrix operations. Therefore, we should include this library to the project to run inference.

To include the CMSIS DSP library to the STM32CubeIDE project, we should create a new folder under the "Drivers/CMSIS" folder and call it as "DSP." Then, we should to go to the path where the firmware installation repository of the STM32CubeIDE is. We should copy the "Include" folder under "STM32Cube_FW_F7_VX.X.X/Drivers/CMSIS/DSP/" and paste it to the "Drivers/CMSIS/DSP/" folder. We should also copy the "Lib" folder under "STM32Cube_FW_F7_VX.X.X/Drivers/CMSIS/" and paste it to the "Drivers/CMSIS" folder. Hence, the file organization for the CMSIS DSP library will be as in Fig. 6.6.

Fig. 6.5 STM32CubeIDE project setup

```
∨  F746NG_CLASSIFIERS_BAYES
   >  Binaries
   >  Includes
   ∨  Core
      ∨  Inc
         >  bayes_cls_config.h
         >  bayes_cls_inference.h
         >  lib_rng.h
         >  lib_serial.h
         >  main.h
         >  stm32f7xx_hal_conf.h
         >  stm32f7xx_it.h
      ∨  Src
         >  bayes_cls_config.c
         >  bayes_cls_inference.c
         >  lib_rng.c
         >  lib_serial.c
         >  main.c
         >  stm32f7xx_hal_msp.c
         >  stm32f7xx_it.c
         >  syscalls.c
         >  sysmem.c
         >  system stm32f7xx.c
```

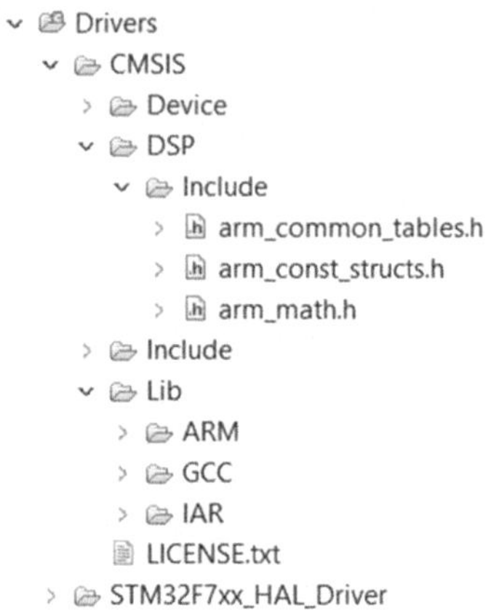

Fig. 6.6 File organization
for the CMSIS DSP library

```
✓ 🖴 Drivers
    ✓ 🗁 CMSIS
        > 🗁 Device
        ✓ 🗁 DSP
            ✓ 🗁 Include
                > 🖹 arm_common_tables.h
                > 🖹 arm_const_structs.h
                > 🖹 arm_math.h
        > 🗁 Include
        ✓ 🗁 Lib
            > 🗁 ARM
            > 🗁 GCC
            > 🗁 IAR
              🖹 LICENSE.txt
    > 🗁 STM32F7xx_HAL_Driver
```

The next step in operation is configuring the project settings for copied folders. Therefore, we should first include the path "/${ProjName}/Drivers/CMSIS/DSP/ Include" to "Include directories" under "Properties − > C/C++ General − > Paths and Symbols − > Includes" of the "Properties" window in STM32CubeIDE. Then, we should go to the "Symbols" tab and add the symbol "ARM_MATH_CM7." Afterward, we should go to the "Properties − > C/C++ Build − > Settings − > Tool Settings − > MCU CGC Linker" and add the file "libarm_cortexM7lfsp_math.a" under the section "Libraries (-l)" by clicking on the green file add icon. Under the "Library search path (-L)" section of the same window, we should add the path "${workspace_loc:/${ProjName}/Drivers/CMSIS/Lib/GCC}" by clicking the same icon. Then, we should click on "Apply and Close" button to apply the configurations to the project.

6.2.3.2 Mbed Studio Project Settings for Deployment

In order to deploy the Bayes classifier to the microcontroller, we should copy the generated header and source files to the root folder of the Mbed Studio project. Besides the generated "bayes_cls_config.h" and "bayes_cls_config.c" files, we should also copy the "bayes_cls_inference.h" and "bayes_cls_inference.c" files to our project root folder. As mentioned in previous section, the function bayes_cls_predict is defined in these files to implement the Bayes classifier using the generated model parameters. This function makes use of the CMSIS DSP library for matrix operations. Therefore, we should include the CMSIS DSP library to the project to run inference.

To compile the CMSIS DSP library, we should change the default Mbed Studio toolchain to GCC_ARM. To do so, we should follow the steps given in Sect. A.3. Then, we should copy the CMSIS DSP library files to our Mbed Studio project. We can create a new folder under our project and name it as "DSP" for copying the header files. We should also copy the "Include" folder to the created DSP folder. The "Include" folder can be found under the path "STM32Cube_FW_F7_VX.XX.X/Drivers/CMSIS/DSP/" in the STM32CubeIDE repository or in STM32CubeF7 GitHub repository of STMicroelectronics. This folder contains three header files as "arm_common_tables.h," "arm_const_structs.h," and "arm_math.h." Then, we should copy the compiled library file "libarm_cortexM7l_math.a" to the root folder of our project. This file can also be found in the STM32CubeIDE repository or in STM32CubeF7 GitHub repository of STMicroelectronics under path "STM32Cube_FW_F7_VX.XX.X/Drivers/CMSIS/Lib/GCC."

6.2.4 Testing the Deployed Classifier on the Microcontroller

To check whether the deployed classifier works as expected, we can use the generated data on PC. Then, we can send it to the STM32 microcontroller. The classifier running on the microcontroller works on this data. Then, we can transfer the classification result back to PC and cross check with the same classifier working in Python on PC. Hence, we can make sure that the generated framework works as expected.

6.2.4.1 STM32CubeIDE Project Settings for Testing

In order to test the deployed classifier on the STM32 microcontroller, we can form a project under STM32CubeIDE. Then, we should add the libraries lib_rng and lib_serial to the project as explained in Sects. 5.1.3.1 and 4.1.2.2, respectively. Afterward, we should add the generated files in the previous section. Hence, we will have the corresponding code block as follows.

```
/* USER CODE BEGIN Includes */
#include "bayes_cls_inference.h"
#include "lib_serial.h"
#include "lib_rng.h"
/* USER CODE END Includes */
```

We should define macros related to the source of the input data, size of the input vector, and size of the output vector. The INPUT macro definition selects whether the input data is obtained from PC or microcontroller at compile time. It can be selected as INPUT_PC or INPUT_MCU, respectively. SIZE_INPUT and SIZE_OUTPUT macros define the number of data inputs and outputs of the model, respectively. We should also define two floating-point arrays to allocate required memory space for inputs and outputs. We define two matrix instances of the structure arm_matrix_instance_f32 to interpret these arrays as matrices. We assign fields in these structures with the required information as the number of columns, number of rows, and memory address of data as follows.

```
/* USER CODE BEGIN 0 */
#define INPUT_PC 1
#define INPUT_MCU 2
#define INPUT INPUT_PC
```

```
#define SIZE_INPUT NUM_FEATURES
#define SIZE_OUTPUT NUM_CLASSES

float input[SIZE_INPUT];
float output[SIZE_OUTPUT];

arm_matrix_instance_f32 mat_input = {.numCols = 1, .numRows=
    NUM_FEATURES, .pData=input};
arm_matrix_instance_f32 mat_output = {.numCols = NUM_CLASSES, .
    numRows=1, .pData=output};
/* USER CODE END 0 */
```

Next, we should form a while loop to acquire data either from PC or microcontroller according to the INPUT macro. Then, we can run the Bayes classifier by calling the function bayes_cls_predict. The first parameter of this function is address of the input matrix structure. The second parameter is the address of the output matrix structure. Finally, we transfer the classification result to PC. We provide the C code corresponding to these definitions in Listing 6.6. Please note that the Python script in Listing 6.7 should already be running on PC for executing this example successfully.

Listing 6.6 While loop for the Bayes classifier

```
/* USER CODE BEGIN WHILE */
while (1)
{
    /* USER CODE END WHILE */

    /* USER CODE BEGIN 3 */
#if (INPUT == INPUT_PC)
        LIB_SERIAL_Receive(input, SIZE_INPUT, TYPE_F32);
#elif (INPUT == INPUT_MCU)
        for (uint32_t i = 0; i < SIZE_INPUT; ++i)
                input[i] = (float)(LIB_RNG_GetRandomNumber() %
                    1000) / 1000.0f;
        LIB_SERIAL_Transmit(input, SIZE_INPUT, TYPE_F32);
#endif
        bayes_cls_predict(&mat_input, &mat_output);
        LIB_SERIAL_Transmit(output, SIZE_OUTPUT, TYPE_F32);
        HAL_Delay(1000);
}
/* USER CODE END 3 */
```

To check whether the deployed classifier works as expected, we need a test setup. We provide the Python script formed for this purpose in Listing 6.7. Here, we first load the previously generated test data. Then, we load our trained Bayes classifier. Afterward, we send the test data to the microcontroller. Next, we perform inference on the PC and microcontroller separately. We finally print both inference results on PC for comparison. Hence, we can make sure that the generated framework works as expected.

Listing 6.7 Python script for comparing Bayes inference results

```
import os.path as osp
import numpy as np
from sklearn2c import BayesClassifier
import py_serial

py_serial.SERIAL_Init("COM4")

test_samples = np.load(osp.join("classification_data","
    cls_test_samples.npy"))
test_labels = np.load(osp.join("classification_data","
    cls_test_labels.npy"))

bayesian = BayesClassifier.load(osp.join("classification_models",
    "bayes_classifier.joblib"))

i = 0
while 1:
    rqType, datalength, dataType = py_serial.
        SERIAL_PollForRequest()
    if rqType == py_serial.MCU_WRITES:
        # INPUT -> FROM MCU TO PC
        inputs = py_serial.SERIAL_Read()

    elif rqType == py_serial.MCU_READS:
        # INPUT -> FROM PC TO MCU
        inputs = test_samples[i:i+1].astype(py_serial.
            SERIAL_GetDType(dataType))
        i = i + 1
        if i >= len(test_samples):
            i = 0
        py_serial.SERIAL_Write(inputs)

    pcout = bayesian.predict(np.reshape(inputs, (1, datalength)))
    rqType, datalength, dataType = py_serial.
        SERIAL_PollForRequest()
    if rqType == py_serial.MCU_WRITES:
        mcuout = py_serial.SERIAL_Read()
        print()
        print("Inputs : " + str(inputs))
        print("PC Output : " + str(pcout))
        print("MCU Output : " + str(mcuout))
        print()
```

6.2.4.2 Mbed Studio Project Settings for Testing

In order to test the deployed Bayes classifier on the STM32 microcontroller, we should add the libraries lib_rng and lib_serial to the Mbed Studio project. To do so, we should copy the files "lib_rng.h," "lib_rng.c," "lib_uart.h," "lib_uart.c," "lib_serial.h," and "lib_serial.c" to our Mbed Studio project folder. Here, we also assume that the steps in Sect. 6.2.3.2 are already done.

The test code for the Mbed Studio project is given in Listing 6.8. This code is similar to the one given for the STM32CubeIDE project. Hence, the explanations there hold here as well. The only difference is that we should include the header file "lib_uart.h" and initialize the UART peripheral unit before the while loop by code in our Mbed Studio project. Please note that the Python script in Listing 6.7 should already be running on PC for executing this example successfully.

Listing 6.8 Bayes classifier test code for Mbed Studio

```
#include "mbed.h"
#include "lib_serial.h"
#include "lib_rng.h"
#include "lib_uart.h"
#include "bayes_cls_inference.h"

#define INPUT_PC 1
#define INPUT_MCU 2
#define INPUT INPUT_PC

#define SIZE_INPUT NUM_FEATURES
#define SIZE_OUTPUT NUM_CLASSES

float input[SIZE_INPUT];
float output[SIZE_OUTPUT];

arm_matrix_instance_f32 mat_input = {.numRows= NUM_FEATURES, .
    numCols = 1,   .pData=input};
arm_matrix_instance_f32 mat_output = {.numRows=1, .numCols =
    NUM_CLASSES,   .pData=output};

int main()
{
    LIB_UART_Init();
#if (INPUT == INPUT_MCU)
    LIB_RNG_Init();
#endif
    while (true)
    {
#if (INPUT == INPUT_PC)
        LIB_SERIAL_Receive(input, SIZE_INPUT, TYPE_F32);
#elif (INPUT == INPUT_MCU)
        for (uint32_t i = 0; i < SIZE_INPUT; ++i)
            input[i] = (float)(LIB_RNG_GetRandomNumber() % 1000)
                / 1000.0f;
        LIB_SERIAL_Transmit(input, SIZE_INPUT, TYPE_F32);
#endif
        bayes_cls_predict(&mat_input, &mat_output);
        LIB_SERIAL_Transmit(output, SIZE_OUTPUT, TYPE_F32);
    HAL_Delay(1000);
    }
}
```

6.3 k-Nearest Neighbor Classifier

k-nearest neighbor (kNN) is the second classifier to be considered in this chapter. It falls into the nonparametric classifier group. We will explain why this is the case in the theoretical background section. Although training the kNN classifier is straightforward, we devote a separate section on this topic for completeness. Then, we will focus on deploying the kNN classifier to the STM32 microcontroller and testing it there.

6.3.1 Theoretical Background

We used a known pdf (Gaussian most of the times) to model likelihood values in the Bayes classifier. This approach simplifies training step of the classifier such that only a few parameters are estimated for this purpose. Besides, inference is based on simple distance calculations between the unknown sample to be classified and class sample mean and covariance values. This approach works well as long as the selected pdf models the actual data distribution in class samples. When this modeling does not work, it will have negative effect on the inference step.

To overcome limitations of the predefined pdf usage in Bayes classifier, another approach is proposed in the literature. Here, class label assignment via maximum posterior probability, as in the Bayes classifier, still holds. However, the posterior pdf is formed by the normalized histogram, introduced in Sect. 5.1.4, directly from the available class training samples. Therefore, if there are M classes, then M normalized histogram functions are formed as posterior pdf. This is the training step of the classifier. In the inference step, the unknown sample posterior probability value is calculated for each class based on the formed pdfs. Then, the unknown sample is assigned to the class with the highest posterior probability value. This is called the Parzen window approach in the literature. We pick the fixed bin size in the normalized histogram formation in this method. As explained in Sect. 5.1.4, this may cause problems. In other words, the bin size may not be set correctly for a given data distribution. Picking a large bin size may lead to crude histogram formation. Picking a small bin size may lead to sparse histogram formation. Hence, the classifier based on this histogram may not work as expected.

To overcome the problems when a fixed bin size is used in normalized histogram formation, another approach is proposed in the literature. Here, the number of samples is fixed for each bin. Hence, the bin size is set adaptively while forming the histogram. The kNN classifier is based on this approach. It is further refined such that the histogram formation step is eliminated in operation. Hence, the training and inference steps are based on training samples directly. To be more specific, training step of the kNN classifier corresponds to keeping training samples from all classes. The inference step is based on finding the closest training samples to the unknown sample to be classified. Then, we assign the unknown sample to the class with maximum number of samples in the closest k neighbor. Hence, the name of the classifier is kNN. There is also a special case for the kNN classifier when the closest neighbor is considered in the inference step. This is called the nearest neighbor (NN) classifier. We assign the unknown sample to the class with the nearest training sample in this classifier.

We can explain working principles of the kNN classifier based on a two-class classification problem. Assume that we have a total of L training samples from two classes C_1 and C_2. When an unknown sample is to be classified, we calculate its distance to the closest k training samples. Hence, we have L distance values as $d_1, d_2, \cdots, d_L$. Then, we sort these values from smallest to largest. Assume that we set k=3. We pick the smallest three distance values as d_a, d_b, d_c. We assign the unknown sample to the class C_1 if the class labels in samples a, b, c have majority (two here) of C_1 value. Otherwise, we assign the unknown sample to the class C_2.

Since distances can be calculated in any dimension, the kNN classifier can be implemented for feature spaces with any dimension. The classifier can also be extended to M classes. The important issue here is that a class should always have majority when the neighbor number is set. As an example, the k value should be odd when we have a two-class problem. When we have a three-class problem, the k value cannot be multiple of three. Hence, the majority in favor of one class will always be satisfied. The NN classifier does not have such constraints in M class classification problem, since only the nearest neighbor is considered in operation.

Let's obtain the decision boundary formed by the NN and kNN classifiers on the two-dimensional random dataset given in Fig. 6.1. To note here, this data is generated by the Python script given in Listing 5.2. Assume that we pick these as training samples for the NN classifier. We can check the feature space at regular intervals and form the corresponding decision boundary as in Fig. 6.7.

As can be seen in Fig. 6.7, the model decides on the output label based on a single sample. The decision boundary represents the closest training sample at each point. Since we pick the NN classifier, it tends to overfit. We can increase the number of neighbors for the model to make better predictions. On the other hand, increasing the number of neighbors more than desired may cause the classifier to underfit. Hence, the classifier always tends to predict the class with highest population. Let's try the above example and form the decision boundary for the kNN classifier with $k = 3$. We provide the formed decision boundary in Fig. 6.8.

As can be seen in Fig. 6.8, we observe less zigzag between two classes compared to the NN case. Hence, the kNN classifier is able to make better predictions, since prediction is made upon a more crowded population. We should add few comments

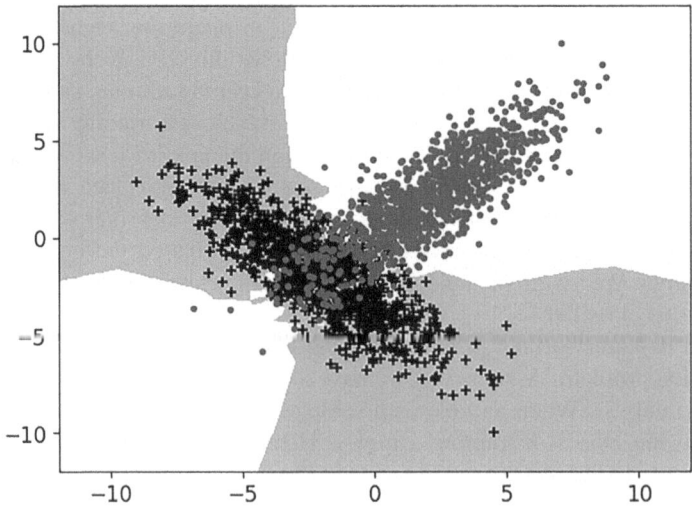

Fig. 6.7 Decision boundary formed by the NN classifier

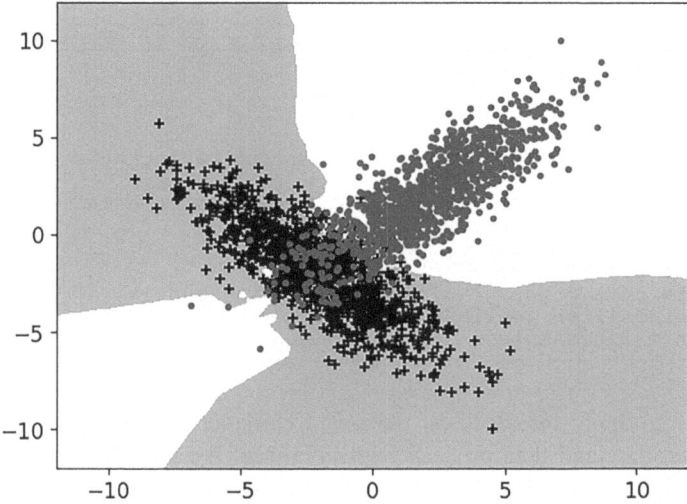

Fig. 6.8 Decision boundary formed by the kNN classifier

on the decision boundary formed by the kNN classifier. As the neighborhood number k is taken a small value, the decision boundary will be more detailed. When the k value becomes bigger, the decision boundary becomes more crude. Therefore, the reader should pick a suitable k value for the problem at hand. As mentioned previously, the decision boundary cannot be represented by an explicit function since kNN is a nonparametric classifier. Hence, it does not have the generalization property. Although the decision boundary is detailed when there is training data around, it becomes less useful as we move away from training samples in the feature space. In fact, the decision boundary only depends on closest training samples in the outer boundary.

6.3.2   Training the Classifier in Python

As in the Bayes classifier, we can form the kNN classifier from labeled training samples at hand using the sklearn2c library. We provide one such example in Listing 6.9. Here, we first load the training data and corresponding labels by the NumPy function load. Afterward, we form the kNN classifier using the class KNNClassifier. This class has the parameter n_neighbors to decide on the neighborhood number in forming the classifier. In our example, we set it as n_neighbors = 5. The formed classifier can be trained by the function train. This function also takes the optional parameter save_path to save the trained classifier to a file. Therefore, we feed the train_samples data with the label set train_labels and save_path parameter in our example. As the operation ends, the trained classifier will be saved as the "knn_classifier.joblib" file under the folder "classification_models."

Listing 6.9 Training the kNN classifier

```
import os.path as osp
import numpy as np
from sklearn2c import KNNClassifier

train_samples = np.load(osp.join("classification_data", "
    cls_train_samples.npy"))
train_labels = np.load(osp.join("classification_data", "
    cls_train_labels.npy"))

knn = KNNClassifier(n_neighbors = 5)
model_save_path = osp.join("classification_models", "
    knn_classifier.joblib")
knn.train(train_samples, train_labels, save_path=model_save_path)
```

In order to apply inference on the saved classifier in Listing 6.9, we should load the test data and then proceed with inference. We provide such an example in Listing 6.10. Here, the test_samples data and test_labels are loaded from the files "cls_test_samples.npy" and "cls_test_labels.npy," respectively. Then, the classifier is loaded by the function load using the saved model path as parameter. Afterward, the kNN classifier predicts the classification result of test_samples using the function predict.

Listing 6.10 Inference with the kNN classifier in Python

```
import os.path as osp
import numpy as np
from sklearn2c import KNNClassifier

test_samples = np.load(osp.join("classification_data","
    cls_test_samples.npy"))
test_labels = np.load(osp.join("classification_data","
    cls_test_labels.npy"))

knn_classifier = KNNClassifier.load(osp.join("
    classification_models", "knn_classifier.joblib"))
predictions = knn_classifier.predict(test_samples)
print(predictions)
```

6.3.3 Deploying the Trained Classifier to the Microcontroller

As explained in previous sections, the kNN classifier needs all training samples to operate. Therefore, deploying the trained classifier to the microcontroller means transferring all training data there. We have 400 training samples per class with two features per sample in the example considered in this chapter. There are two classes in our case. Hence, we should transfer 800 training samples to run inference on the microcontroller. The kNNExporter class allows us to generate C header and source files which are ready to be deployed to the microcontroller. Python script to export the model parameters is given in Listing 6.11.

Listing 6.11 Exporting the kNN classifier

```
import os.path as osp
from sklearn2c import KNNClassifier

model_path = osp.join("classification_models","knn_classifier.
    joblib")
export_path = osp.join("exported_models","classification","
    knn_cls_config")
knn = KNNClassifier.load(model_path)
knn.export(export_path)
```

As we execute the Python script in Listing 6.11, we form the header and source files. The generated C header file for the example considered in this chapter is given in Listing 6.12. This file only keeps the variable names and their size. It will be used by other files to include the corresponding variables from the source file given in Listing 6.13. All variables declared in Listing 6.12 are assigned here.

Listing 6.12 Header file for the kNN classifier configuration

```
#ifndef KNN_CLS_CONFIG_H_INCLUDED
#define KNN_CLS_CONFIG_H_INCLUDED
#define NUM_CLASSES 2
#define NUM_NEIGHBORS 5
#define NUM_FEATURES 2
#define NUM_SAMPLES 1600
extern char* LABELS[NUM_CLASSES];
extern const float DATA[NUM_SAMPLES][NUM_FEATURES];
extern const int DATA_LABELS[NUM_SAMPLES];
#endif
```

Listing 6.13 Source file for the kNN classifier configuration, shortened version

```
#include "knn_cls_config.h"
char* LABELS[NUM_CLASSES] = {0,1};
const float DATA[NUM_SAMPLES][NUM_FEATURES] = {{-2.70255383e
    +00,-4.72532919e+00},...};
const int DATA_LABELS[NUM_SAMPLES] = {0,...};
```

6.3.3.1 STM32CubeIDE Project Settings for Deployment

Now, we have all the parameters required to implement the kNN classifier on the microcontroller. To deploy these parameters to the microcontroller, we should first copy the generated "knn_cls_config.h" and "knn_cls_config.c" files and paste them to the "Core/Inc" and "Core/Src" folders of our STM32CubeIDE project, respectively. Afterward, we should copy the "knn_cls_inference.h" and "knn_cls_inference.c" files from the sklearn2c library and paste them to the "Core/Inc" and "Core/Src" folders of our STM32CubeIDE project, respectively.

6.3.3.2 Mbed Studio Project Settings for Deployment

In order to deploy the kNN classifier to the microcontroller using Mbed Studio, we should copy the generated "knn_cls_config.h" and "knn_cls_config.c" files to the root folder of our project. Afterward, we should copy the "knn_cls_inference.h" and "knn_cls_inference.c" files from the sklearn2c library to our Mbed Studio project. Then, we will be ready to use the classifier on the microcontroller.

6.3.4   Testing the Deployed Classifier on the Microcontroller

To check whether the deployed classifier works as expected, we can use the generated data on PC. Then, we can send it to the STM32 microcontroller. The classifier running on the microcontroller works on this data. Then, we can transfer the classification result back to PC and cross check with the same classifier working in Python on PC. Hence, we can make sure that the generated framework works as expected.

6.3.4.1 STM32CubeIDE Project Settings for Testing

In order to test the deployed classifier on the STM32 microcontroller, we can use the test setup formed for the Bayes classifier in Sect. 6.2.4. We should add the libraries lib_rng and lib_serial to the project as explained in Sects. 5.1.3.1 and 4.1.2.2, respectively. Afterward, we should add the generated files in the previous section to our project. Hence, we will have the corresponding code block as follows.

```
/* USER CODE BEGIN Includes */
#include "lib_serial.h"
#include "lib_rng.h"
#include "knn_cls_inference.h"
/* USER CODE END Includes */
```

We should define macros related to the source of the input data, size of the input vector, and size of the output vector. The INPUT macro definition selects whether the input data is obtained from PC or microcontroller at compile time. It can be selected as INPUT_PC or INPUT_MCU, respectively. SIZE_INPUT and SIZE_OUTPUT macros define the number of data inputs and outputs of the model, respectively. We should define a floating-point array, with size equal to the number of features, for storing the input data. We should also define an integer array, with size equal to the number of classes, for storing output data. Hence, we will have the corresponding code block as follows.

```
/* USER CODE BEGIN 0 */
#define INPUT_PC 1
#define INPUT_MCU 2
#define INPUT    INPUT_PC

#define SIZE_INPUT NUM_FEATURES
#define SIZE_OUTPUT NUM_CLASSES
```

```
float input[SIZE_INPUT];
int output[SIZE_OUTPUT];

/* USER CODE END 0 */
```

Next, we should form a while loop to acquire data either from PC or microcontroller according to the INPUT macro. Then, we can run the kNN classifier by calling the function knn_cls_predict. The first parameter of this function is a pointer to the input array. The second parameter is a pointer to the output array. This output array is filled by the function knn_cls_predict with kNN inference results. Finally, we transfer the classification result to PC. We provide the C code corresponding to these definitions in Listing 6.14. Please note that the Python script in Listing 6.15 should already be running on PC for executing this example successfully.

Listing 6.14 While loop for the kNN classifier

```
/* USER CODE BEGIN WHILE */
while (1)
{
/* USER CODE END WHILE */

/* USER CODE BEGIN 3 */
#if (INPUT == INPUT_PC)
        LIB_SERIAL_Receive(input, SIZE_INPUT, TYPE_F32);
#elif (INPUT == INPUT_MCU)
        for (uint32_t i = 0; i < SIZE_INPUT; ++i)
        input[i] = (float)(LIB_RNG_GetRandomNumber() % 1000) /
            1000.0f;

        LIB_SERIAL_Transmit(input, SIZE_INPUT, TYPE_F32);
#endif

knn_cls_predict(input, output);

LIB_SERIAL_Transmit((void*)&output, SIZE_OUTPUT, TYPE_S32);

HAL_Delay(1000);
}
/* USER CODE END 3 */
```

To check whether the deployed classifier works as expected, we need a test setup. We provide the Python script formed for this purpose in Listing 6.15. Here, we first load the previously generated test data. Then, we load our trained kNN classifier. Afterward, we send the test data to the microcontroller. Next, we perform inference on the PC and microcontroller separately. We finally print both inference results on PC for comparison. Hence, we can make sure that the generated framework works as expected.

Listing 6.15 Python script for comparing kNN inference results

```
import os.path as osp
import numpy as np
```

```
from sklearn2c import KNNClassifier
import py_serial

py_serial.SERIAL_Init("COM4")

test_samples = np.load(osp.join("classification_data","
    cls_test_samples.npy"))
test_labels = np.load(osp.join("classification_data","
    cls_test_labels.npy"))

knn = KNNClassifier.load(osp.join("classification_models", "
    knn_classifier.joblib"))
i = 0

while 1:
    rqType, datalength, dataType = py_serial.
        SERIAL_PollForRequest()
    if rqType == py_serial.MCU_WRITES:
        # INPUT -> FROM MCU TO PC
        inputs = py_serial.SERIAL_Read()

    elif rqType == py_serial.MCU_READS:
        # INPUT -> FROM PC TO MCU
        inputs = test_samples[i:i+1].astype(py_serial.
            SERIAL_GetDType(dataType))
        i = i + 1
        if i >= len(test_samples):
            i = 0
        py_serial.SERIAL_Write(inputs)

    pcout = knn.predict(np.reshape(inputs, (1, datalength)))
    rqType, datalength, dataType = py_serial.
        SERIAL_PollForRequest()
    if rqType == py_serial.MCU_WRITES:
        mcuout = py_serial.SERIAL_Read()
        print()
        print("Inputs : " + str(inputs))
        print("PC Output : " + str(pcout))
        print("MCU Output : " + str(mcuout))
        print()
```

6.3.4.2 Mbed Studio Project Settings for Testing

In order to test the deployed kNN classifier on the STM32 microcontroller, we should add the libraries lib_rng and lib_serial to the Mbed Studio project. To do so, we should copy the files "lib_rng.h," "lib_rng.c," "lib_uart.h," "lib_uart.c," "lib_serial.h," and "lib_serial.c" to our Mbed Studio project folder. Here, we also assume that the steps in Sect. 6.3.3.2 are already done.

The test code for the Mbed Studio project is given in Listing 6.16. This code is similar to the one given for the STM32CubeIDE project. Hence, the explanations there hold here as well. The only difference is that we should include the header file "lib_uart.h" and initialize the UART peripheral before the while loop by code in our Mbed Studio project. Please note that the Python script in Listing 6.15 should already be running on PC for running this example successfully.

Listing 6.16 kNN classifier test code for Mbed Studio

```
#include "mbed.h"
#include "knn_cls_inference.h"
#include "lib_serial.h"
#include "lib_rng.h"
#include "lib_uart.h"

#define INPUT_PC 1
#define INPUT_MCU 2
#define INPUT INPUT_PC

#define SIZE_INPUT NUM_FEATURES
#define SIZE_OUTPUT NUM_CLASSES

float input[SIZE_INPUT];
int output[SIZE_OUTPUT];

int main()
{
    LIB_UART_Init();
#if (INPUT == INPUT_MCU)
    LIB_RNG_Init();
#endif
    while (true)
    {
#if (INPUT == INPUT_PC)
        LIB_SERIAL_Receive(input, SIZE_INPUT, TYPE_F32);
#elif (INPUT == INPUT_MCU)
        for (uint32_t i = 0; i < SIZE_INPUT; ++i)
        input[i] = (float)(LIB_RNG_GetRandomNumber() % 1000) /
            1000.0f;
        LIB_SERIAL_Transmit(input, SIZE_INPUT, TYPE_F32);
#endif
        knn_cls_predict(input, output);
        LIB_SERIAL_Transmit((void*)&output, SIZE_OUTPUT, TYPE_S32
            );
        HAL_Delay(1000);
    }
}
```

6.4 Support Vector Machine Classifier

Support vector machine (SVM) is the third classifier to be considered in this book. It adopts a different approach in forming the decision boundary. We will evaluate this in the theoretical background section next. Afterward, we will provide ways of training the SVM based classifier in Python on PC. This will be followed by deploying the trained classifier to the STM32 microcontroller and testing its performance there.

6.4.1 Theoretical Background

Let's explain the SVM classifier on a simple example by assuming a two-dimensional feature space $\mathbf{x} = [x_1, x_2]$ with two classes C_1 and C_2. We further assume that these two classes are linearly separable. We would like to form a decision boundary to separate C_1 and C_2 by a linear classifier. Hence, the decision boundary becomes $\mathbf{w}^T\mathbf{x} + w_0$ where $\mathbf{w} = [w_1, w_2]$ and w_0 are the classifier parameters to be obtained from the training set. Throughout the derivations, we use Alpaydın's [1] notation and his formalism of the classification problem.

Based on the definition of decision boundary, we will have $\mathbf{w}^T\mathbf{x}[l] + w_0 > 0$ for the sample $\mathbf{x}[l] = [x_1[l], x_2[l]] \in C_1$. Likewise, we will have $\mathbf{w}^T\mathbf{x}[l] + w_0 < 0$ for the sample $\mathbf{x}[l] \in C_2$. We can define the class index $r[l]$ for the training sample such that $r[l] = 1$ for $\mathbf{x}[l] \in C_1$ and $r[l] = -1$ for $\mathbf{x}[l] \in C_2$. Based on these definitions, we can write $r[l](\mathbf{w}^T\mathbf{x}[l] + w_0) > 0$. This inequality holds for all training samples with index $l = 1, \cdots, L$.

We can calculate the distance of the training sample $\mathbf{x}[l]$ to the decision boundary as

$$dist = \frac{|\mathbf{w}^T\mathbf{x}[l] + w_0|}{\|\mathbf{w}\|} \tag{6.18}$$

We know that $\mathbf{w}^T\mathbf{x}[l] + w_0$ is a number. Hence, its absolute value can be represented as $r[l](\mathbf{w}^T\mathbf{x}[l] + w_0)$. Therefore, we can represent Eq. 6.18 as

$$dist = \frac{r[l](\mathbf{w}^T\mathbf{x}[l] + w_0)}{\|\mathbf{w}\|} \tag{6.19}$$

Since we have a linearly separable case, this distance will have at least a positive value for all training samples as

$$\frac{r[l](\mathbf{w}^T\mathbf{x}[l] + w_0)}{\|\mathbf{w}\|} > \rho \tag{6.20}$$

As suggested by Duda *et al.* [7], ρ should be maximized. They also impose the constraint $\rho\|\mathbf{w}\| = 1$ to ensure unique parameters $\mathbf{w}$ and w_0. Therefore, maximizing ρ leads to minimizing $\|\mathbf{w}\|$. As mentioned in Alpaydın [1], we can formalize this as an optimization problem. Solution of the problem gives classifier parameters $\mathbf{w}$ and w_0. In this setting, the margin between the classes will be $2/\|\mathbf{w}\|$ based on the formed decision boundary. Moreover, there will be training samples on both sides of the decision boundary $1/\|\mathbf{w}\|$ away from it. These are support vectors. We can also say that support vectors are the training samples closest to the decision boundary. The derivation is also valid for the N-dimensional feature space. There, the decision boundary will be a hyperplane.

Based on the derivations given in previous paragraphs, finding a decision boundary between two classes is possible as long as they are linearly separable.

If this is not the case, then SVM proposes using functions (called kernels) to map the feature space to a higher dimension such that there exists a separating hyperplane there. This is called the kernel trick. Generally, radial basis function (RBF), polynomial, and sigmoid kernels are used for this purpose.

We should mention that the SVM classifier can be formed only for two classes. Therefore, if more than two classes are to be separated, then a hierarchical approach should be applied. There are mainly two strategies for adapting SVMs to the multi-class classification problem. The first one is the one-to-rest approach. The second one is the one-to-one approach. In both approaches, we consider the multi-class problem as a binary problem. In the one-to-one approach, we break down multiple classes and obtain linear separation for each class pair. In the one-to-rest approach, we pick one class and consider the rest as another class. Assume that we have M classes. We should form M SVM classifiers in the one-to-rest approach. In the one-to-one approach, we should form $M(M-1)/2$ SVM classifiers. Therefore, we obtain multiple binary classification problems in both approaches.

As the SVM is trained, the inference step is straightforward. Let's pick the two-class classification problem. The value of the unknown sample is checked, whether being positive or negative, based on its location to the decision boundary. Then, the class label is assigned accordingly. For the M class case, successive classifiers should be tested to reach the final class label.

Let's obtain the decision boundary formed by the SVM classifier with RBF kernel on the two-dimensional random dataset given in Fig. 6.1. To note here, this data is generated by the Python script given in Listing 5.2. We provide the formed decision boundary in Fig. 6.9. As can be seen in this figure, the SVM classifier smoothly separates two classes. Please note that our decision boundary is not linear due to the kernel trick.

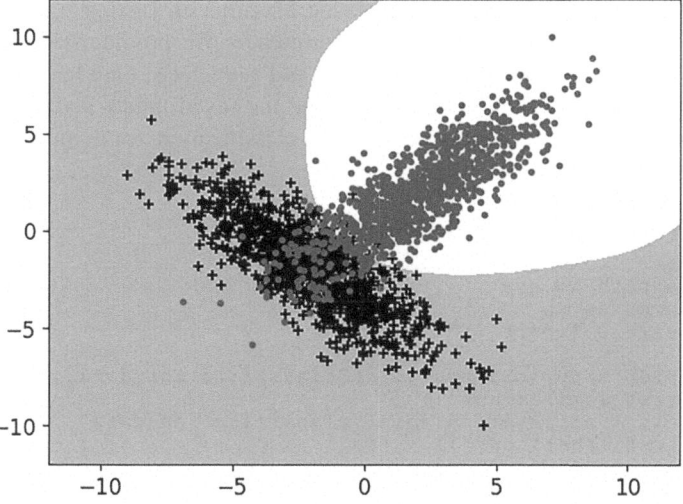

Fig. 6.9 Decision boundary formed by the SVM classifier

6.4.2   Training the Classifier in Python

As in the Bayes classifier, we can form the SVM classifier from the labeled training samples at hand using the sklearn2c library. We provide one such example in Listing 6.17. Here, we first load the training data and corresponding labels by the NumPy function `load`. Afterward, we form the SVM classifier using the class `SVMClassifier`. This class has several parameters in forming the classifier [30]. We use default values for the parameters in our example. The formed classifier can be trained by the function `train`. This function also takes the optional parameter `save_path` to save the trained classifier to a file. Therefore, we feed the `train_samples` data with the label set `train_labels` and `save_path` parameter in our example. As the operation ends, the trained classifier will be saved as the "SVM_classifier.joblib" file under the folder "classification_models."

Listing 6.17 Training the SVM classifier

```
import os.path as osp
import numpy as np
from sklearn2c import SVMClassifier

train_samples = np.load(osp.join("classification_data", "
    cls_train_samples.npy"))
train_labels = np.load(osp.join("classification_data", "
    cls_train_labels.npy"))

svm = SVMClassifier()
model_save_path = osp.join("classification_models", "
    svm_classifier.joblib")
svm.train(train_samples, train_labels, save_path= model_save_path
    )
```

In order to apply inference on the saved classifier in Listing 6.17, we should load the test data and then proceed with inference. We provide such an example in Listing 6.18. Here, the `test_samples` data and `test_labels` are loaded. Then, the classifier is loaded by the function `load` using the saved model path as parameter. Afterward, the SVM classifier predicts the classification result of `test_samples` using the function `predict`.

Listing 6.18 Inference with the SVM classifier in Python

```
import os.path as osp
import numpy as np
from sklearn2c import SVMClassifier

test_samples = np.load(osp.join("classification_data","
    cls_test_samples.npy"))
test_labels = np.load(osp.join("classification_data","
    cls_test_labels.npy"))

svc = SVMClassifier.load(osp.join("classification_models", "
    svm_classifier.joblib"))
predictions = svc.predict(test_samples)
print(predictions)
```

6.4.3   Deploying the Trained Classifier to the Microcontroller

We trained the SVM classifier by given data in Python on PC and saved the classifier in the previous section. The next phase is deploying it to the STM32 microcontroller such that we can use the classifier in an actual classification scenario. Therefore, we should generate the C header and source files of the classifier on PC. To do so, we should first load the trained model using the function load. We should generate the C header and source files automatically by the SVMExporter class function export. In other words, this function automatically generates the C header file that declares configurations of our SVM classifier. It also generates the C source file which stores the information needed to run inference on the microcontroller. We provide a sample Python script to export the classifier parameters in Listing 6.19.

Listing 6.19 Exporting the SVM classifier

```python
import os.path as osp
from sklearn2c import SVMClassifier

model_path = osp.join("classification_models","svm_classifier.
    joblib")
export_path = osp.join("exported_models","classification","
    svm_cls_config")

svm = SVMClassifier.load(model_path)
svm.export(export_path)
```

In Listing 6.19, the function export generates the C header, "svm_cls_config.h," and source file, "svm_cls_config.c." It then saves them to the given export path. Hence, the classifier can be deployed to the microcontroller. The generated C header file for the example considered in this chapter is given in Listing 6.20. This file only keeps variable names and their size. It will be used by other files to include the corresponding source file given in Listing 6.21. All variables declared in Listing 6.20 are assigned here.

Listing 6.20 Header file for the SVM classifier configuration

```c
#ifndef SVM_CLS_CONFIG_H_INCLUDED
#define SVM_CLS_CONFIG_H_INCLUDED
#define NUM_CLASSES 2
#define NUM_INTERCEPTS 1
#define NUM_FEATURES 2
#define NUM_SV 293
enum KernelType{
        LINEAR,
        POLY,
        RBF
};
extern const float coeffs[NUM_CLASSES - 1][NUM_SV];
extern const float SV[NUM_SV][NUM_FEATURES];
extern const float intercepts[NUM_INTERCEPTS];
extern const float w_sum[NUM_CLASSES + 1];
extern const float svm_gamma;
```

```
extern const float coef0;
extern const int degree;
extern const enum KernelType type;
#endif
```

Listing 6.21 Source file for the SVM classifier configuration, shortened version

```
#include "svm_cls_config.h"
const float coeffs[NUM_CLASSES - 1][NUM_SV] = {{-1.,...}};
const float SV[NUM_SV][NUM_FEATURES] = {{-2.70255383e
    +00,-4.72532919e+00},...};
const float intercepts[NUM_INTERCEPTS] = {-0.00951688};
const float w_sum[NUM_CLASSES + 1] = {  0,146,293};
const float svm_gamma = 0.05397466411641588;
const float coef0 = 0.0;
const int degree = 3;
const enum KernelType type = RBF;
```

6.4.3.1 STM32CubeIDE Project Settings for Deployment

Now, we have all the parameters required to implement the SVM classifier on the microcontroller. To deploy these parameters to the microcontroller, we should first copy the generated "svm_cls_config.h" and "svm_cls_config.c" files and paste them to the "Core/Inc" and "Core/Src" folders of our STM32CubeIDE project, respectively. Afterward, we should copy the "svm_cls_inference.h" and "svm_cls_inference.c" files from the sklearn2c library and paste them to the "Core/Inc" and "Core/Src" folders of our STM32CubeIDE project, respectively.

6.4.3.2 Mbed Studio Project Settings for Deployment

In order to deploy the SVM classifier to the microcontroller using Mbed Studio, we should copy the generated "svm_cls_config.h" and "svm_cls_config.c" files to the root folder of our project. Afterward, we should copy the "svm_cls_inference.h" and "svm_cls_inference.c" files from the sklearn2c library to our Mbed Studio project. Then, we will be ready to use the classifier on the microcontroller.

6.4.4 Testing the Deployed Classifier on the Microcontroller

To check whether the deployed classifier works as expected, we can use the generated data on PC. Then, we can send it to the STM32 microcontroller. The classifier running on the microcontroller works on this data. Then, we can transfer the classification result back to PC and cross check with the same classifier working in Python on PC. Hence, we can make sure that the generated framework works as expected.

6.4.4.1 STM32CubeIDE Project Settings for Testing

In order to test the deployed classifier on the STM32 microcontroller, we can form a project under STM32CubeIDE. Then, we should add the libraries lib_rng and lib_serial to the project as explained in Sects. 5.1.3.1 and 4.1.2.2, respectively. Afterward, we should add the generated files in the previous section. Hence, we will have the corresponding code block as follows.

```
/* USER CODE BEGIN Includes */
#include "lib_serial.h"
#include "lib_rng.h"
#include "svm_cls_inference.h"
/* USER CODE END Includes */
```

We should define macros related to the source of the input data, size of the input vector, and size of the output vector. The INPUT macro definition selects whether the input data is obtained from PC or microcontroller at compile time. It can be selected as INPUT_PC or INPUT_MCU, respectively. SIZE_INPUT and SIZE_OUTPUT macros define the number of data inputs and outputs of the model, respectively. We should define a floating-point array, with size equal to the number of features, for storing the input data. We should also define an integer array, with size equal to the number of classes, for storing output data. Hence, we will have the corresponding code block as follows.

```
/* USER CODE BEGIN 0 */
#define INPUT_PC 1
#define INPUT_MCU 2
#define INPUT INPUT_PC

#define SIZE_INPUT NUM_FEATURES
#define SIZE_OUTPUT ((NUM_CLASSES == 2) ? 1 : NUM_CLASSES)

float input[SIZE_INPUT];
float output[SIZE_OUTPUT];
/* USER CODE END 0 */
```

Next, we should form a while loop to acquire data either from PC or microcontroller according to the INPUT macro. Then, we can run the SVM classifier by calling the function svm_cls_score. The first parameter of this function is a pointer to the input array. The second parameter is a pointer to the char array. This output array is filled by the function svm_cls_score with SVM inference results. An alternative for this function is svm_cls_predict. However, the second parameter of this function is integer pointer since it writes the output class label to the pointed address. Finally, we transfer the classification result to PC. We provide the C code corresponding to these definitions in Listing 6.22. Please note that the Python script in Listing 6.23 should already be running on PC for executing this example successfully.

Listing 6.22 While loop for the SVM classifier

```
/* USER CODE BEGIN WHILE */
  while (1)
  {
/* USER CODE END WHILE */

/* USER CODE BEGIN 3 */
/*
 * ORGANIZE INPUTS
 */
#if (INPUT == INPUT_PC)
  LIB_SERIAL_Receive(input, SIZE_INPUT, TYPE_F32);
#elif (INPUT == INPUT_MCU)
  for (uint32_t i = 0; i < SIZE_INPUT; ++i)
  input[i] = (float)(LIB_RNG_GetRandomNumber() % 1000) / 1000.0f;

  LIB_SERIAL_Transmit(input, SIZE_INPUT, TYPE_F32);
#endif
  svm_cls_score(input, output);
  LIB_SERIAL_Transmit((void*)&output, SIZE_OUTPUT, TYPE_F32);
  HAL_Delay(1000);
}
/* USER CODE END 3 */
```

To check whether the deployed classifier works as expected, we need a test setup. We provide the Python script formed for this purpose in Listing 6.23. Here, we first load the previously generated test data. Then, we load our trained SVM classifier. Afterward, we send the test data to the microcontroller. Next, we perform inference on PC and microcontroller separately. We finally print both inference results on the PC for comparison. Hence, we can make sure that the generated framework works as expected.

Listing 6.23 Python script for comparing SVM inference results

```
import os.path as osp
import numpy as np
from sklearn2c import SVMClassifier
import py_serial

py_serial.SERIAL_Init("COM4")

test_samples = np.load(osp.join("classification_data","
    cls_test_samples.npy"))
test_labels = np.load(osp.join("classification_data","
    cls_test_labels.npy"))

svm = SVMClassifier.load(osp.join("classification_models", "
    svm_classifier.joblib"))

i = 0
while 1:
    rqType, datalength, dataType = py_serial.
        SERIAL_PollForRequest()
    if rqType == py_serial.MCU_WRITES:
        # INPUT -> FROM MCU TO PC
        inputs = py_serial.SERIAL_Read()
```

```
    elif rqType == py_serial.MCU_READS:
        # INPUT -> FROM PC TO MCU
        inputs = test_samples[i:i+1].astype(py_serial.
            SERIAL_GetDType(dataType))
        i = i + 1
        if i >= len(test_samples):
            i = 0
        py_serial.SERIAL_Write(inputs)

    pcout = svm.score(np.reshape(inputs, (1, datalength)))
    rqType, datalength, dataType = py_serial.
        SERIAL_PollForRequest()
    if rqType == py_serial.MCU_WRITES:
        mcuout = py_serial.SERIAL_Read()
        print()
        print("Inputs : " + str(inputs))
        print("PC Output : " + str(pcout))
        print("MCU Output : " + str(mcuout))
        print()
```

6.4.4.2 Mbed Studio Project Settings for Testing

In order to test the deployed SVM classifier on the STM32 microcontroller, we should add the libraries lib_rng and lib_serial to the Mbed Studio project. To do so, we should copy the files "lib_rng.h," "lib_rng.c," "lib_uart.h," "lib_uart.c," "lib_serial.h," and "lib_serial.c" to our Mbed Studio project folder. Here, we also assume that the steps in Sect. 6.4.3.2 are already done.

The test code for the Mbed Studio project is given in Listing 6.24. This code is similar to the one given for the STM32CubeIDE project. Hence, the explanations there hold here as well. The only difference is that we should include the header file "lib_uart.h" and initialize the UART peripheral unit before the while loop by code in our Mbed Studio project. Please note that the Python script in Listing 6.23 should already be running on PC for running this example successfully.

Listing 6.24 SVM classifier test code for Mbed Studio

```
#include "mbed.h"
#include "svm_cls_inference.h"
#include "lib_serial.h"
#include "lib_rng.h"
#include "lib_uart.h"
#include <string.h>

#define INPUT_PC 1
#define INPUT_MCU 2
#define INPUT INPUT_PC

#define SIZE_INPUT NUM_FEATURES
#define SIZE_OUTPUT ((NUM_CLASSES == 2) ? 1 : NUM_CLASSES)

float input[SIZE_INPUT];
float output[SIZE_OUTPUT];

int main()
{
```

```
    LIB_UART_Init();
#if (INPUT == INPUT_MCU)
    LIB_RNG_Init();
#endif
    while (true)
    {
#if (INPUT == INPUT_PC)
        LIB_SERIAL_Receive(input, SIZE_INPUT, TYPE_F32);
#elif (INPUT == INPUT_MCU)
        for (uint32_t i = 0; i < SIZE_INPUT; ++i)
            input[i] = (float)(LIB_RNG_GetRandomNumber() % 1000) /
                1000.0f;
        LIB_SERIAL_Transmit(input, SIZE_INPUT, TYPE_F32);
#endif
        svm_cls_score(input, output);
        LIB_SERIAL_Transmit((void*)&output, SIZE_OUTPUT, TYPE_F32);
        HAL_Delay(1000);
    }
}
```

6.5   Decision Tree Classifier

Decision tree is the fourth classifier to be considered in this book. This classifier adopts yet another approach in forming the decision boundary. We will evaluate how this is done in the theoretical background section next. Afterward, we will provide ways of training the decision tree classifier in Python on PC. This will be followed by deploying the trained classifier to the STM32 microcontroller and testing its performance there.

6.5.1   Theoretical Background

The decision tree classifier works by successively splitting data. In this operation, no global parametric model is assumed for the data at hand. Therefore, the decision tree classifier can be categorized under the nonparametric classifier group. Let's focus on the training and inference steps of the classifier.

Data is split successively based on the selected feature and threshold values in training phase of the decision tree classifier. We will use binary decision tree for this purpose in this section. This tree consists of nodes, which split the given data into two disjoint and exhaustive subsets from the previous set or subset. The initial node which keeps the overall training data is called the root. The final nodes which keep subsets obtained at the end of training are called leaves. The aim in splitting the data is to reduce impurity at each successive level compared to the previous one. There are three impurity measures for classification as entropy, Gini index, and misclassification error in the literature. Alpaydın [1] mentions that these three measures produce similar results.

Based on the available training data, we should choose the feature and its threshold value at each split level to create the decision tree. There are different methods in the literature for this purpose. In this book, we do not focus on how this operation is done. We depend on the scikit-learn library implementation of decision tree formation for classification.

As the splits end after successive steps in training the decision tree classifier, several leaf nodes will be formed. Each leaf node consists of a subset of class samples. If a leaf node consists of samples from only one class, then we assign this label to the node as classification output. If a leaf node consists of samples from more than one class, then we assign the label of the majority class samples to the node as classification output. We perform this operation to all leaf nodes to obtain the final decision tree classifier.

The user can decide on the number of splits while forming the decision tree for classification. We would like to have pure leaf nodes at the end of training such that each leaf consists of the same class samples. This can be achieved by increasing the split number. The obtained decision boundary will be detailed in this case. However, this may not be desirable when there is noise in data. Then, pure leaves in fact correspond to overfitting. On the other hand, we will have a crude decision boundary when the split number is low. This may result in classification errors during inference. Therefore, the user should be cautious while setting the split number.

As the training phase of the decision tree classifier is done, it can be represented by a successive set of if-else rules. In other words, successive binary decisions can be used to implement the decision tree classifier. The inference step for the classifier is also realized in the same way. Therefore, when an unknown sample is fed to the classifier, it is tested via successive binary rules. The final reached leaf node class label is fed as the inference result.

Let's obtain the decision boundary formed by the decision tree classifier on the two-dimensional dataset given in Fig. 6.1. To note here, this data is generated by the Python code given in Listing 5.2. We provide the formed decision boundary in Fig. 6.10. As can be seen in this figure, the decision tree classifier separates two classes in a blocky manner. This is due to binary decisions applied during the inference step.

6.5.2 Training the Classifier in Python

As in the Bayes classifier, we can form the decision tree classifier from the labeled training samples at hand using the sklearn2c library. We provide one such example in Listing 6.25. Here, we first load the training data and corresponding labels by the NumPy function `load`. Afterward, we form the decision tree classifier using the class `DTClassifier`. This class has several parameters in forming the classifier [31]. In our example, we used default parameter values. The formed classifier can be trained by the function `train`. This function takes the optional parameter `save_path` to save the trained classifier to a file. Therefore, we feed the `train_samples` data with the label set `train_labels` and `save_path` parameter in our example. As the operation ends, the trained classifier will be saved as the "dtc_classifier.joblib" file under the folder "classification_models."

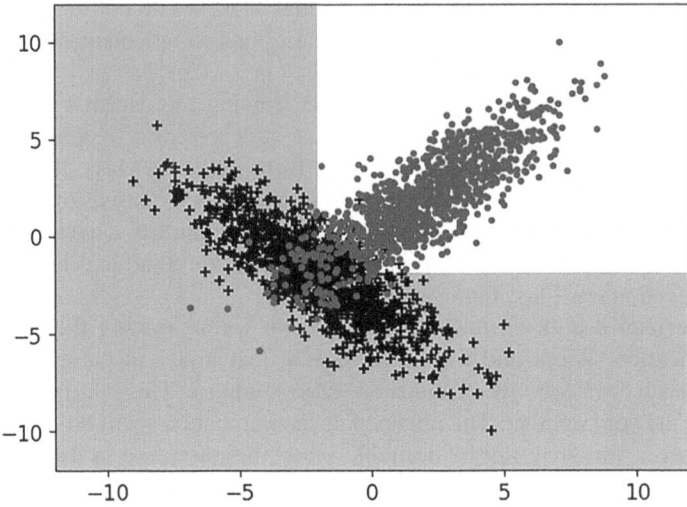

Fig. 6.10 Decision boundary formed by the decision tree classifier

Listing 6.25 Training the decision tree classifier

```
import os.path as osp
import numpy as np
from sklearn2c import DTClassifier

train_samples = np.load(osp.join("classification_data", "
    cls_train_samples.npy"))
train_labels = np.load(osp.join("classification_data", "
    cls_train_labels.npy"))

dtc = DTClassifier()
model_save_path = osp.join("classification_models", "
    dt_classifier.joblib")
dtc.train(train_samples, train_labels, save_path= model_save_path
    )
```

In order to apply inference on the saved classifier in Listing 6.25, we should load the test data and then proceed with inference. We provide such an example in Listing 6.26. Here, the test_samples data and test_labels are loaded from the files "cls_test_samples.npy" and "cls_test_labels.npy," respectively. Then, the classifier is loaded by the function load using the saved model path as parameter. Afterward, the decision tree classifier predicts the classification result of test_samples by using the function predict.

Listing 6.26 Inference with the decision tree classifier in Python

```
import os.path as osp
import numpy as np
from sklearn2c import DTClassifier
```

```
test_samples = np.load(osp.join("classification_data","
    cls_test_samples.npy"))
test_labels = np.load(osp.join("classification_data","
    cls_test_labels.npy"))

dt_classifier = DTClassifier.load(osp.join("classification_models
    ", "dt_classifier.joblib"))
predictions = dt_classifier.predict(test_samples)
print(predictions)
```

6.5.3 Deploying the Trained Classifier to the Microcontroller

We trained the decision tree classifier by given data in Python on PC and saved the trained classifier in the previous section. The next phase is deploying it to the STM32 microcontroller such that we can use the trained classifier in an actual classification scenario. Therefore, we should generate the C header and source files of the trained classifier on PC. To do so, we should first load the trained model using the function load. We should generate the C header and source files automatically by the DTClassifierExporter class function export. In other words, this class automatically generates the C header file that declares configurations of our decision tree classifier. It also generates the C source file which stores the information needed to run inference on the microcontroller. We provide a sample Python script to export the classifier parameters in Listing 6.27.

Listing 6.27 Exporting the decision tree classifier

```
import os.path as osp
from sklearn2c import DTClassifier

model_path = osp.join("classification_models","dt_classifier.
    joblib")
export_path = osp.join("exported_models","classification","
    dt_cls_config")

dtc = DTClassifier.load(model_path)
dtc.export(export_path)
```

In Listing 6.27, the function export generates the C header, "dt_cls_config.h," and source file, "dt_cls_config.c." It then saves them to the given export path. Hence, the classifier can be deployed to the microcontroller. The generated C header file for the example considered in this chapter is given in Listing 6.28. This file only keeps variable names and their size. It will be used by other files to include the corresponding source file given in Listing 6.29. All variables declared in Listing 6.28 are assigned here.

Listing 6.28 Header file for the decision tree classifier configuration

```
#ifndef DT_CLS_CONFIG_H_INCLUDED
#define DT_CLS_CONFIG_H_INCLUDED
#define NUM_NODES 285
```

```
#define NUM_FEATURES 2
#define NUM_CLASSES 2
extern const int LEFT_CHILDREN[NUM_NODES];
extern const int RIGHT_CHILDREN[NUM_NODES];
extern const int SPLIT_FEATURE[NUM_NODES];
extern const float THRESHOLDS[NUM_NODES];
extern const int VALUES[NUM_NODES][NUM_CLASSES];
#endif
```

Listing 6.29 Source file for the decision tree classifier configuration, shortened version

```
#include "dt_cls_config.h"
const int LEFT_CHILDREN[NUM_NODES] = {1,...};
const int RIGHT_CHILDREN[NUM_NODES] = {250,...} ;
const int SPLIT_FEATURE[NUM_NODES] = {0,...};
const float THRESHOLDS[NUM_NODES] = {-0.21506915,...};
const int VALUES[NUM_NODES][NUM_CLASSES] = {{806,794},...};
```

6.5.3.1 STM32CubeIDE Project Settings for Deployment

Now, we have all the parameters required to implement the decision tree classifier on the microcontroller. To deploy these parameters to the microcontroller, we should first copy the generated "dt_cls_config.h" and "dt_cls_config.c" files and paste them to the "Core/Inc" and "Core/Src" folders of our STM32CubeIDE project, respectively. Afterward, we should copy the "dt_cls_inference.h" and "dt_cls_inference.c" files from the sklearn2c library and paste them to the "Core/Inc" and "Core/Src" folders of our STM32CubeIDE project, respectively.

6.5.3.2 Mbed Studio Project Settings for Deployment

In order to deploy the decision tree classifier to the microcontroller using Mbed Studio, we should copy the generated "dt_cls_config.h" and "dt_cls_config.c" files to the root folder of our project. Afterward, we should copy the "dt_cls_inference.h" and "dt_cls_inference.c" files from the sklearn2c library to our Mbed Studio project folder. Then, we will be ready to use the classifier on the microcontroller.

6.5.4   Testing the Deployed Classifier on the Microcontroller

To check whether the deployed classifier works as expected, we can use the generated data on PC. Then, we can send it to the STM32 microcontroller. The classifier running on the microcontroller works on this data. Then, we can transfer the classification result back to PC and cross check with the same classifier working in Python on PC. Hence, we can make sure that the generated framework works as expected.

6.5.4.1 STM32CubeIDE Project Settings for Testing

In order to test the deployed classifier on the STM32 microcontroller, we can form a project under STM32CubeIDE. Then, we should add the libraries lib_rng and lib_serial to the project as explained in Sects. 5.1.3.1 and 4.1.2.2, respectively. Afterward, we should add the generated files in the previous section. Hence, we will have the corresponding code block as follows.

```
/* USER CODE BEGIN Includes */
#include "dt_cls_inference.h"
#include "lib_serial.h"
#include "lib_rng.h"
/* USER CODE END Includes */
```

We should define macros related to the source of the input data, size of the input vector, and size of the output vector. The INPUT macro definition selects whether the input data is obtained from PC or microcontroller at compile time. It can be selected as INPUT_PC or INPUT_MCU, respectively. SIZE_INPUT and SIZE_OUTPUT macros define the number of data inputs and outputs of the model, respectively. We should define a floating-point array, with size equal to the number of features, for storing the input data. We should also define an integer array, with size equal to the number of classes, for storing output data. Hence, we will have the corresponding code block as follows.

```
/* USER CODE BEGIN 0 */
#define INPUT_PC 1
#define INPUT_MCU 2
#define INPUT INPUT_PC

#define SIZE_INPUT NUM_FEATURES
#define SIZE_OUTPUT NUM_CLASSES

float input[SIZE_INPUT];
int output[SIZE_OUTPUT];

/* USER CODE END 0 */
```

Next, we should form a while loop to acquire data either from PC or microcontroller according to the INPUT macro. Then, we can run the decision tree classifier by calling the function dt_cls_predict. The first parameter of this function is a pointer to the input array. The second parameter is the address of the output array. This output array is filled by the function dt_cls_predict with the decision three classifier inference results. Finally, we transfer the classification result to PC. We provide the C code corresponding to these definitions in Listing 6.30. Please note that the Python script in Listing 6.31 should already be running on PC for executing this example successfully.

Listing 6.30 While loop for the decision tree classifier

```
/* USER CODE BEGIN WHILE */
  while (1)
  {
/* USER CODE END WHILE */

/* USER CODE BEGIN 3 */
#if (INPUT == INPUT_PC)
```

```
  LIB_SERIAL_Receive(input, SIZE_INPUT, TYPE_F32);
#elif (INPUT == INPUT_MCU)
  for (uint32_t i = 0; i < SIZE_INPUT; ++i)
    input[i] = (float)(LIB_RNG_GetRandomNumber() % 1000) / 1000.0f;

  LIB_SERIAL_Transmit(input, SIZE_INPUT, TYPE_F32);
#endif
  dt_cls_predict(input, output);

  LIB_SERIAL_Transmit(output, SIZE_OUTPUT, TYPE_S32);
  HAL_Delay(1000);
}
/* USER CODE END 3 */
```

To check whether the deployed classifier works as expected, we need a test setup. We provide the Python script formed for this purpose in Listing 6.31. Here, we first load the previously generated test data. Then, we load our trained decision tree classifier. Afterward, we send the test data to the microcontroller. Next, we perform inference on PC and microcontroller separately. We finally print both inference results on the PC for comparison. Hence, we can make sure that the generated framework works as expected.

Listing 6.31 Python script for comparing decision tree classifier inference results

```python
import os.path as osp
import numpy as np
from sklearn2c import DTClassifier
import py_serial

py_serial.SERIAL_Init("COM4")

test_samples = np.load(osp.join("classification_data","
    cls_test_samples.npy"))
test_labels = np.load(osp.join("classification_data","
    cls_test_labels.npy"))

dtc = DTClassifier.load(osp.join("classification_models", "
    dt_classifier.joblib"))

i = 0
while 1:
    rqType, datalength, dataType = py_serial.
        SERIAL_PollForRequest()
    if rqType == py_serial.MCU_WRITES:
        # INPUT -> FROM MCU TO PC
        inputs = py_serial.SERIAL_Read()

    elif rqType == py_serial.MCU_READS:
        # INPUT -> FROM PC TO MCU
        inputs = test_samples[i:i+1].astype(py_serial.
            SERIAL_GetDType(dataType))
        i = i + 1
        if i >= len(test_samples):
            i = 0
        py_serial.SERIAL_Write(inputs)

    pcout = dtc.predict(np.reshape(inputs, (1, datalength)))
```

```
rqType, datalength, dataType = py_serial.
    SERIAL_PollForRequest()
if rqType == py_serial.MCU_WRITES:
    mcuout = py_serial.SERIAL_Read()
    print()
    print("Inputs : " + str(inputs))
    print("PC Output : " + str(pcout))
    print("MCU Output : " + str(mcuout))
    print()
```

6.5.4.2 Mbed Studio Project Settings for Testing

In order to test the deployed decision tree classifier on the STM32 microcontroller, we should add the libraries lib_rng and lib_serial to the Mbed Studio project. To do so, we should copy the files "lib_rng.h," "lib_rng.c," "lib_uart.h," "lib_uart.c," "lib_serial.h," and "lib_serial.c" to our Mbed Studio project folder. Here, we also assume that the steps in Sect. 6.5.3.2 are already done.

The test code for the Mbed Studio project is given in Listing 6.32. This code is similar to the one given for the STM32CubeIDE project. Hence, the explanations there hold here as well. The only difference is that we should include the header file "lib_uart.h" and initialize the UART peripheral before the while loop by code in our Mbed Studio project. Please note that the Python script in Listing 6.31 should already be running on PC for executing this example successfully.

Listing 6.32 Decision tree classifier test code for Mbed Studio

```
#include "mbed.h"
#include "dt_cls_inference.h"
#include "lib_serial.h"
#include "lib_rng.h"
#include "lib_uart.h"

#define INPUT_PC 1
#define INPUT_MCU 2
#define INPUT INPUT_PC

#define SIZE_INPUT NUM_FEATURES
#define SIZE_OUTPUT NUM_CLASSES

float input[SIZE_INPUT];
int output[SIZE_OUTPUT];

int main()
{
 LIB_UART_Init();
#if (INPUT == INPUT_MCU)
 LIB_RNG_Init();
#endif
 while (true)
 {
#if (INPUT == INPUT_PC)
 LIB_SERIAL_Receive(input, SIZE_INPUT, TYPE_F32);
#elif (INPUT == INPUT_MCU)
 for (uint32_t i = 0; i < SIZE_INPUT; ++i)
 input[i] = (float)(LIB_RNG_GetRandomNumber() % 1000) / 1000.0f;
```

```
LIB_SERIAL_Transmit(input, SIZE_INPUT, TYPE_F32);
#endif
dt_cls_predict(input, output);
LIB_SERIAL_Transmit(output, SIZE_OUTPUT, TYPE_S32);
HAL_Delay(1000);
}
}
```

6.6   Application: Human Activity Recognition via Accelerometer Data

We extracted features from the accelerometer data from the WISDM dataset in Sect. 5.4. As a reminder, this dataset has six classes as walking, jogging, upstairs, downstairs, sitting, and standing. In this section, we train the Bayes classifier to separate these classes based on the raw accelerometer data.

We provide the Python script for human activity recognition in Listing 6.33. We split the data in the WISDM dataset into training and testing sets based on the user ID. We pick the users with ID less than or equal to 28 for training. We pick the rest of users for testing. We benefit from the function create_features to transform the raw data into a format suitable for machine learning. This function segments the time-series data into chunks (each with size TIME_PERIODS and a step size of STEP_DISTANCE) and extracts features from each segment. The Bayes classifier from the sklearn2c library is then instantiated and trained on the training data.

Listing 6.33 Training and testing the Bayes classifier for human activity recognition

```
import os.path as osp
from data_utils import read_data
from feature_utils import create_features
from sklearn import metrics
import sklearn2c
from matplotlib import pyplot as plt

DATA_PATH = osp.join("WISDM_ar_v1.1", "WISDM_ar_v1.1_raw.txt")
TIME_PERIODS = 80
STEP_DISTANCE = 40
data_df = read_data(DATA_PATH)
df_train = data_df[data_df["user"] <= 28]
df_test = data_df[data_df["user"] > 28]

train_segments_df, train_labels = create_features(df_train,
    TIME_PERIODS, STEP_DISTANCE)
test_segments_df, test_labels = create_features(df_test,
    TIME_PERIODS, STEP_DISTANCE)

bayes = sklearn2c.BayesClassifier()
bayes.train(train_segments_df, train_labels)
bayes_preds = bayes.predict(test_segments_df)

conf_matrix = metrics.confusion_matrix(test_labels, bayes_preds)
```

```
cm_display = metrics.ConfusionMatrixDisplay(confusion_matrix =
    conf_matrix, display_labels = bayes.class_names)
cm_display.plot()
cm_display.ax_.set_title("Bayes Classifier Confusion Matrix")
plt.show()

bayes.export("bayes_har_config")
```

We next export the trained Bayes classifier to C configuration files to be used in the microcontroller. We provide the corresponding header and source files in Listings 6.34 and 6.35, respectively. We will provide the complete STM32CubeIDE and Mbed Studio projects in separate folders in the book repository for this application.

Listing 6.34 C header file for trained Bayes classifier for human activity recognition

```
#ifndef BAYES_HAR_CONFIG_H_INCLUDED
#define BAYES_HAR_CONFIG_H_INCLUDED
#define NUM_CLASSES 6
#define NUM_FEATURES 10
#define CASE 3
extern float MEANS[NUM_CLASSES][NUM_FEATURES];
extern const float CLASS_PRIORS[NUM_CLASSES];
extern const float INV_COVS[NUM_CLASSES][NUM_FEATURES][
    NUM_FEATURES];
extern const float DETS[NUM_CLASSES];
#endif
```

Listing 6.35 C source file for trained Bayes classifier for human activity recognition

```
#include "bayes_har_config.h"
float MEANS[NUM_CLASSES][NUM_FEATURES] = {{ 4.54128467e-01,
    8.62727869e+00, 8.48041095e-01, 4.13930911e+01,...}}
const float CLASS_PRIORS[NUM_CLASSES] =
    {0.08280318,0.31996844,0.0503033
    ,0.03965084,0.10627805,0.4009962 };
const float INV_COVS[NUM_CLASSES][NUM_FEATURES][NUM_FEATURES] =
    {{{ 6.67693289e-01, 2.94860563e-01, 1.52868235e
    -01,-9.77798696e-02,...}}}
const float DETS[NUM_CLASSES] = {5.68137130e+12,1.19219336e
    +16,2.99312116e+10,2.05710132e+02, 5.10977915e+13,1.99716386e
    +12};
```

6.7 Application: Keyword Spotting from Audio Signals

We extracted features from the audio data from the spoken digits dataset in Sect. 5.5. As a reminder, this dataset has ten classes as spoken digits from zero to nine. In this section, we train the kNN classifier to separate these classes.

We provide the Python script for keyword spotting in Listing 6.36. We split the data in the spoken digits dataset into training and test sets based on the name of the person who made the recording. If the name yweweler is in the recording path, then that recording is added to the test set. All other recordings are used for training. We use the function mfcc_features to extract MFCC features from both training and test sets. The kNN classifier is then instantiated with three neighbors and trained on MFCC features.

Listing 6.36 Training and testing the kNN classifier for keyword spotting

```
import os
import scipy.signal as sig
from mfcc_func import create_mfcc_features
from sklearn.metrics import confusion_matrix,
    ConfusionMatrixDisplay
import sklearn2c
from matplotlib import pyplot as plt

RECORDINGS_DIR = "recordings"
recordings_list = [(RECORDINGS_DIR, recording_path) for
    recording_path in os.listdir(RECORDINGS_DIR)]

FFTSize = 1024
sample_rate = 8000
numOfMelFilters = 20
numOfDctOutputs = 13
window = sig.get_window("hamming", FFTSize)
test_list = {record for record in recordings_list if "yweweler"
    in record[1]}
train_list = set(recordings_list) - test_list
train_mfcc_features, train_labels = create_mfcc_features(
    train_list, FFTSize, sample_rate, numOfMelFilters,
    numOfDctOutputs, window)
test_mfcc_features, test_labels = create_mfcc_features(test_list,
     FFTSize, sample_rate, numOfMelFilters, numOfDctOutputs,
    window)

knn =sklearn2c.KNNClassifier(n_neighbors = 3)
knn.train(train_mfcc_features, train_labels)
knn_preds = knn.predict(test_mfcc_features)

conf_matrix = confusion_matrix(test_labels, knn_preds)
cm_display = ConfusionMatrixDisplay(confusion_matrix =
    conf_matrix, display_labels = knn.class_names)
cm_display.plot()
cm_display.ax_.set_title("KNN Classifier Confusion Matrix")
plt.show()

knn.export("knn_mfcc_config")
```

We next export the trained kNN classifier to C configuration files to be used in the microcontroller. We provide the corresponding header and source files in Listings 6.37 and 6.38, respectively. We will provide the complete STM32CubeIDE and Mbed Studio projects in separate folders in the book repository for this application.

Listing 6.37 C header file for the trained kNN classifier for keyword spotting

```
#ifndef KNN_MFCC_CONFIG_H_INCLUDED
#define KNN_MFCC_CONFIG_H_INCLUDED
#define NUM_CLASSES 10
#define NUM_NEIGHBORS 3
#define NUM_FEATURES 26
#define NUM_SAMPLES 2500
extern char* LABELS[NUM_CLASSES];
extern const float DATA[NUM_SAMPLES][NUM_FEATURES];
extern const int DATA_LABELS[NUM_SAMPLES];
#endif
```

Listing 6.38 C source file for the trained kNN classifier for keyword spotting

```
#include "knn_mfcc_config.h"
char* LABELS[NUM_CLASSES] = {0.,1.,2.,3.,4.,5.,6.,7.,8.,9.};
const float DATA[NUM_SAMPLES][NUM_FEATURES] = {{ 2.05717278e
    +01,-4.28795815e-04, 2.55274582e+00, 6.97973967e
    -02,...},...};
const int DATA_LABELS[NUM_SAMPLES] = {0,4, ...};
```

6.8 Application: Handwritten Digit Recognition from Digital Images

Assume that we formed an embedded system to recognize handwritten digits in an image. Hence, there is a microcontroller with a digital camera attached to it. When the image of an unknown digit is acquired, features are extracted from it as explained in Chap. 5. We know that there are ten digits. Hence, there are ten classes. In other words, our handwritten digit recognition operation has ten distinct and possible outcomes. The classifier takes the calculated feature values as input and produces the label corresponding to one of the ten digits as output. We should have prior knowledge about the representation of each digit in feature space. This corresponds to training the classifier which requires labeled dataset. In other words, a supervisor should have prepared digit samples with their expected/actual value. During the training phase, a decision boundary is formed between classes. When an unknown digit sample comes, the classifier assigns a class label to the sample based on its location with respect to the decision boundary between classes. We call this step as inference. Hence, the classifier decides on the class of the unknown sample. We use the SVM classifier for this application; nevertheless, reader may choose another classifier.

We provide the Python script for handwritten digit recognition in Listing 6.39. We split the data in the MNIST dataset into training and test sets. We use the function cv2.moments to extract image moments. Then, we use the function cv2.HuMoments to extract Hu moments from computed moments.

Listing 6.39 Training and testing the SVM classifier for handwritten digit recognition

```python
import os
import numpy as np
import cv2
from sklearn.metrics import confusion_matrix,
    ConfusionMatrixDisplay
import sklearn2c
from mnist import load_images, load_labels
from matplotlib import pyplot as plt

train_img_path = os.path.join("MNIST-dataset", "train-images.idx3
    -ubyte")
train_label_path = os.path.join("MNIST-dataset", "train-labels.
    idx1-ubyte")
test_img_path = os.path.join("MNIST-dataset", "t10k-images.idx3-
    ubyte")
test_label_path = os.path.join("MNIST-dataset", "t10k-labels.idx1
    -ubyte")

train_images = load_images(train_img_path)
train_labels = load_labels(train_label_path)
test_images = load_images(test_img_path)
test_labels = load_labels(test_label_path)

train_huMoments = np.empty((len(train_images),7))
test_huMoments = np.empty((len(test_images),7))

for train_idx, train_img in enumerate(train_images):
    train_moments = cv2.moments(train_img, True)
    train_huMoments[train_idx] = cv2.HuMoments(train_moments).
        reshape(7)

for test_idx, test_img in enumerate(test_images):
    test_moments = cv2.moments(test_img, True)
    test_huMoments[test_idx] = cv2.HuMoments(test_moments).
        reshape(7)

svc = sklearn2c.SVMClassifier()
svc.train(train_huMoments, train_labels)
svc_preds = svc.predict(test_huMoments)
cm = confusion_matrix(test_labels, svc_preds, labels=svc.
    class_names)
cm_display = ConfusionMatrixDisplay(confusion_matrix=cm,
    display_labels=svc.class_names)
cm_display.plot()
cm_display.ax_.set_title("SVM Classifier Confusion Matrix")
plt.show()

svc.export("svm_moments_config")
```

We next export the trained SVM classifier to C configuration files to be used in the microcontroller. We provide the corresponding header and source files in Listings 6.40 and 6.41, respectively. We will provide the complete STM32CubeIDE and Mbed Studio projects in separate folders in the book repository for this application.

Listing 6.40 C header file for the trained SVM classifier for handwritten digit recognition

```
#ifndef SVM_MOMENTS_CONFIG_H_INCLUDED
#define SVM_MOMENTS_CONFIG_H_INCLUDED
#define NUM_CLASSES 10
#define NUM_INTERCEPTS 45
#define NUM_FEATURES 7
#define NUM_SV 50770
enum KernelType{
        LINEAR,
        POLY,
        RBF
};
extern const float coeffs[NUM_CLASSES - 1][NUM_SV];
extern const float SV[NUM_SV][NUM_FEATURES];
extern const float intercepts[NUM_INTERCEPTS];
extern const float w_sum[NUM_CLASSES + 1];
extern const float svm_gamma;
extern const float coef0;
extern const int degree;
extern const enum KernelType type;
#endif
```

Listing 6.41 C header file for the trained SVM classifier for handwritten digit recognition

```
#include "svm_moments_config.h"
const float coeffs[NUM_CLASSES - 1][NUM_SV] = {{ 0. , 0. , ...},
    ...};
const float SV[NUM_SV][NUM_FEATURES] = {{ 3.15816167e-01,
    1.75105811e-02, 5.41035080e-04, 2.53788016e-04,}, ...};
const float intercepts[NUM_INTERCEPTS] = {-6.33389052e-01,
    5.45609269e-01, 2.04965002e+00, 6.64191770e-01,...};
const float w_sum[NUM_CLASSES + 1] = {0, 5007,
    6097,12055,18186,24028,29445,35352,39414,44821,50770};
const float svm_gamma = 9.659468371895358;
const float coef0 = 0.0;
const int degree = 3;
const enum KernelType type = RBF;
```

6.9   Summary of the Chapter

We considered classification in this chapter. The aim of classification is to infer type of an unknown sample summarized by its features. Classifier is the machine learning system performing this operation. We started with the performance measures for classifiers in general. Then, we introduced the Bayes, kNN, SVM, and decision tree classifiers. While handling each classifier, we started with its theoretical background. Then, we explored its training on PC in Python language. Afterward, we provided methods of deploying the trained classifier to the microcontroller. As end of chapter applications, we formed classifiers for the accelerometer data, audio signals, and digital images. We benefit from the extracted features in Chap. 5 for this purpose.

Regression

7

7.1   What Is Regression?

Regression aims to find the relationship between two or more variables. Hence, information on one variable can be obtained by the information on other variable(s). We can also frame regression as a function fit problem such that we can fill missing data, filter data, or predict future values based on current and past data. The operator performing regression is called the regressor.

7.1.1   Regression Types

There are four regression methods to be considered in this chapter. We can group them into two categories as parametric and nonparametric as in classifiers. Next, we will explain general characteristics of the two groups. Hence, the reader can understand how they can be implemented in an embedded system.

A parametric regression method can be constructed as a function with explicit parameters to represent the relationship between input and output variables. Let's start with basic definitions. Assume that we have a training set consisting of N observations (samples) $x[n]$ where $n = 1, \cdots , N$ along with their output $y[n]$. Based on this training set, we can construct a function $\hat{y} = f(x)$. The aim here is formalizing the relationship between the variables x and y. Afterward, we can use the function $f(\cdot)$ for regression. In other words, $f(\cdot)$ will be the regressor. Even if parameter calculation step may change among different regression methods, the final regressor will be a function. We will have a linear function of input variables when linear regression is used as we will see in Sect. 7.2. Likewise, we will have a polynomial function of input variables when polynomial regression is used as we will see in Sect. 7.3.

Parametric regression methods have global representation property since we have a function as the regressor. In other words, they can be used to predict future

C. Ünsalan et al., *Embedded Machine Learning with Microcontrollers*,
https://doi.org/10.1007/978-3-031-70912-8_7

values of an input variable. However, parametric regressors have limited modeling capability since they are bound by the fixed model they are based on. Implementing parametric regression methods on an embedded system is another topic. Therefore, we will consider parametric regression method implementation separately in the related sections.

A nonparametric regression method does not have any function to represent the relationship between input and output variables. Instead, the regressor directly depends on the training data. The kNN and decision tree regression methods, to be introduced in Sects. 7.4 and 7.5, fall into the nonparametric regression category. Nonparametric regression methods depend on local information. The advantage of this property is that they have stronger modeling capability since they are not bound by a fixed model. However, nonparametric regression methods do not have global representation property. Therefore, they cannot be used to predict future values of an input variable. Since we do not have a parametric function, implementing nonparametric regressors requires more resources in general. We will consider nonparametric regression method implementation separately in the related sections.

7.1.2 Regression Performance Measures

We should specify metrics to train the regressor and evaluate the final regression model with test data. These metrics should reflect the error between the model output (prediction) and actual output (ground truth). The most popular metric used for this purpose is the mean squared error (MSE).

We can formalize MSE based on observations $x[n]$ where $n = 1, \cdots, N$ along with their output $y[n]$ as follows. Assume that we form the regressor as $\hat{y} = f(x)$. Hence, we will have the regressor output $\hat{y}[n]$ for every observation $x[n]$. We can define MSE as

$$E = \frac{1}{N} \sum_{n=1}^{N} \left(\hat{y}[n] - y[n] \right)^2 \tag{7.1}$$

where N represents the total number of observations. We will use this metric in regression methods to be introduced next.

We should mention one important property here. Decreasing MSE to absolute minimum at all costs may not be desirable. For such cases, we will have the overfit problem such that the regressor may model unnecessary details or noise in training data. Although MSE may be around zero during training, a similar performance may not be obtained during prediction phase for such scenarios. Therefore, the reader should be cautious to avoid such problems.

7.2   Linear Regression

Linear regression is a parametric method used extensively in practice. Therefore, we will consider it in this section. To do so, we will first introduce its theoretical background. Then, we will form the regressor in Python on PC. Afterward, we will deploy the formed regressor to the microcontroller and test its performance there.

7.2.1   Theoretical Background

The relationship between the input and output variables is modeled by a linear function in this regression method. Therefore, we will replace the regressor function $f(\cdot)$ in Sect. 7.1 by $\hat{y} = wx + b$ where w and b are parameters of the regressor. We will estimate these parameters in the training phase. We will perform this operation such that the linear regressor minimizes MSE wrt. given data. To do so, we can rewrite Eq. 7.1 as

$$E = \frac{1}{N} \sum_{n=1}^{N} (wx[n] + b - y[n])^2 \tag{7.2}$$

We can minimize E wrt. w and b by taking its partial derivatives and equating both to zero. Hence, we will have

$$\begin{bmatrix} \frac{\partial E}{\partial w} \\ \frac{\partial E}{\partial b} \end{bmatrix} = \begin{bmatrix} 0 \\ 0 \end{bmatrix} \tag{7.3}$$

We can generalize these operations using matrix notation. Therefore, we can form our matrices as

$$\hat{\mathbf{y}} = \begin{bmatrix} \hat{y}[1] \\ \vdots \\ \hat{y}[N] \end{bmatrix} \quad \mathbf{y} = \begin{bmatrix} y[1] \\ \vdots \\ y[N] \end{bmatrix} \quad \mathbf{x} = \begin{bmatrix} x[1] & 1 \\ \vdots & \vdots \\ x[N] & 1 \end{bmatrix} \quad \mathbf{w} = \begin{bmatrix} w \\ b \end{bmatrix} \tag{7.4}$$

Then, we can rewrite Eq. 7.2 as

$$E = (\hat{\mathbf{y}} - \mathbf{y})^T (\hat{\mathbf{y}} - \mathbf{y}) \tag{7.5}$$

Furthermore, we can write $\hat{\mathbf{y}} = \mathbf{xw}$. Hence, we will have

$$E = \mathbf{w}^T \mathbf{x}^T \mathbf{xw} - \mathbf{w}^T \mathbf{x}^T \mathbf{y} - \mathbf{y}^T \mathbf{xy} + \mathbf{y}^T \mathbf{y} \tag{7.6}$$

Solving Eq. 7.3 wrt. the definition in Eq. 7.6 leads us to

$$\mathbf{w} = (\mathbf{x}^T \mathbf{x})^{-1} \mathbf{x}^T \mathbf{y} \tag{7.7}$$

Hence, we can obtain the linear regressor parameters w and b based on the given training data.

As we obtain parameters for the linear regressor, it will be ready to be used for prediction. Therefore, when an input x is fed to the function $\hat{y} = wx + b$, it will generate the corresponding output $\hat{y}$. We can call this value as the linear regressor output or prediction.

Representing the linear regression in matrix form allows us to generalize the operation to problems with more than one input and output variables. To do so, the reader should form the matrices accordingly. Then, the remaining operations will be the same.

7.2.2 Forming the Regressor in Python

We can implement theoretical operations introduced in the previous section via Python by the sklearn2c library. We provide a working example for linear regression in Listing 7.1. Here, we first load noisy line and sine data points that we generated in Listing 5.3. We load the previously generated data points by the NumPy function load. We indicate input data points for both cases as train_samples. We call the noisy line and sine data as train_line_values and train_sine_values, respectively. Our aim is to predict the actual line and sine data by linear regression. Therefore, we benefit from the object LinearRegressor in the sklearn2c library. To be more specific, we form two objects line_linear_model and sine_linear_model. The next step is fitting the linear regression model to noisy line and sine data separately. Here, the aim is to minimize the MSE between the actual and predicted values. Therefore, we call train method of the LinearRegressor object. This method takes the optional parameter save_path to save the trained regressor to a file. Hence, the trained regressors for the line and sine data are saved as "linear_regressor_line.joblib" and "linear_regressor_sine.joblib" files under the folder "regression_models" as the operation ends.

Listing 7.1 Training the linear regressor and saving the trained model

```
import os.path as osp
import numpy as np
from sklearn2c import LinearRegressor

train_samples = np.load(osp.join("regression_data", "reg_samples.
    npy"))
train_line_values = np.load(osp.join("regression_data", "
    reg_line_values.npy"))
train_sine_values = np.load(osp.join("regression_data", "
    reg_sine_values.npy"))

line_linear_model = LinearRegressor()
linear_save_path = osp.join("regression_models","
    linear_regressor_line.joblib")
line_linear_model.train(train_samples, train_line_values,
    linear_save_path)
```

```
sine_linear_model = LinearRegressor()
linear_save_path = osp.join("regression_models","
    linear_regressor_sine.joblib")
sine_linear_model.train(train_samples, train_sine_values,
    linear_save_path)
```

In order to apply prediction with the saved linear regressors in Listing 7.1, we should again load the noisy line and sine data points and then proceed with prediction. We provide one such example in Listing 7.2. Here, we start with loading the samples, line_values, and sine_values. Then, we load linear regressors for the line and sine data by the function load using the saved model path as parameter. Afterward, we obtain the regressor outputs (predictions) by the function predict. As we obtain predicted values from the models, we assign them to variables line_predictions and sine_predictions for the line and sine data, respectively. Rest of the code plots the input data with its target value as well as predictions of the formed linear regressors.

Listing 7.2 Predictions with the formed linear regressors

```
import os.path as osp
import numpy as np
from matplotlib import pyplot as plt
from sklearn2c import LinearRegressor

samples = np.load(osp.join("regression_data", "reg_samples.npy"))
line_values = np.load(osp.join("regression_data", "
    reg_line_values.npy"))
sine_values = np.load(osp.join("regression_data", "
    reg_sine_values.npy"))

line_model_path = osp.join("regression_models", "
    linear_regressor_line.joblib")
sine_model_path = osp.join("regression_models", "
    linear_regressor_sine.joblib")

linear_regressor_line = LinearRegressor.load(line_model_path)
linear_regressor_sine = LinearRegressor.load(sine_model_path)

line_predictions = linear_regressor_line.predict(samples)
sine_predictions = linear_regressor_sine.predict(samples)

#Plotting functions
fig, (ax1, ax2) = plt.subplots(2, 1)
ax1.plot(samples, line_values, "k.")
ax1.plot(samples, line_predictions, "b")
ax2.plot(samples, sine_values, "k.")
ax2.plot(samples, sine_predictions, "b")
plt.show()
```

We provide the obtained prediction results from Listing 7.2 for the line and sine data in Fig. 7.1. As can be seen in this figure, the linear regressor works well with the line data. Hence, it can predict the underlying line function as in Fig. 7.1a. However, the regressor cannot predict the underlying sine function from the given input-output

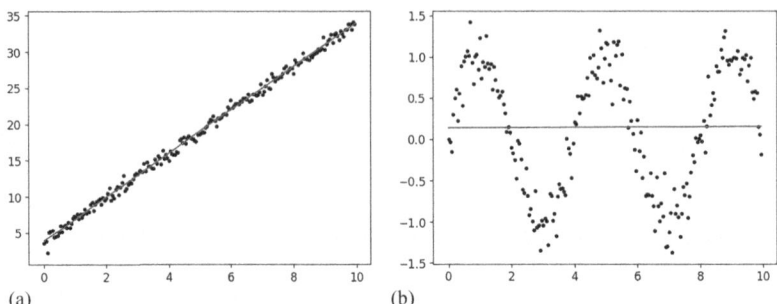

Fig. 7.1 Linear regression example. (**a**) Noisy line data and its linear regression result. (**b**) Noisy sine data and its linear regression result

samples as can be seen in Fig. 7.1b. The main reason is that the sine function is too complex for linear regression. In other words, output of a linear regressor will always be a line (or hyperplane) in N-dimensional space.

7.2.3   Deploying the Formed Regressor to the Microcontroller

We trained linear regressors for the line and sine data in Python on PC and saved the corresponding regression models in the previous section. We proceed with the linear regressor formed for the noisy sine data. The reader can also use the regressor formed for the line data by modifying the corresponding code block. The next phase is deploying the regressor to the STM32 microcontroller. Hence, we can use it in an actual scenario. To do so, we should generate the C header and source files of the trained regressor model on PC. We provide the Python script to export model parameters to C header and source files in Listing 7.3. In this script, we first load the trained linear regressor for the sine data by the function load using the saved model path as parameter. Then, we export the model parameters to two files "linear_reg_config.h" and "linear_reg_config.c" by the function export.

Listing 7.3 Exporting the trained linear regression model

```
import os.path as osp
from sklearn2c import LinearRegressor

# CHOOSE ONE OF THE MODELS
# Parameter linear_reg_config MUST NOT BE CHANGED
# SINCE IT IS INCLUDED IN C INFERENCE FILES

#line_model_path = osp.join("regression_models","
    linear_regressor_line.joblib")
sine_model_path = osp.join("regression_models","
    linear_regressor_sine.joblib")

export_path = osp.join("exported_models","regression","
    linear_reg_config")
```

```
#linear_regressor = LinearRegressor.load(line_model_path)
linear_regressor = LinearRegressor.load(sine_model_path)

linear_regressor.export(export_path)
```

The generated C header file for the example considered in this chapter is given
in Listing 7.4. This file only keeps the variable names and their size. They will be
used by other files to include the corresponding source file given in Listing 7.5. All
variables declared in Listing 7.4 are assigned here.

Listing 7.4 Header file for the linear regression model

```
#ifndef LINEAR_REG_CONFIG_H_INCLUDED
#define LINEAR_REG_CONFIG_H_INCLUDED
#define NUM_FEATURES 1
extern const float COEFFS[NUM_FEATURES];
extern const float OFFSET;
#endif
```

Listing 7.5 Source file for the linear regression model

```
#include "linear_reg_config.h"
const float COEFFS[NUM_FEATURES] = {-0.00016989};
const float OFFSET = 0.14527732756994105;
```

7.2.3.1 STM32CubeIDE Project Settings for Deployment
We should apply the following steps to deploy the trained regressor to the microcontroller in an STM32CubeIDE project. We should first copy the generated header
and source files to the "Core/Inc" and "Core/Src" folders of the STM32CubeIDE
project, respectively. Then, we should copy the files "linear_reg_inference.h"
and "linear_reg_inference.c" from the sklearn2c library to the "Core/Inc" and
"Core/Src" folders of the STM32CubeIDE project, respectively. These files contain
the required C functions to run the linear regressor on the microcontroller.

7.2.3.2 Mbed Studio Project Settings for Deployment
We should apply the following steps to deploy the trained regressor to the microcontroller in an Mbed Studio project. We should first copy the generated header and
source files to the root folder of the Mbed Studio project. Afterward, we should copy
the "linear_reg_inference.h" and "linear_reg_inference.c" files from the sklearn2c
library to the root folder of the Mbed Studio project.

7.2.4 Testing the Deployed Regressor on the Microcontroller

To check whether the deployed regressor works as expected, we can use the generated data on PC. Then, we can send them to the STM32 microcontroller. The regressor running on the microcontroller works on this data. Then, we can transfer the regression result back to PC and cross check with the same regressor working in Python on PC. Hence, we can make sure that the generated framework works as expected.

7.2.4.1 STM32CubeIDE Project Settings for Testing

In order to test the deployed regressor on the STM32 microcontroller, we can form a project under STM32CubeIDE. Then, we should add the libraries lib_rng and lib_serial to the project as explained in Sects. 5.1.3.1 and 4.1.2.2, respectively. Then, we should add the generated files in the previous section to the project. Hence, we will have the corresponding code block as follows.

```
/* USER CODE BEGIN Includes */
#include "linear_reg_inference.h"
#include "lib_serial.h"
#include "lib_rng.h"
/* USER CODE END Includes */
```

We should define macros related to the source of the input data, size of the input vector, and size of the output vector. The INPUT macrodefinition selects whether input data is obtained from PC or microcontroller at compile time. It can be selected as INPUT_PC or INPUT_MCU, respectively. SIZE_INPUT and SIZE_OUTPUT macros define the number of data inputs and outputs of the model, respectively. We should also define two floating-point arrays to allocate required memory space for inputs and outputs. Hence, we will have the corresponding code block as follows.

```
/* USER CODE BEGIN 0 */
#define INPUT_PC 1
#define INPUT_MCU 2
#define INPUT INPUT_PC

#define SIZE_INPUT NUM_FEATURES
#define SIZE_OUTPUT 1

float input[SIZE_INPUT];
float output[SIZE_OUTPUT];
/* USER CODE END 0 */
```

Next, we should form a while loop to acquire data either from PC or microcontroller according to the INPUT macro. Then, we can run the linear regressor by calling the function linear_reg_predict. The first parameter of this function is a pointer to the input array. The second parameter is a pointer to the output array. This output array is filled by the function linear_reg_predict with the prediction results. Finally, we transfer the regression result to PC. We provide the C code corresponding to these definitions in Listing 7.6. Please note that the Python script in Listing 7.7 should already be running on PC for executing this example successfully.

Listing 7.6   Linear regression test code for STM32CubeIDE

```
/* USER CODE BEGIN WHILE */
while (1)
{
/* USER CODE END WHILE */
/* USER CODE BEGIN 3 */
#if (INPUT == INPUT_PC)
 LIB_SERIAL_Receive(input, SIZE_INPUT, TYPE_F32);
#elif (INPUT == INPUT_MCU)
 for (uint32_t i = 0; i < SIZE_INPUT; ++i)
 input[i] = (float)(LIB_RNG_GetRandomNumber() % 1000) / 1000.0f;

 LIB_SERIAL_Transmit(input, SIZE_INPUT, TYPE_F32);
#endif
 linear_reg_predict(input, output);
 LIB_SERIAL_Transmit(output, SIZE_OUTPUT, TYPE_F32);
 HAL_Delay(1000);
}
/* USER CODE END 3 */
```

To check whether the deployed linear regressor works as expected, we need a test setup. We provide the Python script formed for this purpose in Listing 7.7. Here, we first load the previously generated data. Then, we load our trained linear regressor. Afterward, we send the test data to the microcontroller. Next, we perform prediction on PC and microcontroller separately. We finally print both prediction results on PC for comparison. Hence, we can make sure that the generated framework works as expected.

Listing 7.7   Python script for comparing linear regression results

```
import os.path as osp
import numpy as np
from sklearn2c import LinearRegressor
import py_serial

py_serial.SERIAL_Init("COM4")

train_samples = np.load(osp.join("regression_data", "reg_samples.
    npy"))
linear = LinearRegressor().load(osp.join("regression_models","
    linear_regressor_sine.joblib"))

i = 0
while 1:
    rqType, datalength, dataType = py_serial.
        SERIAL_PollForRequest()

    if rqType == py_serial.MCU_WRITES:
        # INPUT -> FROM MCU TO PC
        inputs = py_serial.SERIAL_Read()

    elif rqType == py_serial.MCU_READS:
```

```
# INPUT -> FROM PC TO MCU
inputs = train_samples[i:i+1].astype(py_serial.
    SERIAL_GetDType(dataType))
i = i + 1
if i >= len(train_samples):
    i = 0
py_serial.SERIAL_Write(inputs)

pcout = linear.predict(np.reshape(inputs, (1, datalength)))
rqType, datalength, dataType = py_serial.
    SERIAL_PollForRequest()
if rqType == py_serial.MCU_WRITES:
    mcuout = py_serial.SERIAL_Read()
    print()
    print("Inputs : " + str(inputs))
    print("PC Output : " + str(pcout))
    print("MCU Output : " + str(mcuout))
    print()
```

7.2.4.2 Mbed Studio Project Settings for Testing

In order to test the deployed linear regressor on the STM32 microcontroller, we should add the libraries lib_rng and lib_serial to the Mbed Studio project. To do so, we should copy the files "lib_rng.h," "lib_rng.c," "lib_uart.h," "lib_uart.c," "lib_serial.h," and "lib_serial.c" to our Mbed Studio project folder. Here, we also assume that the steps in Sect. 7.2.3.2 are already done.

The test code for the Mbed Studio project is given in Listing 7.8. This code is similar to the one given for the STM32CubeIDE project. Hence, the explanations there hold here as well. The only difference is that we should include the header file "lib_uart.h" and initialize the UART peripheral unit before the while loop by code in our Mbed Studio project. Please note that the Python script in Listing 7.7 should already be running on PC for executing this example successfully.

Listing 7.8 Linear regression test code for Mbed Studio

```
#include "mbed.h"
#include "linear_reg_inference.h"
#include "lib_serial.h"
#include "lib_rng.h"
#include "lib_uart.h"

#define INPUT_PC 1
#define INPUT_MCU 2
#define INPUT INPUT_PC

#define SIZE_INPUT NUM_FEATURES
#define SIZE_OUTPUT 1

float input[SIZE_INPUT];
float output[SIZE_OUTPUT];

int main()
{
LIB_UART_Init();
```

```
#if (INPUT == INPUT_MCU)
 LIB_RNG_Init();
#endif
while (true)
{
#if (INPUT == INPUT_PC)
 LIB_SERIAL_Receive(input, SIZE_INPUT, TYPE_F32);
#elif (INPUT == INPUT_MCU)
 for (uint32_t i = 0; i < SIZE_INPUT; ++i)
 input[i] = (float)(LIB_RNG_GetRandomNumber() % 1000) / 1000.0f;
  LIB_SERIAL_Transmit(input, SIZE_INPUT, TYPE_F32);
#endif
 linear_reg_predict(input, output);
 LIB_SERIAL_Transmit(output, SIZE_OUTPUT, TYPE_F32);
 HAL_Delay(1000);
 }
}
```

7.3 Polynomial Regression

Although linear regression is easy to implement, it may not model the relationship between the input and output variables sufficiently well. Polynomial regression can be used for this purpose. We will consider it in this section. To do so, we will first introduce the theoretical background. Then, we will form the polynomial regressor in Python on PC. Afterward, we will deploy the formed regressor to the microcontroller and test its performance there.

7.3.1 Theoretical Background

The relationship between the input and output variables is represented by a polynomial in this regression operation. Hence, we will replace the regressor function $f(\cdot)$ in Sect. 7.1 by $\hat{y} = \sum_{k=1}^{K} w_k x^k + b$. As in linear regression, we will estimate the parameters w_k and b by minimizing MSE wrt. given data for the training step. We will benefit from the matrix formalism introduced in Sect. 7.3 for this purpose. However, we should update matrix definitions.

We can explain theoretical background on training the polynomial regression by a second-order polynomial such that $\hat{y} = w_2 x^2 + w_1 x + b$. Hence, we can model the relationship between the input and output variables as

$$\mathbf{x} = \begin{bmatrix} x^2[1] & x[1] & 1 \\ \vdots & \vdots & \\ x^2[N] & x[N] & 1 \end{bmatrix} \quad \mathbf{w} = \begin{bmatrix} w_2 \\ w_1 \\ b \end{bmatrix} \tag{7.8}$$

Derivations in Sect. 7.2.1 will hold since we have the same matrix form as in linear regression. Hence, we will obtain the polynomial regressor parameters w_2,

w_1, and b by Eq. 7.7. As we obtain parameters for the polynomial regressor, it will be ready to be used for prediction. Therefore, when an input x is fed to the function $\hat{y} = w_2 x^2 + w_1 x + b$, it will generate the corresponding output $\hat{y}$. We can call this value as the polynomial regressor output or prediction.

The same derivations hold when we use a polynomial with degree more than two. Only the matrices in Eq. 7.7 should be updated accordingly. Moreover, we can extend the polynomial regression operations to problems with more than one input and output as in linear regression. The reader should only update the corresponding matrices. Then, the remaining operations will be the same.

7.3.2 Forming the Regressor in Python

As in Sect. 7.2.2, we can implement theoretical operations introduced in the previous section via Python using the sklearn2c library. We provide a working example for polynomial regression in Listing 7.9. Here, we first apply the data loading steps as in Listing 7.1. Our aim is to predict the actual line and sine data by polynomial regression. Therefore, we benefit from the object `PolynomialRegressor` in the sklearn2c library. To be more specific, we form two objects `line_poly_model` and `sine_poly_model`. The next step is fitting the polynomial regression model to noisy line and sine data separately. Here, the aim is to minimize the MSE between the actual and predicted values.

The difference between linear and polynomial regression is that we can use the Nth power of features and their products to form new features in the latter approach. For example, if we have two features x_1 and x_2, then the second-order polynomials will be x_1^2, $x_1 x_2$, x_2^2. We also include the features to these terms, as if they are first-order polynomials. Eventually, we obtain polynomials up to degree two. sklearn2c library has a simple interface to create such features. In Listing 7.9, we form second-order features within the polynomial regressor by the parameter `deg = 2` in the object `PolynomialRegressor`. Afterward, we follow the same steps as in Listing 7.1. Finally, the trained regressors for the line and sine data are saved as "poly_regressor_line.joblib" and "poly_regressor_sine.joblib" files under the folder "regression_models."

Listing 7.9 Training the polynomial regressor and saving the trained model

```
import os.path as osp
import numpy as np
from sklearn2c import PolynomialRegressor

train_samples = np.load(osp.join("regression_data", "reg_samples.
    npy"))
train_line_values = np.load(osp.join("regression_data", "
    reg_line_values.npy"))
train_sine_values = np.load(osp.join("regression_data", "
    reg_sine_values.npy"))

line_poly_model = PolynomialRegressor(deg = 2)
```

```
poly_save_path = osp.join("regression_models","
    poly_regressor_line.joblib")
line_poly_model.train(train_samples, train_line_values,
    poly_save_path)

sine_poly_model = PolynomialRegressor(deg = 2)
poly_save_path = osp.join("regression_models","
    poly_regressor_sine.joblib")
sine_poly_model.train(train_samples, train_sine_values,
    poly_save_path)
```

In order to apply prediction with the saved polynomial regressors in Listing 7.9, we should again load the noisy line and sine data points and then proceed with prediction. We provide one such example in Listing 7.10. Here, we start with loading the samples, line_values, and sine_values. Then, we load the polynomial regressors for the line and sine data by the function load using the saved model path as parameter. Afterward, we obtain the regressor outputs (predictions) by the function predict. As we obtain predicted values from the models, we assign them to variables line_predictions and sine_predictions for the line and sine data, respectively. Rest of the code plots the input data with its target value as well as prediction of the polynomial regressors.

Listing 7.10 Predictions with the polynomial regressor

```
import os.path as osp
import numpy as np
from matplotlib import pyplot as plt
from sklearn2c import PolynomialRegressor

samples = np.load(osp.join("regression_data", "reg_samples.npy"))
line_values = np.load(osp.join("regression_data", "
    reg_line_values.npy"))
sine_values = np.load(osp.join("regression_data", "
    reg_sine_values.npy"))

line_model_path = osp.join("regression_models", "
    poly_regressor_line.joblib")
sine_model_path = osp.join("regression_models", "
    poly_regressor_sine.joblib")

poly_regressor_line = PolynomialRegressor.load(line_model_path)
poly_regressor_sine = PolynomialRegressor.load(sine_model_path)

line_predictions = poly_regressor_line.predict(samples)
sine_predictions = poly_regressor_sine.predict(samples)

# Plotting
fig, (ax1, ax2) = plt.subplots(2, 1)
ax1.plot(samples, line_values, "k.")
ax1.plot(samples, line_predictions, "b")
ax2.plot(samples, sine_values, "k.")
ax2.plot(samples, sine_predictions, "b")
plt.show()
```

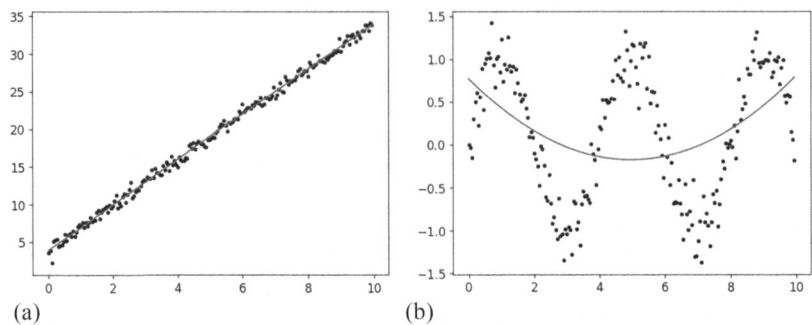

Fig. 7.2 Polynomial regression example with the usage of second-order polynomial in operation. (**a**) Noisy line data and its polynomial regression result. (**b**) Noisy sine data and its polynomial regression result

We provide the obtained predictions from Listing 7.10 for the line and sine data in Fig. 7.2. As can be seen in this figure, the second-order polynomial based regressor works well with the line data. Hence, it can predict the underlying line function as in Fig. 7.2a. However, it cannot predict the underlying sine function from the given input-output samples as can be seen in Fig. 7.2b. Again, the main reason is that the sine function is too complex for the second-order polynomial based regressor. Therefore, we should increase the polynomial order.

We can increase the polynomial order in Listing 7.9 by the code line `PolynomialRegressor(deg = 6)` for the noisy line and sine data. Then, we can retrain the models. We provide the obtained results for the sixth-order polynomial in Fig. 7.3. As can be seen in this figure, the sixth-order polynomial based regressor successfully predicts the sine function and is not affected by noise in data. However, the regressor is now affected by the noise term in the line data. In other words, it also learns noise and predicts it as well as line data. Therefore, we can say that the second-order polynomial regression is not sufficient to estimate the sine function and underfits to it. The sixth-order polynomial overfits to data and estimates noise too. That is why higher-order polynomial regression is not always the best option.

7.3.3   Deploying the Formed Regressor to the Microcontroller

We trained polynomial regressors for the line and sine data in Python on PC and saved the corresponding regression models in the previous section. We proceed with the polynomial regressor formed for the noisy sine data as in Sect. 7.2.3. The reader can also use the regressor formed for the line data by modifying the corresponding code block. The next phase is deploying the regressor to the STM32 microcontroller. Hence, we can use it in an actual scenario. To do so, we should generate the C header and source files of the trained regressor model on PC. We provide the Python script to export model parameters to C header and source files in Listing 7.11. In this script,

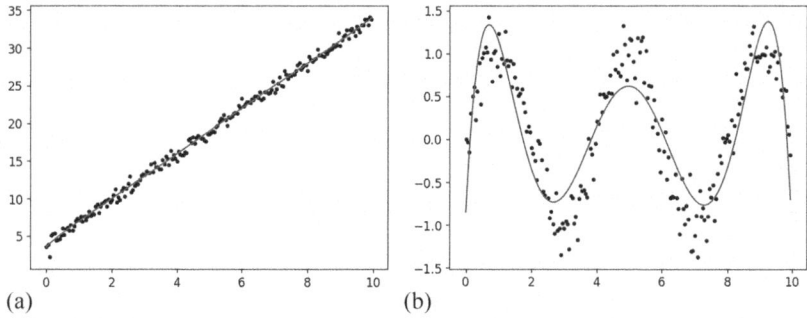

Fig. 7.3 Polynomial regression example with the usage of sixth-order polynomial in operation. (**a**) Noisy line data and its polynomial regression result. (**b**) Noisy sine data and its polynomial regression result

we first load the trained polynomial regressor for the sine data by the function `load` using the saved model path as parameter. Then, we export the model parameters to two files "poly_reg_config.h" and "poly_reg_config.c" by the function `export`.

Listing 7.11 Exporting the trained polynomial regression model

```
import os.path as osp
from sklearn2c import PolynomialRegressor

# CHOOSE ONE OF THE MODELS
# Parameter poly_reg_config MUST NOT BE CHANGED
# SINCE IT IS INCLUDED IN C INFERENCE FILES

#line_model_path = osp.join("regression_models","
    poly_regressor_line.joblib")
sine_model_path = osp.join("regression_models","
    poly_regressor_sine.joblib")

export_path = osp.join("exported_models","regression","
    poly_reg_config")

#poly_regressor = PolynomialRegressor.load(line_model_path)
poly_regressor = PolynomialRegressor.load(sine_model_path)

poly_regressor.export(export_path)
```

The generated C header file for the example considered in this chapter is given in Listing 7.12. This file only keeps the variable names and their size. They will be used by other files to include the corresponding source file given in Listing 7.13. All variables declared in Listing 7.12 are assigned here.

Listing 7.12 Header file for the polynomial regression model

```
#ifndef POLY_REG_CONFIG_H_INCLUDED
#define POLY_REG_CONFIG_H_INCLUDED
#define NUM_FEATURES 6
#define NUM_INPUTS 1
extern const float COEFFS[NUM_FEATURES];
extern const float OFFSET;
extern char *feature_names[NUM_FEATURES];
#endif
```

Listing 7.13 Source file for the polynomial regression model

```
#include "poly_reg_config.h"
const float COEFFS[NUM_FEATURES] = { 7.13940067e+00,-8.03689477e
    +00, 3.33981721e+00,-6.36406464e-01, 5.64113211e
    -02,-1.88486684e-03};
const float OFFSET = -0.7198341978970282;
char *feature_names[NUM_FEATURES] = {"0","0*0","0*0*0","0*0*0*0",
    "0*0*0*0*0","0*0*0*0*0*0"};
```

7.3.3.1 STM32CubeIDE Project Settings for Deployment

We should apply the following steps to deploy the trained regressor to the microcontroller in an STM32CubeIDE project. We should first copy the generated header and source files in the previous section to the "Core/Inc" and "Core/Src" folders of the STM32CubeIDE project, respectively. Then, we should copy the files "poly_reg_inference.h" and "poly_reg_inference.c" from the sklearn2c library to the "Core/Inc" and "Core/Src" folders of the STM32CubeIDE project, respectively. These files contain the required C functions to run the polynomial regressor on the microcontroller.

7.3.3.2 Mbed Studio Project Settings for Deployment

We should apply the following steps to deploy the trained regressor to the microcontroller in an Mbed Studio project. We should first copy the generated header and source files in the previous section to the root folder of the Mbed Studio project. Afterward, we should copy the "poly_reg_inference.h" and "poly_reg_inference.c" files from the sklearn2c library to the root folder of the Mbed Studio project.

7.3.4 Testing the Deployed Regressor on the Microcontroller

To check whether the deployed regressor works as expected, we can use the generated data on PC. Then, we can send them to the STM32 microcontroller. The regressor running on the microcontroller works on this data. Then, we can transfer the regression result back to PC and cross check with the same regressor working in Python on PC. Hence, we can make sure that the generated framework works as expected.

7.3.4.1 STM32CubeIDE Project Settings for Testing

In order to test the deployed regressor on the STM32 microcontroller, we can use the test setup formed for the linear regressor in Sect. 7.2.4. Then, we should add the generated files in the previous section to our project. Hence, we will have the corresponding code block as follows.

```
/* USER CODE BEGIN Includes */
#include ''poly_reg_inference.h''
#include ''lib_serial.h''
#include "lib_rng.h"
/* USER CODE END Includes */
```

We can use the macrodefinitions formed for linear regression in Sect. 7.2.4 in the test setup here as well. Hence, we will have the following C code snippet for polynomial regression. The reader should pay attention to the code line #define SIZE_INPUT NUM_INPUTS below. This code line was #define SIZE_INPUT NUM_FEATURES for linear regression. There, the input size was equal to the number of features. However, the input size is not equal to number of features in polynomial regression. Therefore, we had to define the input size as such.

```
/* USER CODE BEGIN 0 */
#define INPUT_PC 1
#define INPUT_MCU 2
#define INPUT   INPUT_PC

#define SIZE_INPUT NUM_INPUTS
#define SIZE_OUTPUT 1

float input[SIZE_INPUT];
float output[SIZE_OUTPUT];
/* USER CODE END 0 */
```

Next, we should form a while loop to acquire data either from PC or microcontroller according to the INPUT macro. Then, we can run the polynomial regressor by the function poly_reg_predict. The first parameter of this function is a pointer to the input array. The second parameter is a pointer to the output array. This output array is filled by the function poly_reg_predict with prediction results. Finally, we transfer the regression result to PC. We provide the C code corresponding to these definitions in Listing 7.14. Please note that the Python script in Listing 7.15 should already be running on PC for executing this example successfully.

Listing 7.14 Polynomial regression test code for STM32CubeIDE

```
/* USER CODE BEGIN WHILE */
while (1)
{
```

```
/* USER CODE END WHILE */

/* USER CODE BEGIN 3 */
#if (INPUT == INPUT_PC)
 LIB_SERIAL_Receive(input, SIZE_INPUT, TYPE_F32);
#elif (INPUT == INPUT_MCU)
 for (uint32_t i = 0; i < SIZE_INPUT; ++i)
 input[i] = (float)(LIB_RNG_GetRandomNumber() % 1000) / 1000.0f;
 LIB_SERIAL_Transmit(input, SIZE_INPUT, TYPE_F32);
#endif
poly_reg_predict(input, output);
LIB_SERIAL_Transmit(output, SIZE_OUTPUT, TYPE_F32);
HAL_Delay(1000);
}
/* USER CODE END 3 */
```

To check whether the deployed polynomial regressor works as expected, we need a test setup. We provide the Python script formed for this purpose in Listing 7.15. Here, we first load the previously generated data. Then, we load our trained polynomial regressor. Afterward, we send the test data to the microcontroller. Next, we perform inference on PC and microcontroller separately. We finally print both inference results on PC for comparison. Hence, we can make sure that the generated framework works as expected.

Listing 7.15 Python script for comparing polynomial regression results

```
import os.path as osp
import numpy as np
from sklearn2c import PolynomialRegressor
import py_serial

py_serial.SERIAL_Init("COM4")
train_samples = np.load(osp.join("regression_data", "reg_samples.
    npy"))
poly = PolynomialRegressor.load(osp.join("regression_models","
    poly_regressor_sine.joblib"))

i = 0
while 1:
    rqType, datalength, dataType = py_serial.
        SERIAL_PollForRequest()

    if rqType == py_serial.MCU_WRITES:
        # INPUT -> FROM MCU TO PC
        inputs = py_serial.SERIAL_Read()

    elif rqType == py_serial.MCU_READS:
        # INPUT -> FROM PC TO MCU
        inputs = train_samples[i.i+1].astype(py_serial.
            SERIAL_GetDType(dataType))
        i = i + 1
        if i >= len(train_samples):
            i = 0
        py_serial.SERIAL_Write(inputs)

    pcout = poly.predict(inputs)
```

```
    rqType, datalength, dataType = py_serial.
        SERIAL_PollForRequest()
    if rqType == py_serial.MCU_WRITES:
        mcuout = py_serial.SERIAL_Read()
        print()
        print("Inputs : " + str(inputs))
        print("PC Output : " + str(pcout))
        print("MCU Output : " + str(mcuout))
        print()
```

7.3.4.2 Mbed Studio Project Settings for Testing

In order to test the deployed polynomial regressor on the STM32 microcontroller, we should add the libraries lib_rng and lib_serial to the Mbed Studio project. To do so, we should copy the files "lib_rng.h," "lib_rng.c," "lib_uart.h," "lib_uart.c," "lib_serial.h," and "lib_serial.c" to our Mbed Studio project folder. Here, we also assume that the steps in Sect. 7.3.3.2 are already done.

The test code for the Mbed Studio project is given in Listing 7.16. This code is similar to the one given for the STM32CubeIDE project. Hence, the explanations there hold here as well. The only difference is that we should include the header file "lib_uart.h" and initialize the UART peripheral unit before the while loop by code in our Mbed Studio project. Please note that the Python script in Listing 7.15 should already be running on PC for executing this example successfully.

Listing 7.16 Polynomial regression test code for Mbed Studio

```
#include "mbed.h"
#include "poly_reg_inference.h"
#include "lib_serial.h"
#include "lib_rng.h"
#include "lib_uart.h"

#define INPUT_PC 1
#define INPUT_MCU 2
#define INPUT INPUT_PC

#define SIZE_INPUT NUM_INPUTS
#define SIZE_OUTPUT 1

float input[SIZE_INPUT];
float output[SIZE_OUTPUT];

int main()
{
LIB_UART_Init();
#if (INPUT == INPUT_MCU)
  LIB_RNG_Init();
#endif
while (true)
{
#if (INPUT == INPUT_PC)
  LIB_SERIAL_Receive(input, SIZE_INPUT, TYPE_F32);
#elif (INPUT == INPUT_MCU)
  for (uint32_t i = 0; i < SIZE_INPUT; ++i)
  input[i] = (float)(LIB_RNG_GetRandomNumber() % 1000) / 1000.0f;
  LIB_SERIAL_Transmit(input, SIZE_INPUT, TYPE_F32);
#endif
```

```
poly_reg_predict(input, output);
LIB_SERIAL_Transmit(output, SIZE_OUTPUT, TYPE_F32);
HAL_Delay(1000);
}
}
```

7.4   kNN Regression

kNN regression is the first nonparametric approach to be evaluated in this chapter. To do so, we start with introducing its theoretical background. Then, we will form the kNN regressor in Python on PC. Afterward, we will deploy the formed regressor to the microcontroller and test its performance there.

7.4.1   Theoretical Background

Working principles of the kNN regressor is similar to the kNN classifier introduced in Sect. 6.3. However, the regressor has its own properties. Therefore, we can summarize working principles of the kNN regressor as follows. Assume that we would like to form the regressor output for a given input. First, we obtain the k nearest neighbors of this input from the training set. Then, the regressor output is calculated as the average of these values. We can call this as both the training and prediction step since the kNN regressor directly depends on training data while forming its output.

As in the kNN classifier, the k value is set by the user in regression. The higher the value of k, the more crude is the regression output. The limiting case for k is 1. Here, the nearest value is obtained from the training set. Then, it is fed as the regressor output or prediction.

Calculating the average value for the regressor output can be done in two different ways as uniform and weighted. All selected neighbors are assigned equal weight while calculating the average in the uniform case. Selected neighbors have nonuniform weight assigned to them in the weighted case. One such weight assignment can be taken as inverse of the distance between the regressor output and its neighbor from the training set. Hence, nearby neighbors have more weight compared to far away neighbors.

7.4.2   Forming the Regressor in Python

As in Sect. 7.2.2, we can implement theoretical operations introduced in the previous section via Python by using the sklearn2c library. We provide a working example for kNN regression in Listing 7.17. Here, we first apply the data generation steps as in Listing 7.1. The next step is fitting the kNN regression model to noisy line and sine data separately. Our aim is to predict the actual line and sine data by kNN regression. Therefore, we benefit from the object KNNRegressor in the sklearn2c library. To be more specific, we form two objects line_knn_model and sine_knn_model. While doing so, we set the neighborhood parameter for the kNN regression as one by the code line KNNRegressor(n_neighbors =1) for both cases. Afterward, we follow the same steps as in Listing 7.1. Finally, the trained regressors for the line and sine data are saved as "knn_regressor_line.joblib" and "knn_regressor_sine.joblib" files under the folder "regression_models."

Listing 7.17   Training the kNN regressor and saving the trained model

```
import os.path as osp
import numpy as np
from sklearn2c import KNNRegressor

train_samples = np.load(osp.join("regression_data", "reg_samples.
    npy"))
train_line_values = np.load(osp.join("regression_data", "
    reg_line_values.npy"))
train_sine_values = np.load(osp.join("regression_data", "
    reg_sine_values.npy"))

line_knn_model = KNNRegressor(n_neighbors = 1)
knn_save_path = osp.join("regression_models","knn_regressor_line.
    joblib")
line_knn_model.train(train_samples, train_line_values,
    knn_save_path)

sine_knn_model = KNNRegressor(n_neighbors = 1)
knn_save_path = osp.join("regression_models","knn_regressor_sine.
    joblib")
sine_knn_model.train(train_samples, train_sine_values,
    knn_save_path)
```

In order to apply prediction with the saved kNN regressors in Listing 7.17, we should again load the noisy line and sine data points and then proceed with prediction. We provide such an example in Listing 7.18. Here, we start with loading the samples, line_values, and sine_values. Then, we load the kNN regressors for the line and sine data by the function load using the saved model path as parameter. Afterward, we obtain the regressor outputs (predictions) by the function predict. As we obtain predicted values from the models, we assign them to variables line_predictions and sine_predictions for the line and sine data, respectively. Rest of the code plots the input data with its target values as well as predictions of the kNN regressors.

Listing 7.18 Predictions with the kNN regressor

```
import os.path as osp
import numpy as np
from sklearn2c import KNNRegressor
from matplotlib import pyplot as plt

samples = np.load(osp.join("regression_data", "reg_samples.npy"))
line_values = np.load(osp.join("regression_data", "
    reg_line_values.npy"))
sine_values = np.load(osp.join("regression_data", "
    reg_sine_values.npy"))

line_model_path = osp.join("regression_models", "
    knn_regressor_line.joblib")
sine_model_path = osp.join("regression_models", "
    knn_regressor_sine.joblib")

knn_regressor_line = KNNRegressor.load(line_model_path)
knn_regressor_sine = KNNRegressor.load(sine_model_path)

line_predictions = knn_regressor_line.predict(samples)
sine_predictions = knn_regressor_sine.predict(samples)

fig, (ax1, ax2) = plt.subplots(2, 1)
ax1.plot(samples, line_values, "k.")
ax1.plot(samples, line_predictions, "b")
ax2.plot(samples, sine_values, "k.")
ax2.plot(samples, sine_predictions, "b")
plt.show()
```

We provide the obtained results from Listing 7.18 for the line and sine data in Fig. 7.4. As can be seen in this figure, the kNN regressor provides highly quantized predictions. The main reason for this result is that the regressor output depends on the nearest neighbor. Thus, the model tends to overfit to data. Therefore, we should increase the neighborhood number.

We can increase the neighborhood number in the kNN regressor in Listing 7.17 by the code line `KNeighborsRegressor(n_neighbors=5)` for the noisy line and sine

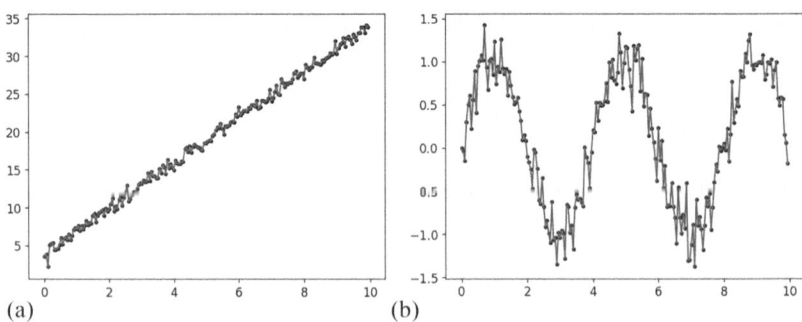

(a)   (b)

Fig. 7.4 kNN regression example with the usage of one neighbor in operation. (**a**) Noisy line data and its kNN regression result. (**b**) Noisy sine data and its kNN regression result

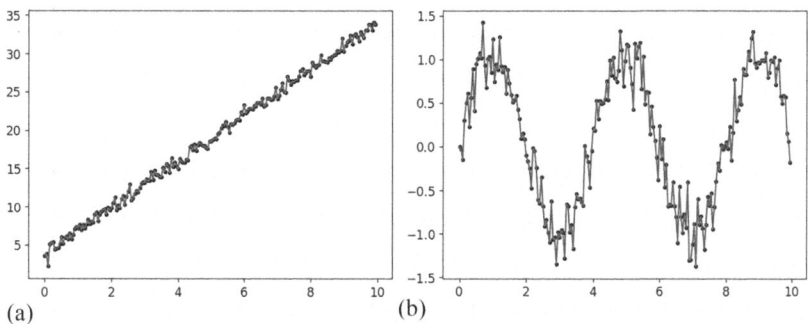

Fig. 7.5 kNN regression example with the usage of five neighbors in operation. (**a**) Noisy line data and its kNN regression result. (**b**) Noisy sine data and its kNN regression result

data. Then, we can retrain the models. We provide the obtained results in Fig. 7.5. As can be seen in this figure, the regression result is smoother compared to previous results.

7.4.3   Deploying the Formed Regressor to the Microcontroller

We trained kNN regressors for the line and sine data in Python on PC and saved the corresponding regression models in the previous section. We proceed with the kNN regressor formed for the noisy sine data as in Sect. 7.2.3. The reader can also use the regressor formed for the line data by modifying the corresponding code block. The next phase is deploying the regressor to the STM32 microcontroller. Hence, we can use it in an actual scenario. To do so, we should generate the C header and source files of the trained regressor model on PC. We provide the Python script to export model parameters to C header and source files in Listing 7.19. In this script, we first load the trained kNN regressor for the sine data by the function load using the saved model path as parameter. Then, we export the model parameters to two files "knn_reg_config.h" and "knn_reg_config.c" by the function export.

Listing 7.19 Exporting the trained kNN regression model

```
import os.path as osp
from sklearn2c import KNNRegressor

# CHOOSE ONE OF THE MODELS
# Parameter knn_reg_config MUST NOT BE CHANGED
# SINCE IT IS INCLUDED IN C INFERENCE FILES

#line_model_path = osp.join("regression_models","
    knn_regressor_line.joblib")
sine_model_path = osp.join("regression_models","
    knn_regressor_sine.joblib")

export_path = osp.join("exported_models","regression","
    knn_reg_config")

#knn_regressor = KNNRegressor.load(line_model_path)
```

```
knn_regressor = KNNRegressor.load(sine_model_path)

knn_regressor.export(export_path)
```

The generated C header file for the example considered in this chapter is given in Listing 7.20. This file only keeps the variable names and their size. They will be used by other files to include the corresponding source file given in Listing 7.21. All variables declared in Listing 7.20 are assigned here.

Listing 7.20 Header file for the kNN regression configuration

```
#ifndef KNN_REG_CONFIG_H_INCLUDED
#define KNN_REG_CONFIG_H_INCLUDED
#define NUM_NEIGHBORS 5
#define NUM_FEATURES 1
#define NUM_SAMPLES 200
extern const float DATA[NUM_SAMPLES][NUM_FEATURES];
extern const float DATA_VALUES[NUM_SAMPLES];
#endif
```

Listing 7.21 Source file for the kNN regression configuration

```
#include "knn_reg_config.h"
const float DATA[NUM_SAMPLES][NUM_FEATURES] = {{0.}, {0.05},...,
    {9.9 },{9.95}};
const float DATA_VALUES[NUM_SAMPLES] = {{ 0.07155747}, { 0.190616
    },..., { 0.1335265 }, { 0.32602236}};
```

7.4.3.1 STM32CubeIDE Project Settings for Deployment

We should apply the following steps to deploy the trained regressor to the microcontroller in an STM32CubeIDE project. We should first copy the generated header and source files in the previous section to the "Core/Inc" and "Core/Src" folders of the STM32CubeIDE project, respectively. Then, we should copy the files "knn_reg_inference.h" and "knn_reg_inference.c" from the sklearn2c library to the "Core/Inc" and "Core/Src" folders of the STM32CubeIDE project, respectively. These files contain the required C functions to run the kNN regressor on the microcontroller.

7.4.3.2 Mbed Studio Project Settings for Deployment

We should apply the following steps to deploy the trained regressor to the microcontroller in an Mbed Studio project. We should first copy the generated header and source files in the previous section to the root folder of the Mbed Studio project. Afterward, we should copy the "knn_reg_inference.h" and "knn_reg_inference.c" files from the sklearn2c library to the root folder of the Mbed Studio project.

7.4.4 Testing the Deployed Regressor on the Microcontroller

To check whether the deployed regressor works as expected, we can use the generated data on PC. Then, we can send them to the STM32 microcontroller. The regressor running on the microcontroller works on this data. Then, we can transfer the regression result back to PC and cross check with the same regressor working in Python on PC. Hence, we can make sure that the generated framework works as expected.

7.4.4.1 STM32CubeIDE Project Settings for Testing
In order to test the deployed regressor on the STM32 microcontroller, we can use the test setup formed for the linear regressor in Sect. 7.2.4. Then, we should add the generated files in the previous section to our project. Hence, we will have the corresponding code block as follows.

```
/* USER CODE BEGIN Includes */
#include "knn_reg_inference.h"
#include "lib_serial.h"
#include "lib_rng.h"
/* USER CODE END Includes */
```

We can use the same macrodefinitions formed for linear regression in Sect. 7.2.4 in the test setup here as well. Hence, we will have the following C code for kNN regression.

```
/* USER CODE BEGIN 0 */
#define INPUT_PC 1
#define INPUT_MCU 2
#define INPUT INPUT_PC

#define SIZE_INPUT NUM_FEATURES
#define SIZE_OUTPUT 1

float input[SIZE_INPUT];
float output[SIZE_OUTPUT];
/* USER CODE END 0 */
```

Next, we should form a while loop to acquire data either from PC or microcontroller according to the INPUT macro. Then, we can run the kNN regressor by the function knn_reg_predict. The first parameter of this function is a pointer to the input array. The second parameter is a pointer to the output array. This output array is filled by the function knn_reg_predict with prediction results. Finally, we transfer the regression result to PC. We provide the C code corresponding to these definitions in Listing 7.22. Please note that the Python script in Listing 7.23 should already be running on PC for executing this example successfully.

Listing 7.22 kNN regression test code for STM32CubeIDE

```
/* USER CODE BEGIN WHILE */
while (1)
{
/* USER CODE END WHILE */

/* USER CODE BEGIN 3 */
#if (INPUT == INPUT_PC)
 LIB_SERIAL_Receive(input, SIZE_INPUT, TYPE_F32);
#elif (INPUT == INPUT_MCU)
 for (uint32_t i = 0; i < SIZE_INPUT; ++i)
 input[i] = (float)(LIB_RNG_GetRandomNumber() % 1000) / 1000.0f;
 LIB_SERIAL_Transmit(input, SIZE_INPUT, TYPE_F32);
#endif
knn_reg_predict(input, output);
LIB_SERIAL_Transmit(output, SIZE_OUTPUT, TYPE_F32);
HAL_Delay(1000);
}
/* USER CODE END 3 */
```

To check whether the deployed kNN regressor works as expected, we need a test setup. We provide the Python script formed for this purpose in Listing 7.23. Here, we first load the previously generated data. Then, we load our trained kNN regressor. Afterward, we send the test data to the microcontroller. Next, we perform inference on PC and microcontroller separately. We finally print both inference results on PC for comparison. Hence, we can make sure that the generated framework works as expected.

Listing 7.23 Python script for comparing kNN regression results

```
import os.path as osp
import numpy as np
from sklearn2c import KNNRegressor
import py_serial

py_serial.SERIAL_Init("COM4")
train_samples = np.load(osp.join("regression_data", "reg_samples.
    npy"))
knn = KNNRegressor().load(osp.join("regression_models","
    knn_regressor_sine.joblib"))

i = 0
while 1:
    rqType, datalength, dataType = py_serial.
        SERIAL_PollForRequest()

    if rqType == py_serial.MCU_WRITES:
        # INPUT -> FROM MCU TO PC
        inputs = py_serial.SERIAL_Read()

    elif rqType == py_serial.MCU_READS:
        # INPUT -> FROM PC TO MCU
```

```
            inputs = train_samples[i:i+1].astype(py_serial.
                SERIAL_GetDType(dataType))
            i = i + 1
            if i >= len(train_samples):
                i = 0
            py_serial.SERIAL_Write(inputs)

        pcout = knn.predict(inputs)
        rqType, datalength, dataType = py_serial.
            SERIAL_PollForRequest()
        if rqType == py_serial.MCU_WRITES:
            mcuout = py_serial.SERIAL_Read()
            print()
            print("Inputs : " + str(inputs))
            print("PC Output : " + str(pcout))
            print("MCU Output : " + str(mcuout))
            print()
```

7.4.4.2 Mbed Studio Project Settings for Testing

In order to test the deployed kNN regressor on the STM32 microcontroller, we should add the libraries lib_rng and lib_serial to the Mbed Studio project. To do so, we should copy the files "lib_rng.h," "lib_rng.c," "lib_uart.h," "lib_uart.c," "lib_serial.h," and "lib_serial.c" to our Mbed Studio project folder. Here, we also assume that the steps in Sect. 7.4.3.2 are already done.

The test code for the Mbed Studio project is given in Listing 7.24. This code is similar to the one given for the STM32CubeIDE project. Hence, the explanations there hold here as well. The only difference is that we should include the header file "lib_uart.h" and initialize the UART peripheral unit before the while loop by code in our Mbed Studio project. Please note that the Python script in Listing 7.23 should already be running on PC for executing this example successfully.

Listing 7.24 kNN regression test code for Mbed Studio

```
#include "mbed.h"
#include "knn_reg_inference.h"
#include "lib_serial.h"
#include "lib_rng.h"
#include "lib_uart.h"

#define INPUT_PC 1
#define INPUT_MCU 2
#define INPUT INPUT_PC

#define SIZE_INPUT NUM_FEATURES
#define SIZE_OUTPUT 1

float input[SIZE_INPUT];
float output[SIZE_OUTPUT];

int main()
{
LIB_UART_Init();
#if (INPUT == INPUT_MCU)
```

```
LIB_RNG_Init();
#endif
while (true)
{
#if (INPUT == INPUT_PC)
 LIB_SERIAL_Receive(input, SIZE_INPUT, TYPE_F32);
#elif (INPUT == INPUT_MCU)
 for (uint32_t i = 0; i < SIZE_INPUT; ++i)
 input[i] = (float)(LIB_RNG_GetRandomNumber() % 1000) / 1000.0f;
 LIB_SERIAL_Transmit(input, SIZE_INPUT, TYPE_F32);
#endif
knn_reg_predict(input, output);
LIB_SERIAL_Transmit(output, SIZE_OUTPUT, TYPE_F32);
HAL_Delay(1000);
}
}
```

7.5   Decision Tree Regression

Decision tree regression has a nonparametric approach to model the relationship
between the input and output variables. We will consider it in this section. To do so,
we will first introduce the theoretical background for decision tree regression. Then,
we will form the regressor in Python on PC. Afterward, we will deploy the formed
regressor to the microcontroller and test its performance there.

7.5.1   Theoretical Background

Decision tree regression resembles the decision tree classifier introduced in
Sect. 6.5. Therefore, we explain it in line with the definitions given there. While
doing so, we will emphasize the differences between the regressor and classifier.
Let's get started.

As in the classifier case, the decision tree regressor works by successively
splitting data. In this operation, no global parametric model is assumed for the
data at hand. Therefore, the decision tree regressor can be categorized under the
nonparametric regression group. Let's focus on the training and prediction steps of
the regressor.

Data is split successively based on the selected feature and threshold values in
training phase of the decision tree regression. We will use binary decision tree for
this purpose. This tree consists of nodes, which split the given data into two disjoint
and exhaustive subsets from the previous set or subset. The initial node which keeps
the overall training data is called the root. The final nodes which keep subsets
obtained at the end of training are called leaves. The aim in splitting the data is to
reduce impurity at each successive level compared to the previous one. The impurity
measure used in regression is different from the ones used in classification. What we
can represent is as follows. During splitting of data, the aim is having smaller sum of

MSE values over all nodes in the current level, compared to all nodes in the previous level.

MSE value for each node is calculated between the output of that node compared to data in that node. The usual practice for calculating the output value for a node is by calculating the average of data in that branch. As a result, MSE for that node turns out to be variance of data in that node [1]. Hence, the sum of node variances is minimized during branching.

Based on the available training data, we should choose the used feature and its threshold value at each split level to form the decision tree. There are different methods in literature for this purpose. In this book, we do not focus on how this operation is done. We depend on the scikit-learn library implementation of decision tree formation for regression.

As the splits end after successive steps in training the decision tree regressor, several leaf nodes will be formed. Each leaf node will consist of several sample values. We assign average value of these samples as the regression result for that leaf node. We perform this operation for all leaf nodes to obtain the final decision tree regressor.

The user can decide on the number of splits while forming the decision tree for regression. We would like to have pure leaf nodes at the end of training. Hence, each leaf node consists of the same or similar sample values. This can be achieved by increasing the split number. We will have a detailed input-output representation as we increase the split number. However, this may not be desirable when there is noise in data. Here, pure leaves in fact correspond to overfitting. On the other hand, we will have a crude input-output representation when the split number is low. This may not be desirable during the prediction phase. Therefore, the user should be cautious while setting the split number.

As training of the decision tree regressor is done, it can be represented by a successive set of if-else rules. In other words, successive binary decisions can be used to implement the decision tree regressor. The inference step for the regressor is also realized in the same way. Therefore, when an input is fed to the regressor, it is tested via successive binary rules. The final reached leaf node value is fed as the regressor output or prediction.

7.5.2 Forming the Regressor in Python

As in Sect. 7.2.2, we can implement theoretical operations introduced in the previous section via Python by the sklearn2c library. We provide a working example for decision tree regression in Listing 7.25. Here, we first apply the data generation steps as in Listing 7.1. The next step is fitting the decision tree regression model to noisy line and sine data separately. Our aim is to predict the actual line and sine data by decision tree regression. Therefore, we benefit from the object DTRegressor in the sklearn2c library. To be more specific, we form two objects line_dtr_model and sine_dtr_model. While doing so, we set the depth parameter for the decision tree regression as max_depth = 2 by the code line DTRegressor (max_depth =2) for both cases. Afterward, we follow the same steps as in Listing 7.1. Finally, the trained regressors for the line and sine data are saved as "dt_regressor_line.joblib" and "dt_regressor_sine.joblib" files under the folder "regression_models."

Listing 7.25 Training the decision tree regressor and saving the trained model

```
import os.path as osp
import numpy as np
from sklearn2c import DTRegressor

train_samples = np.load(osp.join("regression_data", "reg_samples.
    npy"))
train_line_values = np.load(osp.join("regression_data", "
    reg_line_values.npy"))
train_sine_values = np.load(osp.join("regression_data", "
    reg_sine_values.npy"))

line_dtr_model = DTRegressor(max_depth =2)
dtr_save_path = osp.join("regression_models", "dt_regressor_line.
    joblib")
line_dtr_model.train(train_samples, train_line_values,
    dtr_save_path)

sine_dtr_model = DTRegressor(max_depth =2)
dtr_save_path = osp.join("regression_models", "dt_regressor_sine.
    joblib")
sine_dtr_model.train(train_samples, train_sine_values,
    dtr_save_path)
```

In order to apply prediction with the saved decision tree regressors in Listing 7.25, we should again load the noisy line and sine data points then proceed with prediction. We provide one such example in Listing 7.26. Here, we start with loading the samples, line_values, and sine_values. Then, we load the decision tree regressors for the line and sine data by the function load using the saved model path as parameter. Afterward, we obtain the regressor outputs (predictions) by the function predict. As we obtain predicted values from the models, we assign them to variables line_predictions and sine_predictions for the line and sine data, respectively. Rest of the code plots the input data with its target value as well as predictions of decision tree regressors.

Listing 7.26 Predictions with the decision tree regressor

```
import os.path as osp
import numpy as np
from sklearn2c import DTRegressor
from matplotlib import pyplot as plt

samples = np.load(osp.join("regression_data", "reg_samples.npy"))
line_values = np.load(osp.join("regression_data", "
    reg_line_values.npy"))
sine_values = np.load(osp.join("regression_data", "
    reg_sine_values.npy"))
```

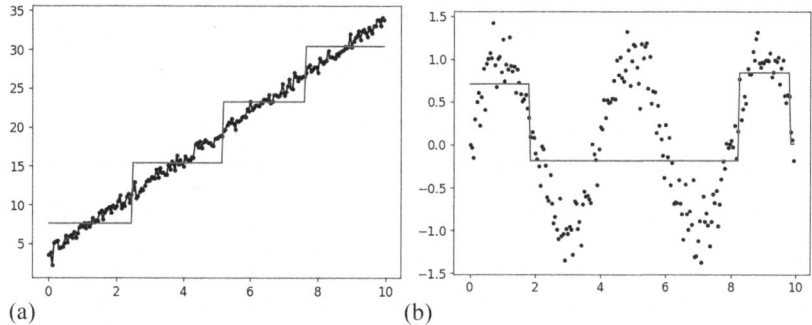

Fig. 7.6 Decision tree regression example with the usage of maximum depth two in operation. (**a**) Noisy line data and its decision tree regression result. (**b**) Noisy sine data and its decision tree regression result

```
line_model_path = osp.join("regression_models", "
    dt_regressor_line.joblib")
sine_model_path = osp.join("regression_models", "
    dt_regressor_sine.joblib")

dt_regressor_line = DTRegressor.load(line_model_path)
dt_regressor_sine = DTRegressor.load(sine_model_path)

line_predictions = dt_regressor_line.predict(samples)
sine_predictions = dt_regressor_sine.predict(samples)

fig, (ax1, ax2) = plt.subplots(2, 1)
ax1.plot(samples, line_values, "k.")
ax1.plot(samples, line_predictions, "b")
ax2.plot(samples, sine_values, "k.")
ax2.plot(samples, sine_predictions, "b")
plt.show()
```

We provide the obtained predictions from Listing 7.26 for the line and sine data in Fig. 7.6. As can be seen in this figure, when the maximum depth is low (two here), the prediction results are not encouraging. Therefore, we should increase the maximum depth in the decision tree regressor.

We can increase the maximum depth of the decision tree regression in Listing 7.25 by the code line DTRegressor(max_depth=10) for the noisy line and sine data. Then, we can retrain the models. We provide the obtained results in Fig. 7.7. As can be seen in this figure, the obtained prediction results are fairly good. However, we can observe that the regressor overfits to the data at hand in this setting.

7.5.3   Deploying the Formed Regressor to the Microcontroller

We trained decision tree regressors for the line and sine data in Python on PC and saved the corresponding regression models in the previous section. We proceed with the decision tree regressor formed for the noisy sine data as in Sect. 7.2.3. The reader

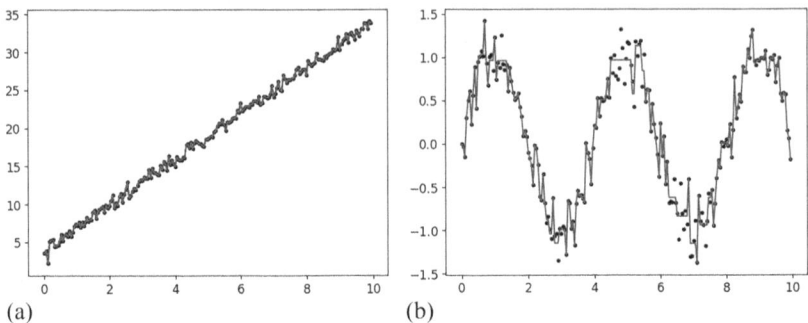
(a)   (b)

Fig. 7.7 Decision tree regression example with the usage of maximum depth ten in operation. (**a**) Noisy line data and its decision tree regression result. (**b**) Noisy sine data and its decision tree regression result

can also use the regressor formed for the line data by modifying the corresponding code block. The next phase is deploying the regressor to the STM32 microcontroller. Hence, we can use it in an actual scenario. To do so, we should generate the C header and source files of the trained regressor model on PC. We provide the Python script to export model parameters to C header and source files in Listing 7.27. In this script, we first load the trained decision tree regressor for the sine data by the function load using the saved model path as parameter. Then, we export the model parameters to two files "dt_reg_config.h" and "dt_reg_config.c" by the function export.

Listing 7.27 Exporting the trained decision tree regression model

```
import os.path as osp
from sklearn2c import DTRegressor

# CHOOSE ONE OF THE MODELS
# Parameter dt_reg_config MUST NOT BE CHANGED
# SINCE IT IS INCLUDED IN C INFERENCE FILES

#line_model_path = osp.join("regression_models","
    dt_regressor_line.joblib")
sine_model_path = osp.join("regression_models","dt_regressor_sine
    .joblib")

export_path = osp.join("exported_models","regression","
    dt_reg_config")

#dt_regressor = DTRegressor.load(line_model_path)
dt_regressor = DTRegressor.load(sine_model_path)

dt_regressor.export(export_path)
```

The generated C header file for the example considered in this chapter is given in Listing 7.28. This file only keeps the variable names and their size. They will be used by other files to include the corresponding source file given in Listing 7.29. All variables declared in Listing 7.28 are assigned here.

Listing 7.28 Header file for the decision tree regression configuration

```
#ifndef DT_REG_CONFIG_H_INCLUDED
#define DT_REG_CONFIG_H_INCLUDED
#define NUM_FEATURES 1
#define NUM_NODES 399
extern const int LEFT_CHILDREN[NUM_NODES];
extern const int RIGHT_CHILDREN[NUM_NODES];
extern const int SPLIT_FEATURE[NUM_NODES];
extern const float THRESHOLDS[NUM_NODES];
extern const float VALUES[NUM_NODES];
#endif
```

Listing 7.29 Source file for the decision tree regression configuration

```
#include "dt_reg_config.h"
const int LEFT_CHILDREN[NUM_NODES] = {  1,   2,   3,..., 397, -1,
    -1};
const int RIGHT_CHILDREN[NUM_NODES] = { 72, 13, 10,..., 398, -1,
    -1} ;
const int SPLIT_FEATURE[NUM_NODES] = { 0, 0, 0,..., 0,-2,-2};
const float THRESHOLDS[NUM_NODES] = { 1.77499998, 0.27500001,
    0.175,..., 9.92499971,-2. ,-2.};
const float VALUES[NUM_NODES] = { 0.14443214, 0.7366521 ,
    0.21800425,..., 0.22977443, 0.1335265, 0.32602236};
```

7.5.3.1 STM32CubeIDE Project Settings for Deployment

We should apply the following steps to deploy the trained regressor to the microcontroller in an STM32CubeIDE project. We should first copy the generated header and source files in the previous section to the "Core/Inc" and "Core/Src" folders of the STM32CubeIDE project, respectively. Then, we should copy the files "dt_reg_inference.h" and "dt_reg_inference.c" from the sklearn2c library to the "Core/Inc" and "Core/Src" folders of the STM32CubeIDE project, respectively. These files contain the required C functions to run the decision tree regressor on the microcontroller.

7.5.3.2 Mbed Studio Project Settings for Testing

We should apply the following steps to deploy the trained regressor to the microcontroller in an Mbed Studio project. We should first copy the generated header and source files in the previous section to the root folder of the Mbed Studio project. Afterward, we should copy the "dt_reg_inference.h" and "dt_reg_inference.c" files from the sklearn2c library to the root folder of the Mbed Studio project.

7.5.4 Testing the Deployed Regressor on the Microcontroller

To check whether the deployed regressor works as expected, we can use the generated data on PC. Then, we can send them to the STM32 microcontroller. The regressor running on the microcontroller works on this data. Then, we can transfer the regression result back to PC and cross check with the same regressor working in Python on PC. Hence, we can make sure that the generated framework works as expected.

7.5.4.1 STM32CubeIDE Project Settings for Testing

In order to test the deployed regressor on the STM32 microcontroller, we can use the test setup formed for the linear regressor in Sect. 7.2.4. Then, we should add the generated files in the previous section to our project. Hence, we will have the corresponding code block as follows.

```
/* USER CODE BEGIN Includes */
#include "dt_reg_inference.h"
#include "lib_serial.h"
#include "lib_rng.h"
/* USER CODE END Includes */
```

We can use the same macrodefinitions formed for linear regression in Sect. 7.2.4 in the test setup here as well. Hence, we will have the following C code for decision tree regression.

```
/* USER CODE BEGIN 0 */
#define INPUT_PC 1
#define INPUT_MCU 2
#define INPUT INPUT_PC

#define SIZE_INPUT NUM_FEATURES
#define SIZE_OUTPUT 1

float input[SIZE_INPUT];
float output[SIZE_OUTPUT];
/* USER CODE END 0 */
```

Next, we should form a while loop to acquire data either from PC or microcontroller according to the INPUT macro. Then, we can run the decision tree regressor by calling the function dt_reg_predict. The first parameter of this function is a pointer to the input array. The second parameter is a pointer to the output array. This output array is filled by the function dt_reg_predict with prediction results. Finally, we transfer the regression result to PC. We provide the C code corresponding to these definitions in Listing 7.30. Please note that the Python script in Listing 7.23 should already be running on PC for executing this example successfully.

Listing 7.30 Decision tree regression test code for STM32CubeIDE

```c
/* USER CODE BEGIN WHILE */
while (1)
{
/* USER CODE END WHILE */

/* USER CODE BEGIN 3 */
#if (INPUT == INPUT_PC)
 LIB_SERIAL_Receive(input, SIZE_INPUT, TYPE_F32);
#elif (INPUT == INPUT_MCU)
 for (uint32_t i = 0; i < SIZE_INPUT; ++i)
 input[i] = (float)(LIB_RNG_GetRandomNumber() % 1000) / 1000.0f;
 LIB_SERIAL_Transmit(input, SIZE_INPUT, TYPE_F32);
#endif
dt_reg_predict(input, output);
LIB_SERIAL_Transmit(output, SIZE_OUTPUT, TYPE_F32);
HAL_Delay(1000);
}
/* USER CODE END 3 */
```

To check whether the deployed decision tree regressor works as expected, we need a test setup. We provide the Python script formed for this purpose in Listing 7.31. Here, we first load the previously generated data. Then, we load our trained decision tree regressor. Afterward, we send the test data to the microcontroller. Next, we perform inference on PC and microcontroller separately. We finally print both inference results on PC for comparison. Hence, we can make sure that the generated framework works as expected.

Listing 7.31 Python script for comparing decision tree regression results

```python
import os.path as osp
import numpy as np
from sklearn2c import DTRegressor
import py_serial

py_serial.SERIAL_Init("COM6")
train_samples = np.load(osp.join("regression_data", "reg_samples.
    npy"))
dtr = DTRegressor().load(osp.join("regression_models", "
    dt_regressor_sine.joblib"))

i = 0
while 1:
    rqType, datalength, dataType = py_serial.
        SERIAL_PollForRequest()

    if rqType == py_serial.MCU_WRITES:
        # INPUT -> FROM MCU TO PC
        inputs = py_serial.SERIAL_Read()

    elif rqType == py_serial.MCU_READS:
        # INPUT -> FROM PC TO MCU
        inputs = train_samples[i : i + 1].astype(py_serial.
            SERIAL_GetDType(dataType))
        i = i + 1
        if i >= len(train_samples):
```

```
      i = 0
  py_serial.SERIAL_Write(inputs)

pcout = dtr.predict(inputs)
rqType, datalength, dataType = py_serial.
    SERIAL_PollForRequest()
if rqType == py_serial.MCU_WRITES:
    mcuout = py_serial.SERIAL_Read()
    print()
    print("Inputs : " + str(inputs))
    print("PC Output : " + str(pcout))
    print("MCU Output : " + str(mcuout))
    print()
```

7.5.4.2 Mbed Studio Project Settings for Testing

In order to test the deployed decision tree regressor on the STM32 microcontroller, we should add the libraries lib_rng and lib_serial to the Mbed Studio project. To do so, we should copy the files "lib_rng.h," "lib_rng.c," "lib_uart.h," "lib_uart.c," "lib_serial.h," and "lib_serial.c" to our Mbed Studio project folder. Here, we also assume that the steps in Sect. 7.5.3.2 are already done.

The test code for the Mbed Studio project is given in Listing 7.32. This code is similar to the one given for the STM32CubeIDE project. Hence, the explanations there hold here as well. The only difference is that we should include the header file "lib_uart.h" and initialize the UART peripheral unit before the while loop by code in our Mbed Studio project. Please note that the Python script in Listing 7.31 should already be running on PC for executing this example successfully.

Listing 7.32 Decision tree regression test code for Mbed Studio

```
#include "mbed.h"
#include "dt_reg_inference.h"
#include "lib_serial.h"
#include "lib_rng.h"
#include "lib_uart.h"

#define INPUT_PC 1
#define INPUT_MCU 2
#define INPUT INPUT_PC

#define SIZE_INPUT NUM_FEATURES
#define SIZE_OUTPUT 1

float input[SIZE_INPUT];
float output[SIZE_OUTPUT];

int main()
{
LIB_UART_Init();
#if (INPUT == INPUT_MCU)
 LIB_RNG_Init();
#endif
while (true)
{
#if (INPUT == INPUT_PC)
 LIB_SERIAL_Receive(input, SIZE_INPUT, TYPE_F32);
```

```
#elif (INPUT == INPUT_MCU)
  for (uint32_t i = 0; i < SIZE_INPUT; ++i)
  input[i] = (float)(LIB_RNG_GetRandomNumber() % 1000) / 1000.0f;
  LIB_SERIAL_Transmit(input, SIZE_INPUT, TYPE_F32);
#endif
dt_reg_predict(input, output);
LIB_SERIAL_Transmit(output, SIZE_OUTPUT, TYPE_F32);
HAL_Delay(1000);
}
}
```

7.6 Application: Estimating Future Temperature Values

We can use the regression methods introduced in this chapter to predict future temperature values based on previous information. Therefore, we train a model in Python on PC. Then, we embed the model to the microcontroller.

7.6.1 The SML2010 Dataset

We benefit from the SML2010 dataset in this application. This dataset can be used to analyze the indoor climate in terms of temperature and humidity values. These are among the most used physical variables in multiple applications in different working areas, such as health, electronic circuits, and precision agriculture.

The dataset is available in [25]. It consists of indoor temperature, outdoor temperature, indoor relative humidity, and outdoor relative humidity values collected in 15-minute intervals over two months, specifically between March and April 2012. Data was acquired using temperature and humidity sensors. In this application, we use its filtered version consisting of indoor temperature, outdoor temperature, and humidity values. The reader can download the filtered dataset from [11].

7.6.2 Training the Regression Model in Python on PC

The first step in forming the regressor for predicting future temperature values is forming and training the model. We picked the linear regressor in this application for this purpose. However, other regressors considered in this chapter can also be used as well. We provide the Python script for our linear regressor in Listing 7.33.

Listing 7.33 Python script for linear regressor formation, training, and testing

```python
import numpy as np
import pandas as pd
from joblib import dump
from sklearn.model_selection import train_test_split
from sklearn.linear_model import LinearRegression
```

```python
from sklearn.metrics import mean_absolute_error
from matplotlib import pyplot as plt

df = pd.read_csv('temperature_dataset.csv')
y = df['Room_Temp'][::4]
prev_values_count = 5

X = pd.DataFrame()
for i in range(prev_values_count, 0, -1):
    X['t-' + str(i)] = y.shift(i)

X = X[prev_values_count:]
y = y[prev_values_count:]

X_train, X_test, y_train, y_test = train_test_split(X, y,
    test_size=0.2, random_state=0)

linear_model = LinearRegression()
linear_model.fit(X_train, y_train)

y_train_predicted = linear_model.predict(X_train)
y_test_predict = linear_model.predict(X_test)

fig, ax = plt.subplots(1,1)
ax.plot(y_test.to_numpy(), label = "Actual values")
ax.plot(y_test_predict, label = "Predicted values")
plt.legend()
plt.show()

mae_train = np.sqrt(mean_absolute_error(y_train,
    y_train_predicted))
mae_test = np.sqrt(mean_absolute_error(y_test, y_test_predict))

print(f"Training set MAE: {mae_train}\n")
print(f"Test set MAE:{mae_test}")

dump(linear_model, 'temperature_prediction_lin.joblib')
```

In Listing 7.33, we first read the dataset using the Pandas function read_csv. Originally, the data sampling period is 15 minutes as explained in Sect. 7.6. We assume that temperature changes can be better observed in one hour intervals.Therefore, we resample the dataset accordingly. In our application, we use the Room_Temp values for regression. Therefore, we form a new DataFrame with previous five temperature records. While doing so, we remove the first five rows containing NaN values. Afterward, we split the dataset into training and test sets by the scikit-learn library function train_test_split. Finally, we fit the linear regression model to the training set.

7.6.3   Implementation on the Microcontroller

We next export the trained regressor C configuration files to be used in the microcontroller. To do so, we follow the same steps explained in relevant sections of this chapter. We will provide the complete STM32CubeIDE and Mbed Studio projects in separate folders in the book repository for this application.

7.7 Summary of the Chapter

We considered regression in this chapter. The aim in regression is forming a relationship between one or more input and output variables. Hence, information on one variable can be obtained by using the information on other variables. To fully explain the regression concept, we started with its definition. Then, we considered linear, polynomial, kNN, and decision tree regression methods. While handling each regression method, we covered its theoretical background. Then, we explored its formation in Python language on PC. Afterward, we evaluated methods to deploy the formed regressor to the microcontroller. We formed a regressor to estimate future temperature values as end of chapter application.

Clustering

8

8.1 What Is Clustering?

Clustering aims to group feature values based on their inherent characteristics. While doing so, labeled data is not used. Therefore, the training step in clustering is different from the classification and regression methods introduced in previous chapters. To be more specific, the user feeds feature values and number of clusters to be obtained during training. The clustering algorithm groups feature values based on predefined rules and provides cluster centers as output.

Clustering algorithms find wide range of applications in big data analysis. We will not cover them in this book. Instead, we will focus on clustering applications suitable for embedded systems. One such application is the fall detection system. We can represent this system as a subpart of the human activity recognition problem introduced in Chap. 6. There, we had more than one class to represent a specific activity type. To do so, we had labeled data such that the classifier learns each class characteristics from them.

We can reformulate the human activity recognition problem as fall detection and form an embedded system for this purpose. Our system will be based on a clustering algorithm. Hence, normal human activity is taken as one cluster and the fall action is taken as another cluster. Let's explain general working principles of the fall detection system based on a generic clustering algorithm. There should be a sensor (or sensors) placed on the observed person to acquire data continuously. We extract the corresponding feature values from the acquired data. The clustering algorithm should learn inherent characteristics of the normal and fall actions from these. We can call this as the training or learning phase of the system. One way of doing this is letting the system collect sensor data for a certain time duration. Or the total number of samples to be collected can be set. In this scenario, data is collected and then used to form clusters. Hence, we have a batch clustering operation. Cluster formation can also be done on the fly such that each feature value is processed and clusters evolve accordingly. In this scenario, it may be beneficiary to have initial

© The Author(s), under exclusive license to Springer Nature Switzerland AG 2025
C. Ünsalan et al., *Embedded Machine Learning with Microcontrollers*,
https://doi.org/10.1007/978-3-031-70912-8_8

locations for the normal and fall clusters in the feature space. Therefore, we should
have idea about the possible cluster center locations before the operation starts.

As the training phase ends, the embedded fall detection system will be ready for
inference. Hence, the system will decide on whether the person has fallen or not
at a given time instant. Here, we can check the distance of the observed feature
value from the normal and fall clusters. The closest cluster can be picked to indicate
the corresponding cluster label. If the observed feature value falls into the fall
cluster, then the system can send a warning signal to a remote location via wireless
communication or SMS message.

8.2   k-Means Clustering

k-means is the first clustering algorithm to be considered in this chapter. This
algorithm depends on a simple yet elegant idea. Therefore, it is well-known and
used in practice extensively. In this section, we will start with explaining the k-
means clustering algorithm from a theoretical perspective. Then, we will consider
implementation details of the two versions of the algorithm, as batch and online, on
PC and microcontroller.

8.2.1   Theoretical Background

General form of the k-means clustering algorithm is the Gaussian mixture model
(GMM) which aims to model an unknown pdf as mixture of Gaussian functions [7].
To do so, each Gaussian is represented by its mean vector and covariance matrix.
Besides, each Gaussian also has a weight to be used in forming the mixture. These
value are estimated from given feature values by the expectation maximization
method [6]. k-means clustering algorithm simplifies this approach by assuming
identity covariance matrix for Gaussians. Only cluster mean vector values are
estimated. Hence, the name of the algorithm is k-means.

We can form the k-means clustering algorithm in two different ways. The first
way is batch sample processing. The second way is online sample processing.
Let's start with the batch k-means algorithm. This algorithm assumes that data
samples are available before the operation. We first assign random cluster centers
(centroids). Then, we calculate the distance between all unlabeled samples with
these centroids. Afterward, we assign each sample to the closest centroid and
recalculate cluster centroids. We iterate this process until centroid locations do not
change or a maximum number of iterations are reached. We provide the batch k-
means clustering algorithm in Algorithm 2.

The inference step in batch k-means clustering algorithm works as follows. When
a new sample comes, we assign it to the cluster with the closest centroid location.
Cluster centroids are fixed in the inference step of the batch k-means clustering
algorithm. Hence, once set, they do not change afterward.

Algorithm 2 The batch k-means clustering algorithm

1: **procedure** KMEANS(X, K)
2:     Initialize empty clusters $C = \{c_1, c_2, \ldots, c_K\}$
3:     Initialize cluster means $M = \{0, 0, \ldots, 0\}$
4:     Initialize cluster sums $S = \{0, 0, \ldots, 0\}$
5:     Initialize cluster sample counts $T = \{0, 0, \ldots, 0\}$
6:     Randomly select K points from X as initial centroids $M_1, M_2, \ldots, M_K$
7:     **while** until centroids converge **do**
8:         **for** $i \leftarrow 1$ to N **do**
9:             **for** $k \leftarrow 1$ to K **do**
10:                 Compute the distance between sample X_i and cluster centroid M_k. $d(i, k)$
11:                 Find the closest cluster $\hat{k} \leftarrow \arg\min_k d(i, k)$
12:             **end for**
13:             Increment $T_{\hat{k}}$. $T_{\hat{k}} \leftarrow T_{\hat{k}} + 1$
14:             Add sample to cluster sum $S_{\hat{k}}$. $S_{\hat{k}} \leftarrow S_{\hat{k}} + X_i$
15:             Recompute the cluster centroids $M_k = \frac{\sum s_{\hat{k}}}{T_{\hat{k}}}$
16:         **end for**
17:     **end while**
18:     Assign the new sample x to the closest cluster
19: **end procedure**

In the online k-means clustering algorithm, the idea is almost the same as in batch clustering. The only difference is that data samples come on the fly. Hence, we should make calculations as a new sample comes to be clustered. In other words, we always update cluster centroids in the online k-means clustering algorithm. We provide the online k-means clustering algorithm in Algorithm 3.

When we compare the batch and online clustering algorithms in Algorithms 2 and 3, we can see that they have the same steps for processing the data at hand. The difference between them occurs when a new sample comes. We do not update cluster centroids in the batch k-means clustering algorithm. Thus, new samples do not affect the assigned cluster label of upcoming samples. The online k-means clustering algorithm updates cluster centroids based on the new sample. This mechanism makes the online k-means clustering algorithm robust to shifts in data distribution. However, it becomes more vulnerable to outliers since they may shift centroids dramatically.

8.2.2 Forming the Clustering Algorithm in Python

We formed k-means clustering functions under the sklearn2c library. We provide the usage of functions on an actual example in Listing 8.1. Here, we first load the data we used for classification tasks previously by the NumPy function `load`. Then, we form the k-means clustering object by the class `Kmeans`. This class is compatible with the scikit-learn `sklearn.cluster.KMeans` class and accepts the same arguments. The cluster number is set as eight by default. If the reader wants to set a specific cluster number such as two, then we should use the code line `kmeans = Kmeans(n_clusters =2, random_state = 42, n_init="auto")`.

Algorithm 3 The online k-means clustering algorithm

1: **procedure** KMEANS(X, K)
2:     Initialize empty clusters $C = \{c_1, c_2, \ldots, c_K\}$
3:     Initialize cluster means $M = \{0, 0, \ldots, 0\}$
4:     Initialize cluster sums $S = \{0, 0, \ldots, 0\}$
5:     Initialize cluster sample counts $T = \{0, 0, \ldots, 0\}$
6:     Randomly select K points from X as initial centroids $M_1, M_2, \ldots, M_K$
7:     **while** until centroids converge **do**
8:         **for** x in X **do**
9:             **for** $k \leftarrow 1$ to K **do**
10:                 Compute the distance between sample X_i and cluster centroid M_k. $d(x, k)$
11:                 Find the closest cluster $\hat{k} \leftarrow \arg\min_k d(x, k)$
12:             **end for**
13:             Increment $T_{\hat{k}}$. $T_{\hat{k}} \leftarrow T_{\hat{k}} + 1$
14:             Add sample to cluster sum $S_{\hat{k}}$. $S_{\hat{k}} \leftarrow S_{\hat{k}} + X_i$
15:             Recompute the cluster centroids $M_k = \frac{\sum S_{hatk}}{T_{\hat{k}}}$
16:         **end for**
17:     **end while**
18:     **for** $k \leftarrow 1$ to K **do**
19:         Compute the distance between new sample x and cluster centroids M_k. $d(x, k)$
20:         Find closest cluster $\hat{k} \leftarrow \arg\min_k d(x, k)$
21:         Increment $T_{\hat{k}}$. $T_{\hat{k}} \leftarrow T_{\hat{k}} + 1$
22:         Add sample to cluster sum $S_{\hat{k}}$. $S_{\hat{k}} \leftarrow S_{\hat{k}} + x$
23:         Recompute the cluster centroids $M_k = \frac{\sum S_{\hat{k}}}{T_{\hat{k}}}$
24:     **end for**
25: **end procedure**

Listing 8.1 k-means clustering in Python

```python
import os.path as osp
import numpy as np
from sklearn2c.clustering import Kmeans

train_samples = np.load(osp.join("classification_data", "
    cls_train_samples.npy"))
train_labels = np.load(osp.join("classification_data", "
    cls_train_labels.npy"))

kmeans = Kmeans(random_state = 42, n_init="auto")
kmeans_model_dir = osp.join("clustering_models", "
    kmeans_clustering.joblib")
kmeans.train(train_samples, save_path=kmeans_model_dir)
```

In our particular example, in Listing 8.1, we set random_state = 42 for the sake of reproducibility. The formed clustering object can be trained by the function train. This method forms the cluster centers and assigns cluster labels to each data point. The train function also takes optional parameter save_path to save the trained classifier to a file. Therefore, we feed the train_samples data with the label set train_labels and save_path parameter in our example. As the operation ends, the

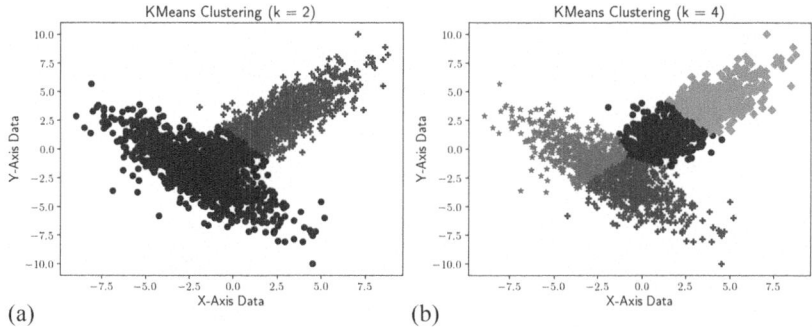

Fig. 8.1 k-means clustering example with two and four clusters. (**a**) Two clusters. (**b**) Four clusters

trained clustering algorithm is saved as the "kmeans_clustering.joblib" file under the folder "clustering_models."

We provide the obtained clustering results with two cluster centers in Fig. 8.1a. As can be seen in this figure, the data at hand is clustered fairly well. We also provide the obtained clustering results on the same dataset with four cluster centers in Fig. 8.1b. As can be seen in this figure, the k-means algorithm with four clusters splits the data redundantly. Therefore, choosing the right number of clusters is important while running the k-means clustering algorithm.

8.2.3   Deploying the Formed Clustering Algorithm to the Microcontroller

As in previous chapters, we should transfer the k-means clustering algorithm to the microcontroller. Hence, we can perform inference there. To do so, we should generate the C header and source files of the k-means clustering algorithm on PC. We provide the Python script formed for this purpose in Listing 8.2. In this script, we first load the trained k-means clustering algorithm by the function load using the saved model path as parameter. Then, we export the model parameters to two files "kmeans_clus_config.h" and "kmeans_clus_config.c" by the function export.

Listing 8.2 Exporting the trained k-means clustering algorithm parameters

```
import os.path as osp
from sklearn2c.clustering import Kmeans

kmeans_model_dir = osp.join("clustering_models", "
    kmeans_clustering.joblib")
kmeans_config_dir = osp.join("exported_models", "clustering", "
    kmeans_clus_config")

kmeans = Kmeans.load(kmeans_model_dir)
kmeans.export(kmeans_config_dir)
```

The generated C header file for the example considered in this chapter is given in Listing 8.3. This file only keeps the variable names and their size. They will be used by other files to include the corresponding source file given in Listing 8.4. All variables declared in Listing 8.3 are assigned here.

Listing 8.3 Header file for the k-means clustering algorithm

```
#ifndef KMEANS_CLUS_CONFIG_H_INCLUDED
#define KMEANS_CLUS_CONFIG_H_INCLUDED
#define NUM_CLUSTERS 8
#define NUM_FEATURES 2
extern int num_samples_per_cluster[NUM_CLUSTERS];
extern float centroids[NUM_CLUSTERS][NUM_FEATURES];
#endif
```

Listing 8.4 Source file for the k-means clustering algorithm

```
#include "kmeans_clus_config.h"
int num_samples_per_cluster[NUM_CLUSTERS] =
    {200,283,102,298,142,296, 81,198};
float centroids[NUM_CLUSTERS][NUM_FEATURES] =
    {{-3.38653056,-0.31088688},..., {-0.39354354,-3.60976311}};
```

8.2.3.1 STM32CubeIDE Project Settings for Deployment

We should apply the following steps to deploy the trained clustering algorithm to the microcontroller in an STM32CubeIDE project. We should first copy the generated header and source files in the previous section to the "Core/Inc" and "Core/Src" folders of the STM32CubeIDE project, respectively. Then, we should copy the files "kmeans_clus_inference.h" and "kmeans_clus_inference.c" from the sklearn2c library to the "Core/Inc" and "Core/Src" folders of the STM32CubeIDE project, respectively. These files contain the required C function to run the k-means clustering algorithm on the microcontroller.

8.2.3.2 Mbed Studio Project Settings for Deployment

We should apply the following steps to deploy the trained clustering algorithm to the microcontroller in an Mbed Studio project. We should first copy the generated header and source files in the previous section to the root folder of the Mbed Studio project. Afterward, we should copy the "kmeans_clus_inference.h" and "kmeans_clus_inference.c" files from the sklearn2c library to the root folder of the Mbed Studio project.

8.2.4 Testing the Deployed Clustering Algorithm on the Microcontroller

To check whether the deployed clustering algorithm works as expected, we can use the generated data on PC. Then, we can send them to the STM32 microcontroller.

The clustering algorithm running on the microcontroller works on this data. Then, we can transfer the clustering result back to PC and cross check with the same clustering algorithm working in Python on PC. Hence, we can make sure that the generated framework works as expected.

8.2.4.1 STM32CubeIDE Project Settings for Testing

In order to test the deployed clustering algorithm on the STM32 microcontroller, we can form a project under STM32CubeIDE. Then, we should add the libraries lib_rng and lib_serial to the project as explained in Sects. 5.1.3.1 and 4.1.2.2, respectively. Then, we should add the generated files in the previous section to the project. Hence, we will have the corresponding code block as follows.

```
/* USER CODE BEGIN Includes */
#include "kmeans_clus_inference.h"
#include "lib_serial.h"
#include "lib_rng.h"
/* USER CODE END Includes */
```

We should define macros related to the source of the input data, size of the input vector, and size of the output vector. The INPUT macrodefinition selects whether input data is obtained from PC or microcontroller at compile time. It can be selected as INPUT_PC or INPUT_MCU, respectively. SIZE_INPUT and SIZE_OUTPUT macros define the number of data inputs and outputs of the model, respectively. We should also define two floating-point arrays to allocate required memory space for inputs and outputs. Hence, we will have the corresponding code block as follows.

```
/* USER CODE BEGIN 0 */
#define INPUT_PC 1
#define INPUT_MCU 2
#define INPUT INPUT_PC

#define SIZE_INPUT NUM_INPUTS
#define SIZE_OUTPUT 1

float input[SIZE_INPUT];
float output[SIZE_OUTPUT];
/* USER CODE END 0 */
```

Next, we should form a while loop to acquire data either from PC or micro-controller according to the INPUT macro. Then, we can run the k-means clustering algorithm by the function kmeans_clus_predict. The first parameter of this function is a pointer to the input array. The second parameter is a pointer to the output array. This output array is filled by the function kmeans_clus_predict with prediction results. The third parameter is a Boolean type which decides whether the clustering will be offline or online. The reader can set it to 0 and 1 for offline and online clustering, respectively. Finally, we transfer the clustering result to PC. We provide the C code corresponding to these definitions in Listing 8.5. Please note that the Python script in Listing 8.6 should already be running on PC for executing this example successfully.

Listing 8.5 k-means clustering algorithm test code for STM32CubeIDE

```
/* USER CODE BEGIN WHILE */
while (1)
{
/* USER CODE END WHILE */

/* USER CODE BEGIN 3 */
#if (INPUT == INPUT_PC)
 LIB_SERIAL_Receive(input, SIZE_INPUT, TYPE_F32);
#elif (INPUT == INPUT_MCU)
 for (uint32_t i = 0; i < SIZE_INPUT; ++i)
 input[i] = (float)(LIB_RNG_GetRandomNumber() % 1000) / 1000.0f;
LIB_SERIAL_Transmit(input, SIZE_INPUT, TYPE_F32);
#endif
kmeans_clus_predict(input, output, 0);
LIB_SERIAL_Transmit(output, SIZE_OUTPUT, TYPE_F32);
HAL_Delay(1000);
}
/* USER CODE END 3 */
```

To check whether the deployed clustering algorithm works as expected, we need a test setup. We provide the Python script formed for this purpose in Listing 8.6. Here, we first load the previously generated data. Then, we load our trained clustering algorithm. Afterward, we send the test data to the microcontroller. Next, we perform inference on PC and microcontroller separately. We finally print both inference results on PC for comparison. Hence, we can make sure that the generated framework works as expected.

Listing 8.6 Python script for comparing k-means clustering results

```
import os.path as osp
import numpy as np
from sklearn2c.clustering import Kmeans
import py_serial

py_serial.SERIAL_Init("COM4")

test_samples = np.load(osp.join("classification_data","
    cls_test_samples.npy"))
test_labels = np.load(osp.join("classification_data","
    cls_test_labels.npy"))

kmeans = Kmeans.load(osp.join("clustering_models", "
    kmeans_clustering.joblib"))

i = 0
while 1:
    rqType, datalength, dataType = py_serial.
        SERIAL_PollForRequest()
    if rqType == py_serial.MCU_WRITES:
        # INPUT -> FROM MCU TO PC
        inputs = py_serial.SERIAL_Read()

    elif rqType == py_serial.MCU_READS:
        # INPUT -> FROM PC TO MCU
```

```
    inputs = test_samples[i:i+1].astype(py_serial.
        SERIAL_GetDType(dataType))
    i = i + 1
    if i >= len(test_samples):
        i = 0
    py_serial.SERIAL_Write(inputs)

pcout = kmeans.predict(np.reshape(inputs, (1, datalength)).
    astype(np.double))
rqType, datalength, dataType = py_serial.
    SERIAL_PollForRequest()
if rqType == py_serial.MCU_WRITES:
    mcuout = py_serial.SERIAL_Read()
    print()
    print("Inputs : " + str(inputs))
    print("PC Output : " + str(pcout))
    print("MCU Output : " + str(mcuout))
    print()
```

8.2.4.2 Mbed Studio Project Settings for Testing

In order to test the deployed k-means clustering algorithm on the STM32 microcontroller, we should add the libraries lib_rng and lib_serial to the Mbed Studio project. To do so, we should copy the files "lib_rng.h," "lib_rng.c," "lib_uart.h," "lib_uart.c," "lib_serial.h," and "lib_serial.c" to our Mbed Studio project folder. Here, we also assume that the steps in Sect. 8.2.3.2 are already done.

The test code for the Mbed Studio project is given in Listing 8.7. This code is similar to the one given for the STM32CubeIDE project. Hence, the explanations there hold here as well. The only difference is that we should include the header file "lib_uart.h" and initialize the UART peripheral unit before the while loop by code in our Mbed Studio project. Please note that the Python script in Listing 8.6 should already be running on PC for executing this example successfully.

Listing 8.7 k-means clustering test code for Mbed Studio

```
#include "mbed.h"
#include "kmeans_clus_inference.h"
#include "lib_serial.h"
#include "lib_rng.h"
#include "lib_uart.h"

#define INPUT_PC 1
#define INPUT_MCU 2
#define INPUT INPUT_PC

#define SIZE_INPUT NUM_FEATURES
#define SIZE_OUTPUT 1

float input[SIZE_INPUT];
float output[SIZE_OUTPUT];

int main()
{
LIB_UART_Init();
#if (INPUT == INPUT_MCU)
```

```
 LIB_RNG_Init();
#endif
while (true)
{
#if (INPUT == INPUT_PC)
 LIB_SERIAL_Receive(input, SIZE_INPUT, TYPE_F32);
#elif (INPUT == INPUT_MCU)
 for (uint32_t i = 0; i < SIZE_INPUT; ++i)
  input[i] = (float)(LIB_RNG_GetRandomNumber() % 1000) / 1000.0f;
 LIB_SERIAL_Transmit(input, SIZE_INPUT, TYPE_F32);
#endif
kmeans_clus_predict(input, output, 0);
LIB_SERIAL_Transmit(output, SIZE_OUTPUT, TYPE_F32);
HAL_Delay(1000);
}
}
```

8.3   Density-Based Spatial Clustering of Applications with Noise

Density-based spatial clustering of applications with noise (DBSCAN) is the second clustering algorithm to be considered in this chapter. Clusters are formed based on the radius of neighborhood and number of neighbors parameters in this algorithm. We will start with explaining DBSCAN from a theoretical perspective in this section. Then, we will consider its implementation details on the microcontroller.

8.3.1   Theoretical Background

Let D be a set of points. We would like to cluster them with the DBSCAN algorithm. We can explain the algorithm based on the definitions in the original paper [8]. We should add the core point definition to simplify the explanation. A point in D with at least $minPts$ points in its ϵ distance is a core point. Each core point should be in one cluster or group of core points can form a cluster. Based on these constraints, we can form clusters. Beforehand, we should make two definitions as directly density reachable and density reachable.

A point q is directly density reachable from a core point p if it is within ϵ distance to it. The point q can be a core point or not. If it is not a core point, then it will be at the outer boundary (edge) of the formed cluster. If q is a core point, then such points can form a chain within the cluster. We will use this formation in the density reachable definition as follows. A point q is density reachable from a core point p if there is a chain of points $p_1, \cdots, p_n, p_1 = p, p_n = q$ such that p_{i+1} is directly density reachable from p_i.

Based on the definitions in the previous paragraph, we can form clusters in the DBSCAN algorithm as follows. If p is a core point, then it forms a cluster with all its density reachable points. This operation is done iteratively to cover all core points

in D and form final clusters. Nonclustered points in D are called noise. Hence, they
are the points not density reachable by any core point. We provide the pseudocode
for the DBSCAN algorithm in Algorithm 4 [29].

Algorithm 4 The DBSCAN algorithm

1: **procedure** DBSCAN($X, \epsilon, minPts$)
2:    **for** x **in** X **do** // Iterate over every point
3:        **if** $L(x) \neq$ undefined **then**
4:            **continue**
5:        **end if**
6:        $N \leftarrow RangeQuery(X, distDunc, p, \epsilon)$
7:        **if** $|N| < minPts$ **then**
8:            $L(x) \leftarrow$ **Noise**
9:            **continue**
10:        **end if**
11:        $c \leftarrow c + 1$
12:        $L(x) \leftarrow c$
13:        $S \leftarrow N \; x$
14:        **for** s **in** S **do**
15:            **if** $L(s) =$ Noise **then**
16:                $L(s) \leftarrow c$
17:            **end if**
18:            **if** $L(s) =$ Undefined **then**
19:                **continue**
20:            **end if**
21:            $N \leftarrow RangeQuery(X, distFunc, s, \epsilon)$
22:            $L(s) \leftarrow c$
23:            **if** $|N| < minPts$ **then** continue // Core-point check
24:            **end if**
25:            $S \leftarrow S \cup N$
26:        **end for**
27:    **end for**
28: **end procedure**

Here, the function RangeQuery returns the neighborhood of a sample point based
on the given distance function distFunc. Pseudocode of this function is given in
Algorithm 5.

Algorithm 5 The RangeQuery function

1: **function** RANGEQUERY($X, distFunc, s, \epsilon$)
2:    $N \leftarrow []$
3:    **for** x **in** X **do**
4:        **if** $distFunc(s, x) \leq \epsilon$ **then**
5:            $N \leftarrow N \cup x$
6:        **end if**
7:    **end for** **return** N
8: **end function**

The DBSCAN algorithm has certain advantages compared to the k-means clustering algorithm. DBSCAN does not require the number of clusters a priori as opposed to k-means. Moreover, it is robust to outliers. DBSCAN finds a cluster based on the neighborhood information. Hence, it can find clusters regardless of data shape. On the other hand, DBSCAN is not deterministic around cluster borders. It is sensitive to differences in different features, as $minPts$ and ϵ parameters may not work for all feature values.

8.3.2   Forming the Clustering Algorithm in Python

We formed DBSCAN functions under the sklearn2c library. We provide the usage of functions on an actual example in Listing 8.8. Here, we first load the data we used for classification tasks previously by the NumPy function load. Then, we form the DBSCAN clustering object by the class Dbscan. This class is compatible with the scikit-learn sklearn.cluster.DBSCAN class and accepts the same arguments. Here, $minPts$ and ϵ in the previous sections ares represented by min_samples and eps, respectively. Therefore, the reader can set specific values for the DBSCAN clustering algorithm such as Dbscan(eps=0.5, min_samples=5). Default values for these parameters are eps=0.5 and min_samples=5.

Listing 8.8 DBSCAN clustering in Python

```
import numpy as np
import os.path as osp
from sklearn2c.clustering import Dbscan

train_samples = np.load(osp.join("classification_data", "
    cls_train_samples.npy"))
train_labels = np.load(osp.join("classification_data", "
    cls_train_labels.npy"))

dbscan = Dbscan(eps = 1)
model_save_path = osp.join("clustering_models", "
    dbscan_clustering.joblib")
dbscan.train(train_samples, model_save_path)
```

In our particular example, in Listing 8.8, the formed clustering object can be trained by the function train. This function also takes optional parameter save_path to save the trained classifier to a file. Therefore, we feed the train_samples data with the label set train_labels and save_path parameter in our example. As the operation ends, the trained clustering method is saved as the "dbscan_clustering.joblib" file under the folder "clustering_models."

We applied the DBSCAN algorithm on the same dataset used in the previous section with three different parameter sets. The aim here is to emphasize their effect on the final clustering result. Therefore, we set the eps parameter to 1, 0.5, and 1 where eps determines the maximum distance between two points to be considered in the same neighborhood. The min_samples parameter is set to 5, 5, 3 respectively,

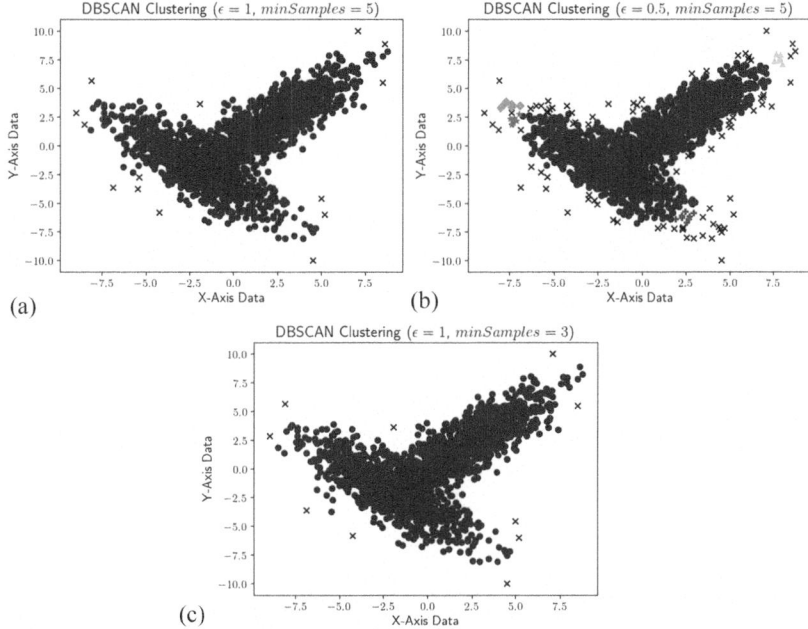

Fig. 8.2 DBSCAN clustering example with different parameters. (**a**) eps=1 and min_samples = 5. (**b**) eps=0.5 and min_samples = 5. (**c**) eps=1 and min_samples = 3

specifying the minimum number of samples required to form a dense region. The clusters obtained through DBSCAN represent groups of data points with similar density. We provide the obtained results in Fig. 8.2.

In Fig. 8.2, points marked as "x" represent outliers. Other colored points represent the clusters inferred by DBSCAN. Please note that DBSCAN tends to find more clusters and outliers when we decrease the eps parameter from 1 to 0.5. On the other hand, outliers in Fig. 8.2a are included in the large cluster when we decrease the min_samples parameter from 5 to 3 as in Fig. 8.2c. Therefore, the reader should select the DBSCAN parameters with care.

8.3.3 Deploying the Formed Clustering Algorithm to the Microcontroller

As in the previous section, we should transfer the DBSCAN clustering algorithm parameters to the microcontroller. Hence, we can perform inference there. To do so, we should generate the C header and source files of the DBSCAN clustering algorithm on PC. We provide the Python script formed for this purpose in Listing 8.9. In this script, we first load the trained DBSCAN clustering algorithm by the function load using the saved model path as parameter. Then, we export the model parameters to two files "dbscan_clus_config.h" and "dbscan_clus_config.c" by the function export.

Listing 8.9 Exporting the trained DBSCAN clustering algorithm parameters

```
import os.path as osp
from sklearn2c.clustering import Dbscan

dbscan_model_dir = osp.join("clustering_models", "
    dbscan_clustering.joblib")
dbscan_config_dir = osp.join("exported_models", "clustering", "
    dbscan_clus_config")

dbscan = Dbscan.load(dbscan_model_dir)
dbscan.export(dbscan_config_dir)
```

The generated C header file for the example considered in this chapter is given in Listing 8.10. This file only keeps the variable names and their size. They will be used by other files to include the corresponding source file given in Listing 8.11. All variables declared in Listing 8.10 are assigned here.

Listing 8.10 Header file for the DBSCAN clustering algorithm

```
#ifndef DBSCAN_CLUS_CONFIG_H_INCLUDED
#define DBSCAN_CLUS_CONFIG_H_INCLUDED
#define NUM_CORE_POINTS 1572
#define NUM_FEATURES 2
#define NUM_CLUSTERS 2
#define EPS 1
extern float CORE_POINTS[NUM_CORE_POINTS][NUM_FEATURES];
extern int LABELS[NUM_CORE_POINTS];
#endif
```

Listing 8.11 Source file for the DBSCAN clustering algorithm

```
#include "dbscan_clus_config.h"
float CORE_POINTS[NUM_CORE_POINTS][NUM_FEATURES] = {{-2.70255383e
    +00,-4.72532919e+00},..., {-2.07285632e+00,-2.22146536e+00}};
int LABELS[NUM_CORE_POINTS] = {0,..., 0};
```

8.3.3.1 STM32CubeIDE Project Settings for Deployment

We should apply the following steps to deploy the trained clustering algorithm to the microcontroller in an STM32CubeIDE project. We should first copy the generated header and source files in the previous section to the "Core/Inc" and "Core/Src" folders of the STM32CubeIDE project, respectively. Then, we should copy the files "dbscan_clus_inference.h" and "dbscan_clus_inference.c" from the sklearn2c library to the "Core/Inc" and "Core/Src" folders of the STM32CubeIDE project, respectively. These files contain the required C function to run the DBSCAN clustering algorithm on the microcontroller.

8.3.3.2 Mbed Studio Project Settings for Deployment

We should apply the following steps to deploy the trained clustering algorithm to the microcontroller in an Mbed Studio project. We should first copy the generated header and source files in the previous section to the root folder of the Mbed Studio project. Afterward, we should copy the "dbscan_clus_inference.h" and "dbscan_clus_inference.c" files from the sklearn2c library to the root folder of the Mbed Studio project.

8.3.4 Testing the Deployed Clustering Algorithm on the Microcontroller

To check whether the deployed clustering algorithm works as expected, we can use the generated data on PC. Then, we can send them to the STM32 microcontroller. The clustering algorithm running on the microcontroller works on this data. Then, we can transfer the clustering result back to PC and cross check with the same clustering algorithm working in Python on PC. Hence, we can make sure that the generated framework works as expected.

8.3.4.1 STM32CubeIDE Project Settings for Testing

In order to test the deployed clustering algorithm on the STM32 microcontroller, we can form a project under STM32CubeIDE. Then, we should add the libraries lib_rng and lib_serial to the project as explained in Sects. 5.1.3.1 and 4.1.2.2, respectively. Then, we should add the generated files in the previous section to the project. Hence, we will have the corresponding code block as follows.

```
/* USER CODE BEGIN Includes */
#include "dbscan_clus_inference.h"
#include "lib_serial.h"
#include "lib_rng.h"
/* USER CODE END Includes */
```

We should define macros related to the source of the input data, size of the input vector, and size of the output vector. The INPUT macrodefinition selects whether input data is obtained from PC or microcontroller at compile time. It can be selected as INPUT_PC or INPUT_MCU, respectively. SIZE_INPUT and SIZE_OUTPUT macros define the number of data inputs and outputs of the model, respectively. We should also define one floating-point array and one signed integer array to allocate required memory space for inputs and outputs. Hence, we will have the corresponding code block as follows.

```
/* USER CODE BEGIN 0 */
#define INPUT_PC 1
#define INPUT_MCU 2
#define INPUT    INPUT_PC

#define SIZE_INPUT NUM_FEATURES
```

```
#define SIZE_OUTPUT 1

float input[SIZE_INPUT];
int32_t output[SIZE_OUTPUT];
/* USER CODE END 0 */
```

Next, we should form a while loop to acquire data either from PC or microcontroller according to the INPUT macro. Then, we can run the DBSCAN clustering algorithm by the function dbscan_clus_predict. The first parameter of this function is a pointer to the input array. The second parameter is a pointer to the output array. This output array is filled by the function dbscan_clus_predict with prediction results. Finally, we transfer the clustering result to PC. We provide the C code corresponding to these definitions in Listing 8.12. Please note that the Python script in Listing 8.13 should already be running on PC for executing this example successfully.

Listing 8.12 DBSCAN clustering algorithm test code for STM32CubeIDE

```
/* USER CODE BEGIN WHILE */
while (1)
{
/* USER CODE END WHILE */

/* USER CODE BEGIN 3 */
#if (INPUT == INPUT_PC)
  LIB_SERIAL_Receive(input, SIZE_INPUT, TYPE_F32);
#elif (INPUT == INPUT_MCU)
  for (uint32_t i = 0; i < SIZE_INPUT; ++i)
  input[i] = (float)(LIB_RNG_GetRandomNumber() % 1000) / 1000.0f;
  LIB_SERIAL_Transmit(input, SIZE_INPUT, TYPE_F32);
#endif
dbscan_clus_predict(input, output);
LIB_SERIAL_Transmit(output, SIZE_OUTPUT, TYPE_S32);
HAL_Delay(1000);
}
/* USER CODE END 3 */
```

To check whether the deployed clustering algorithm works as expected, we need a test setup. We provide the Python script formed for this purpose in Listing 8.13. Here, we first load the previously generated data. Then, we load our trained clustering algorithm. Afterward, we send the test data to the microcontroller. Next, we perform inference on PC and microcontroller separately. We finally print both inference results on PC for comparison. Hence, we can make sure that the generated framework works as expected.

Listing 8.13 Python script for comparing DBSCAN clustering results

```
import os.path as osp
from sklearn2c.clustering import Dbscan
import py_serial
import numpy as np
```

```
py_serial.SERIAL_Init("COM4")

test_samples = np.load(osp.join("classification_data","
    cls_test_samples.npy"))
test_labels = np.load(osp.join("classification_data","
    cls_test_labels.npy"))
MODELS_DIR = "clustering_models"

dbscan_model_dir = osp.join(MODELS_DIR, "dbscan_clustering.joblib
    ")
dbscan = Dbscan.load(dbscan_model_dir)

i = 0
while 1:
    rqType, datalength, dataType = py_serial.
        SERIAL_PollForRequest()
    if rqType == py_serial.MCU_WRITES:
        # INPUT -> FROM MCU TO PC
        inputs = py_serial.SERIAL_Read()

    elif rqType == py_serial.MCU_READS:
        # INPUT -> FROM PC TO MCU
        inputs = test_samples[i:i+1].astype(py_serial.
            SERIAL_GetDType(dataType))
        i = i + 1
        if i >= len(test_samples):
            i = 0
        py_serial.SERIAL_Write(inputs)

    pcout = dbscan.predict(np.reshape(inputs, (1, datalength)).
        astype(np.double))
    rqType, datalength, dataType = py_serial.
        SERIAL_PollForRequest()
    if rqType == py_serial.MCU_WRITES:
        mcuout = py_serial.SERIAL_Read()
        print()
        print("Inputs : " + str(inputs))
        print("PC Output : " + str(pcout))
        print("MCU Output : " + str(mcuout))
        print()
```

8.3.4.2 Mbed Studio Project Settings for Testing

In order to test the deployed DBSCAN clustering algorithm on the STM32 micro-controller, we should add the libraries lib_rng and lib_serial to the Mbed Studio project. To do so, we should copy the files "lib_rng.h," "lib_rng.c," "lib_uart.h," "lib_uart.c," "lib_serial.h," and "lib_serial.c" to our Mbed Studio project folder. Here, we also assume that the steps in Sect. 8.3.3.2 are already done.

The test code for the Mbed Studio project is given in Listing 8.14. This code is similar to the one given for the STM32CubeIDE project. Hence, the explanations there hold here as well. The only difference is that we should include the header file "lib_uart.h" and initialize the UART peripheral unit before the while loop by code in our Mbed Studio project. Please note that the Python script in Listing 8.13 should already be running on PC for executing this example successfully.

Listing 8.14 DBSCAN clustering test code for Mbed Studio

```
#include "mbed.h"
#include "dbscan_clus_inference.h"
#include "lib_serial.h"
#include "lib_rng.h"
#include "lib_uart.h"

#define INPUT_PC 1
#define INPUT_MCU 2
#define INPUT INPUT_PC

#define SIZE_INPUT NUM_FEATURES
#define SIZE_OUTPUT 1

float input[SIZE_INPUT];
int32_t output[SIZE_OUTPUT];

int main()
{
LIB_UART_Init();
#if (INPUT == INPUT_MCU)
  LIB_RNG_Init();
#endif
while (true)
{
#if (INPUT == INPUT_PC)
  LIB_SERIAL_Receive(input, SIZE_INPUT, TYPE_F32);
#elif (INPUT == INPUT_MCU)
  for (uint32_t i = 0; i < SIZE_INPUT; ++i)
  input[i] = (float)(LIB_RNG_GetRandomNumber() % 1000) / 1000.0f;
  LIB_SERIAL_Transmit(input, SIZE_INPUT, TYPE_F32);
#endif
dbscan_clus_predict(input, output);
LIB_SERIAL_Transmit(output, SIZE_OUTPUT, TYPE_S32);
HAL_Delay(1000);
}
}
```

8.4   Application: Fall Detection System

This application focuses on anomaly detection based on human activity data obtained from accelerometer. Thus, the WISDM dataset introduced in Sect. 5.4 is used for this purpose. Preprocessing of the dataset is done using the code snippets in Listings 5.9 and 5.10. Once the feature extraction is done, we form a DBSCAN object under the sklearn2c library. Here, the epsilon parameter is set as 8. The minSamples parameter is set as 5. When the DBSCAN is run, samples labeled as −1 are decided as outliers. Hence, we can call them as anomaly in operation. We provide the Python script formed for this purpose in Listing 8.15. The script also produces C header and source files to run the generated DBSCAN algorithm on the microcontroller.

Listing 8.15 Forming and exporting the DBSCAN algorithm to run on the WISDM dataset

```
import os.path as osp
from data_utils import read_data
from feature_utils import create_features
from sklearn2c import Dbscan

DATA_PATH = osp.join("WISDM_ar_v1.1", "WISDM_ar_v1.1_raw.txt")
TIME_PERIODS = 80
STEP_DISTANCE = 40
data_df = read_data(DATA_PATH)
df_train = data_df[data_df["user"] <= 28]
df_test = data_df[data_df["user"] > 28]

train_segments_df, train_labels = create_features(df_train,
    TIME_PERIODS, STEP_DISTANCE)
dbscan = Dbscan(eps = 8, min_samples = 5)
dbscan.train(train_segments_df)
dbscan.predict(train_segments_df)
dbscan.export("dbscan_clus_export")
```

8.5   Application: Image Quantization

In this application, we apply the online k-means clustering algorithm for color image quantization. Color quantization is a process on images where distinct colors are mapped to smaller set of representative colors. Hence, the number of distinct colors in the image is reduced. The k-means clustering algorithm constitutes clusters based on RGB values of pixels. The number of distinct colors after quantization is equal to the number of clusters formed by the k-means clustering algorithm. We provide the Python script formed for this purpose in Listing 8.16. The script also produces C header and source files to run the generated k-means clustering algorithm on the microcontroller. After exporting the created model to the microcontroller, the user should run the k-means inference function on the microcontroller by iterating over all image pixels.

Listing 8.16 Color image quantization using the k-means clustering algorithm

```
import cv2
from sklearn2c import Kmeans

img = cv2.imread("im_rgb.jpg")

img_flat = img.reshape(-1,3)
kmeans = Kmeans()

kmeans.train(img_flat)
kmeans.export("image_quantization")
```

8.6   Summary of the Chapter

We introduced the k-means and DBSCAN clustering algorithms in this chapter.
These algorithms have different working principles. Hence, they provide different
clusters for the same dataset at hand. In order to understand the working principles
of each clustering algorithm, we first covered its theoretical background. Then,
we explored its formation in Python on PC. Afterward, we introduced methods to
deploy the formed clustering algorithm to the STM32 microcontroller. As end of
chapter applications, we provided solution to two real-life problems via clustering
as fall detection and image quantization.

The TensorFlow Platform and Keras API

9

9.1 Introduction to the TensorFlow Platform and Keras API

We will extensively use the TensorFlow platform and Keras API on PC while developing neural network models in the following chapters. Therefore, we will start with basic definitions in this section. This will be followed by installation of TensorFlow in Python under PC. Finally, we will consider using TensorFlow under Google Colab in cloud.

9.1.1 Basic Definitions

As of writing this book, there were two popular and widely used platforms in machine learning and neural network applications. These were PyTorch developed by Meta (Facebook AI Research Lab) and TensorFlow developed by Google. We picked the latter in this book since it has a version called TensorFlow Lite which can be used on microcontrollers.

TensorFlow is formed such that arithmetic operations and function realizations can be done on data with different dimensions in the same way. Hence, the same structure can be used for data with different dimensions. This simplifies life for us. By the way, we will be using the TensorFlow version 2.13 under Python in this book. TensorFlow can be used in forming and training neural network models. We will explain what a neuron means and how a neural network model can be formed in Chaps. 10 and 11 in detail. Here, we can briefly explain them as follows: a neuron as a nonlinear processing unit. A neural network is a structure formed from neurons in layered form. Each layer consists of more than one neuron. Neurons in one layer will be connected to others in the next or previous layer. TensorFlow allows forming a neural network structure at lowest level. This may be unnecessary for some applications. Fortunately, the Keras API under TensorFlow simplifies these operations. Since we are using TensorFlow version 2.13 in this book, Keras is included in TensorFlow by default.

9.1.2   Installation to PC

We will be using a PC with Windows operating system in this book. Therefore, we will explain the TensorFlow installation process for it. Please check the official TensorFlow website for other operating systems. This website is also extremely useful due to the available tutorials and function explanations.

Installing TensorFlow can be done easily by using the command `pip install tensorflow`. The TensorFlow website also recommends that pip should be upgraded to the most recent version beforehand. This can be done by the command `pip install --upgrade pip`. We strongly suggest the reader to check the official TensorFlow website for possible updates in the future.

As we install TensorFlow to our PC, we can check whether all operations have been done correctly by opening a new Python file and adding the below code to it. If the code works correctly, then we should be seeing the TensorFlow version printed on the command window. Hence, we can proceed. Otherwise, we should check the error and its possible solutions.

```
import tensorflow as tf
print("TensorFlow version:", tf.__version__)
```

TensorFlow allows using GPU to speed up operations. To do so, a supported GPU card should be available on the PC. The user should also install extra packages to benefit from the GPU. We direct the reader to the official TensorFlow website for this purpose.

9.1.3   Using TensorFlow Under Google Colab in Cloud

Google Colab allows us to use TensorFlow in cloud. Hence, the reader does not deal with the TensorFlow installation process. The other advantage of Google Colab is that it offers GPU usage for a limited time free of charge. Hence, the reader can benefit from the computation advantage of GPUs while training and testing neural network models. Google Colab allows writing Python code under Jupyter notebook. This has the advantage of adding readable comments and observing partial outputs directly on a web browser. Moreover, the code written on Colab can be downloaded to local PC in Python or Jupyter notebook form. We direct the reader to the Google Colab user guide for all these operations.

9.2 Constant and Variable Declarations

We can start explaining TensorFlow with constant and variable declarations. Therefore, we will first consider constants. Then, we will introduce variables in this section. For more information on constant and variable declarations, please visit [46] and [47], respectively.

9.2.1 Constant Declaration

TensorFlow has a flexible structure to represent constant values. Therefore, we can handle a scalar, vector, matrix, and tensor (N-dimensional matrix) in the same way. Let's consider each option in detail.

9.2.1.1 Scalar

We can define a scalar value under TensorFlow by the command tf.constant. This scalar constant can be one of the types mentioned in [48]. We can use the **print** function to observe how the constant has been kept. Let's give an example for the constant declaration. Assume that we define two constants a and b in integer and float forms, respectively, as in Listing 9.1. As we execute the code, it will define and print these scalars.

Listing 9.1 Constant scalar assignment in TensorFlow

```
import tensorflow as tf

a = tf.constant(2)
b = tf.constant(5.0)

print("Scalar (1 entry):\n %s \n" % a)
print("Scalar (1 entry):\n %s \n" % b)
```

9.2.1.2 Vector

We can define a vector under TensorFlow by the command tf.constant. This vector constant can be one of the types mentioned in [48]. If no dtype is specified, then it is inferred from the value by default. We can use the **print** function to observe how the constant has been kept. Let's give an example for the constant vector declaration. We can define the Vector as in Listing 9.2. As we execute the code, it will define and print an integer vector with three entries.

Listing 9.2 Constant vector assignment in TensorFlow

```
import tensorflow as tf

Vector = tf.constant([5,6,2])

print("Vector (3 entries) :\n %s \n" % Vector)
```

9.2.1.3 Matrix

We can define a matrix under TensorFlow by the command `tf.constant`. This matrix constant can be one of the types mentioned in [48]. If no `dtype` is specified, then it is inferred from the value by default. We can use the **print** function to observe how the constant has been kept. Let's give an example for the constant matrix declaration. We can define the `Matrix` as in Listing 9.3. As we execute the code, it will define and print an integer matrix with three by three entries.

Listing 9.3   Constant matrix assignment in TensorFlow

```
import tensorflow as tf

Matrix = tf.constant([[1,2,3],[2,3,4],[3,4,5]])

print("Matrix (3x3 entries):\n %s \n" % Matrix)
```

9.2.1.4 Tensor

We can define a tensor under TensorFlow by the command `tf.constant`. This tensor constant can be one of the types mentioned in [48]. If no `dtype` is specified, then it is inferred from the value by default. We can use the **print** function to observe how the constant has been kept. Let's give an example for the constant tensor declaration. We can define the `Tensor` as in Listing 9.4. As we execute the code, it will define and print an integer tensor with three by three by three entries.

Listing 9.4   Constant tensor assignment in TensorFlow

```
import tensorflow as tf

Tensor = tf.constant( [ [[1,2,3],[2,3,4],[3,4,5]] ,
    [[4,5,6],[5,6,7],[6,7,8]] , [[7,8,9],[8,9,10],[9,10,11]] ] )

print("Tensor (3x3x3 entries) :\n %s \n" % Tensor)
```

9.2.2   Variable Declaration

We can define allowed variable types under TensorFlow using the function `tf
.Variable` as in Listing 9.5. Since the type of these variables is not defined, TensorFlow inferred them from the initial value. As we execute the code, it will define and print a float and integer variable with three entries.

Listing 9.5 Variable assignment in TensorFlow

```
import tensorflow as tf

v1 = tf.Variable([1.,2.,3.])
v2 = tf.Variable([0,0,0])

print("Variable v1 is", v1)
print("Variable v2 is", v2)
```

9.3 Arithmetic Operations

As we define constants and variables, we can apply arithmetic operations on them. We will consider these operations in this section.

9.3.1 Addition and Subtraction

We can apply addition and subtraction operations on constants and variables with the functions `tf.add` and `tf.subtract`, respectively. We can also use the addition and subtraction operations, $+$ and $-$, respectively, to perform these operations. We provide a sample code to show how these operations are performed in Listing 9.6.

Listing 9.6 Addition and subtraction operations in TensorFlow

```
import tensorflow as tf

a = tf.Variable([[1, 2, 3], [2, 3, 4], [3, 4, 5]])
b = tf.Variable([[4, 5, 6], [5, 6, 7], [6, 7, 8]])

add1 = tf.add(a, b)
add2 = a + b

print("Addition with the add function: ", add1)
print("Addition with the + operator: ", add2)

sub1 = tf.subtract(a, b)
sub2 = a - b

print("Subtraction with the subtract function: ", sub1)
print("Subtraction with the - operator: ", sub2)
```

In Listing 9.6, the values a and b can be taken from any scalar, vector, matrix, or tensor pair defined in the previous section. We will get the results in the same way. This example indicates the standardization under TensorFlow.

9.3.2   Element-Wise Multiplication

We can apply element-wise multiplication to scalars, vectors, matrices or tensors with the function `tf.multiply`. We can also use the multiplication operator $*$ to perform this operation. We provide a sample code to show how these operations are performed in Listing 9.7.

Listing 9.7 Element-wise multiplication in TensorFlow

```
import tensorflow as tf
a = tf.Variable([[1, 2, 3], [2, 3, 4], [3, 4, 5]])
b = tf.Variable([[4, 5, 6], [5, 6, 7], [6, 7, 8]])

mul1 = tf.multiply(a, b)
mul2 = a * b

print("Element-wise multiplication with the multiply function: ",
      mul1)
print("Element-wise multiplication with the * operator: ", mul2)
```

As we execute the code in Listing 9.7, we can observe that element-wise multiplication has been done between two tensors. The same code can be modified to apply element-wise multiplication between a scalar and vector, two vectors, and two matrices. To do so, the reader should place constant declarations in Sect. 9.2.1.

9.3.3   Vector and Matrix Multiplications

TensorFlow has the function `tf.matmul` to perform vector, matrix, or tensor multiplication between two constants or variables. We provide a sample code to multiply two vectors and matrices separately in Listing 9.8. As can be seen in this code, we can easily multiply two vectors and matrices defined in TensorFlow as long as their dimensions match.

Listing 9.8 Vector and matrix multiplications

```
import tensorflow as tf
a = tf.Variable([[1], [2]])
b = tf.Variable([[3, 4]])

vecmul = tf.matmul(a, b)

print("Vector multiplication: ", vecmul)

c = tf.Variable([[1, 2, 3], [2, 3, 4], [3, 4, 5]])
d = tf.Variable([[4, 5, 6], [5, 6, 7], [6, 7, 8]])

matmul = tf.matmul(c, d)

print("Matrix multiplication: ", matmul)
```

All the mentioned arithmetic operations introduced in this section indicate the standardization under TensorFlow. Hence, all tensor operations can be performed easily. Nevertheless, TensorFlow's capabilities are not limited with tensor operations.

9.4 Activation Functions

In Chap. 10, we will see that each neuron has a nonlinear activation function. Researchers proposed several such functions for this purpose and TensorFlow consists of the most used ones. We summarize four activation functions next.

9.4.1 Hard Limiter

The hard limiter activation function is defined as

$$\varphi(x) = \begin{cases} 0, & x \leq 0 \\ 1, & x > 0 \end{cases} \quad (9.1)$$

The input-output plot of the function is given in Fig. 9.1. We will provide the implementation of this activation function in the following chapters whenever needed.

9.4.2 Sigmoid

The sigmoid activation function is defined as

$$\varphi(x) = \frac{1}{1 + e^{-x}} \quad (9.2)$$

Fig. 9.1 Hard limiter activation function

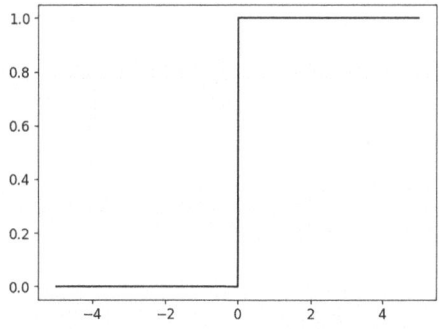

Fig. 9.2 Sigmoid activation
function

Fig. 9.3 ReLU activation
function

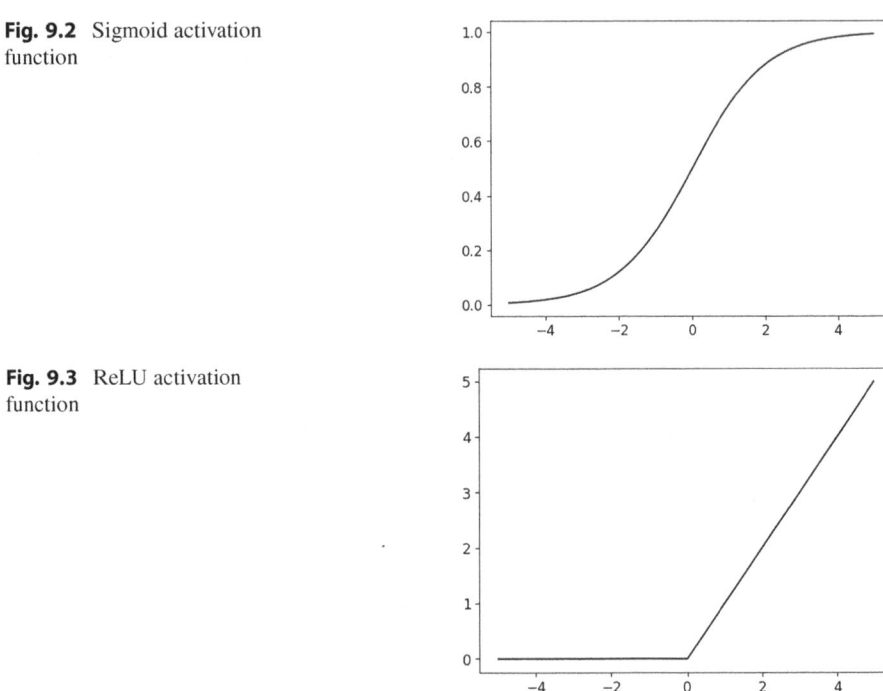

The input-output plot of the function is given in Fig. 9.2. TensorFlow has this
function represented as `tf.keras.activations.sigmoid`.

9.4.3   Rectified Linear Unit

The rectified linear unit (ReLU) activation function is defined as

$$\varphi(x) = \begin{cases} 0, & x \leq 0 \\ x, & x > 0 \end{cases} \tag{9.3}$$

ReLU can also be defined as $\varphi(x) = \max(0, x)$. The input-output plot of the
function is given in Fig. 9.3. TensorFlow has this function represented as `tf.keras.activations.relu`.

9.4.4   Softmax

Softmax is a special function used extensively at the output layer of a neural network
on classification tasks. It normalizes the output values. The definition of the softmax
function is

Fig. 9.4 Softmax function

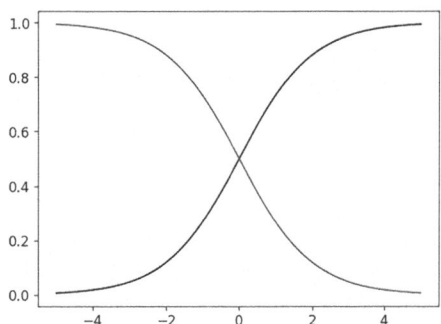

$$\varphi(x_i) = \frac{e^{x_i}}{\sum_{j=1}^{J} e^{x_j}}, \text{ for } i = 1, \ldots, J \tag{9.4}$$

In Eq. 9.4, we did not take entries of the sum term in denominator as input. This value will be used as the normalizing constant for all inputs x_i for $i = 1, \cdots, J$. The input-output plot of the function for $J = 2$ is given in Fig. 9.4. TensorFlow has this function represented as `tf.keras.activations.softmax`.

9.5   NumPy and TensorFlow

NumPy `ndarray` arrays and `tf.tensor` objects are compatible. Therefore, Tensor-Flow operations automatically convert `ndarray` to tensors. NumPy operations also automatically convert tensors to NumPy `ndarray`. A NumPy array can be converted to a tensor by the function `tf.convert_to_tensor`. Besides, a tensor can be converted to NumPy `ndarray` via its `.numpy` method. We provide a sample code to show conversions between NumPy and TensorFlow objects in Listing 9.9.

Listing 9.9 TensorFlow and NumPy conversions

```
import tensorflow as tf

T = tf.constant([[1,2,3],[4,5,6]])

print("Tensor T is ", T)

numpy_arr = T.numpy()
print("Numpy ndarray num_T value is ", numpy_arr)

N= tf.convert_to_tensor(numpy_arr)

print("Convert back to TF Tensor: ", N)
```

As can be seen in Listing 9.9, we first create a constant matrix using TensorFlow. Then, we convert it to a NumPy array. Finally, we convert the same matrix to

TensorFlow tensor. Therefore, we show that NumPy and TensorFlow data types can be used interchangeably. TensorFlow to NumPy conversion also yields more human-readable format for TensorFlow output.

9.6   Application: TensorFlow Datasets and Loading Data

TensorFlow has available datasets to be used in operations. One way of downloading and loading these datasets is installing the `tensorfow_datasets` package by running the script `pip install tensorflow-datasets` in the terminal window. Afterward, we can list available datasets under TFDS repository and download the appropriate dataset for our needs. The function `tfds.load` downloads the datasets into the home folder under TensorFlow dataset directory. `ds` is a TensorFlow `Dataset` object. Therefore, we can iterate over it to get a single element or batches by calling the function `batch`. Each element is a Python `dict` object by default. On the other hand, we can load it as a tuple by adding the parameter `as_supervised = True`. We provide such a usage in Listing 9.10. For more information on this topic, please see [49].

Listing 9.10 Load dataset with TensorFlow

```
import tensorflow_datasets as tfds

tfds.list_builders()

ds_train = tfds.load('mnist', split='train') # Returns dict
#ds = tfds.load('mnist', split='train', as_supervised=True) #
    Returns tuple

for example in ds_train:
        print(list(example.keys()))
        image = example["image"]
        label = example["label"]
        print(image.shape, label)
```

The reader can also load data by using the function `load_data` under Keras. We provide a sample code in Listing 9.11 to load the MNIST dataset this way. As the code is executed, the MNIST dataset is imported as a Python variable. Originally, the function `load_data` returns a tuple of tuples. One tuple consists of training samples and corresponding labels. The other tuple consists of test samples and their corresponding labels. In the code, we directly assign these to variables `(x_train, y_train), (x_test, y_test)`.

Listing 9.11 Load dataset with Keras

```
import tensorflow as tf

(x_train, y_train), (x_test, y_test) = tf.keras.datasets.mnist.
    load_data()
```

9.7 Application: Load and Store Neural Network Models

We will be dealing with neural network models in the following chapters. It will take time to train them. Hence, once a model has been trained, we can save it to be used in the future. Likewise, we can load an available model. TensorFlow allows us save and load a model by its available functions. There are two ways to save TensorFlow models. The first way is saving weights of an existing model. The second way is saving the model with its computation graph. We provide a sample code to perform these operations in Listing 9.12. If we use the Keras H5 format, then it will generate a single file called "model.h5" in the current working directory. If we use the TensorFlow SavedModel format, then it will generate a folder with the computation graph and associated weights.

Listing 9.12 Saving a model in TensorFlow

```
import keras

input = keras.Input(shape=(32,))
output = keras.layers.Dense(1)(input)
model = keras.Model(input, output)
model.compile(optimizer="adam", loss="mean_squared_error")

# Saves model in Keras h5 format
model.save("model.h5")

# Saves model in TF SavedModel format
model.save("model")
```

After saving a model with the Keras H5 format, we can easily load it by calling the function `keras_model = keras.models.load_model("model.h5")`. If we saved the model as TensorFlow SavedModel, then we can use the folder name of the model instead as `keras_model = keras.models.load_model("model")`. For more information on this topic, please see [50].

9.8 Application: Converting Neural Network Models from Other Platforms

We picked TensorFlow as the platform to be used for neural network applications throughout the book. However, the reader may be working on other platforms such as PyTorch and MATLAB. Therefore, we provide ways of converting a neural network model formed in these platforms to TensorFlow format. To do so, we will benefit from the ONNX format. Let's start with explaining it.

9.8.1   The ONNX Format

The open neural network exchange (ONNX) is an open format that defines a common set of operators and common file format for machine and deep learning models. It is designed to represent models in a way that is independent of the training framework. This allows training the model in one framework and then using it in another framework for inference. Therefore, ONNX defines an extensible computation graph model, as well as definitions of built-in operators, and standard data types.

9.8.2   From PyTorch to TensorFlow

This book is based on the TensorFlow platform for neural networks and TensorFlow Lite to embed the generated models to the microcontroller. Fortunately, the ONNX format allows us to embed a PyTorch model to the microcontroller. We will explain such an operation can be done in this section.

We should install the following libraries using the Python package manager (pip) to our current development environment. The torch and torchvision libraries will be used to load pretrained models from the PyTorch repository. ONNX will be used as the intermediate format. onnx_graphsurgeon and sng4onnx are dependencies of the onnx2tf package which will be used to convert an ONNX model to TensorFlow.

```
pip install torch torchvision onnx
pip install onnx_graphsurgeon --index-url https://pypi.ngc.nvidia
    .com
pip install sng4onnx
pip install onnx2tf
```

We provide a conversion example from PyTorch to TensorFlow in Listing 9.13. In this script, we start by importing torch.onnx to convert the PyTorch model to ONNX format. We use torchvision to load the Pytorch model and pretrained weights on the ImageNet dataset. PyTorch is based on dynamic computation graphs. This means the graph structure is created on the fly when the model is fed with an input data. Therefore, we should create a dummy variable to feed the model and retrieve its structure. Also, PyTorch uses the NCHW format in contrast to the NHWC format used in TensorFlow. Here, N represents and the number of samples. C, H, and W represent the number of channels and height and width of an image, respectively. Therefore, we create the dummy input variable based on this data format with N = 1, C = 3, H = 224, and W = 224 in the code. The function torch.randn creates a random tensor with given dimensions with values in the interval [0, 1]. In other words, a single normalized RGB image with size (224,224) is created by this function. In this example, we use the MobileNet3-small model. To note here, all

Torchvision models and their corresponding weights are available at [24]. We load the model using `torchvision.models.mobilenet_v3_small(weights = pretrained)` with its corresponding ImageNet weights. Then, we export the model in ONNX format using the `dummy_input` and output model filename.

Listing 9.13 Converting PyTorch model to ONNX format

```
import torch.onnx
import torchvision

dummy_input = torch.randn(1, 3, 224, 224)
pretrained = "MobileNet_V3_Small_Weights.IMAGENET1K_V1"
model = torchvision.models.mobilenet_v3_small(weights =
    pretrained)
torch.onnx.export(model, dummy_input, "mobilenetv3small.onnx")
```

In Listing 9.14, we use the exported ONNX model and convert it to the corresponding TensorFlow model. To do so, all we need to do is importing the onnx2tf library and calling the `convert` function from it. Here, the required arguments are the input ONNX model path and output folder path for the TensorFlow model. The function produces the model in the `SavedModel` format as well as 16- and 32-bit float TensorFlow Lite models. Output type of the created TensorFlow model can also be chosen. To create a Keras model with HDF5 format, we should set `output_h5=True` in the function. Likewise, we should set `output_keras_v3=True` to create a Keras model with Keras V3 format.

Listing 9.14 Converting the ONNX model to TensorFlow and TensorFlow Lite model

```
import onnx2tf

onnx2tf.convert(
    input_onnx_file_path="mobilenetv3small.onnx",
    output_folder_path="mobilenetv3small_tf",
    output_h5=True,
    non_verbose=True,
)
```

9.8.3 From MATLAB to TensorFlow

The ONNX format can also be used to convert a neural network model generated under MATLAB to TensorFlow format. The reader can follow the specific steps under MATLAB to convert the model to ONNX format for this purpose. Then, we can apply the operations based on the onnx2tf library in the previous section to obtain the corresponding TensorFlow model.

9.9   Summary of the Chapter

TensorFlow and Keras API are the two platforms to extensively use in the following chapters on neural networks. Therefore, we introduced them in this chapter. In other words, the neural network-based methods in the following chapters are extensively based on multidimensional array operations. TensorFlow provides a platform to perform all these operations. The Keras API simplifies life for use while benefiting from TensorFlow. Therefore, we considered both platforms in this chapter. To do so, we started with their installation. Then, we explored constant and variable declarations in TensorFlow. Afterward, we defined arithmetic operations on these. Then, we considered activation functions under TensorFlow. We also provided the link between the NumPy library and TensorFlow. We considered loading data as the first end of chapter application. Next, we introduced loading and storing neural network models under TensorFlow as the second end of chapter application. Finally, we evaluated converting neural network models from other platforms to TensorFlow format as the third end of chapter application.

Fundamentals of Neural Networks

10

10.1 Background

In order to understand the single neuron, we will first consider its structure in this section. Then, we will provide its mathematical definition. These will help us in forming the single neuron (and neural networks in general) in TensorFlow and Keras.

10.1.1 Structure of the Single Neuron

We can represent the single neuron schematically as in Fig. 10.1. In this figure, x_k for $k = 1, \cdots, K$ stand for the inputs. w_k for $k = 1, \cdots, K$ are the corresponding weight terms. b is the bias term. $\varphi(\cdot)$ is the activation function and y is output of the neuron.

The neuron structure in Fig. 10.1 can be summarized as follows. The neuron has K inputs. Each input is multiplied by a weight term. Then, multiplications are summed up. The bias term is added to the sum. The overall sum is fed to the activation function to obtain output y of the neuron. Possible activation functions to be used in this operation have been given in Sect. 9.4.

10.1.2 Mathematical Definition of the Single Neuron

We can mathematically represent the single neuron by its inputs, weights, bias term, activation function, and output as

$$y = \varphi \left(\sum_{k=1}^{K} w_k x_k + b \right) \tag{10.1}$$

© The Author(s), under exclusive license to Springer Nature Switzerland AG 2025   229
C. Ünsalan et al., *Embedded Machine Learning with Microcontrollers*,
https://doi.org/10.1007/978-3-031-70912-8_10

Fig. 10.1 The single neuron

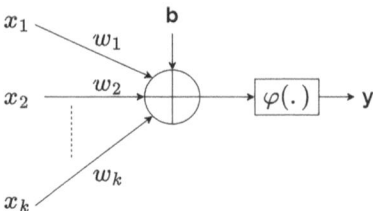

The representation in Eq. 10.1 allows the single neuron to be trained (by adjusting its weights and bias term). Hence, it can be used as a classifier or regressor. We will consider these options in Sects. 10.5 and 10.6, respectively. A single neuron may not be sufficient in forming a good classifier or regressor. Instead, using several neurons in layered form will lead to a more powerful structure for this purpose. We will see this option in detail in Chap. 11.

We can observe input-output relationship of the single neuron when we have two inputs. To do so, let's assume that the activation function of the neuron is sigmoid. We provide four input-output relationships by manually setting weights w_1, w_2, and bias term b as in Fig. 10.2.

As can be seen in Fig. 10.2, we can change the output of the neuron by modifying w_1 and w_2. The bias term, b, shifts the output in x_1-x_2 axis. We will threshold the output y to form a classifier from the single neuron. We will show that the formed decision boundary will be a straight line for the two-class classifier case in Sect. 10.5. We can use the single neuron, without thresholding its output, as a regressor. Assume that we would like to represent the linear relationship $y = 3x + 4$. As we select the single neuron weight $w_1 = 3$, bias term $b = 4$, and activation function as ReLU, we can form a regressor as long as the input value is greater than $-4/3$. We will consider forming a regressor via single neuron in Sect. 10.6.

10.2   Forming the Single Neuron in TensorFlow

TensorFlow allows forming the single neuron from constant, variable, and activation function definitions introduced in Chap. 9. In this section, we will consider this operation in two stages as class definition and parameter initialization. Then, we will show how output of the single neuron can be obtained for a given input.

10.2.1 Class Definition and Parameter Initialization

We can define the neuron as an object with input, weight, bias, activation function, and output terms. To do so, we can form the class definition as in Listing 10.1.

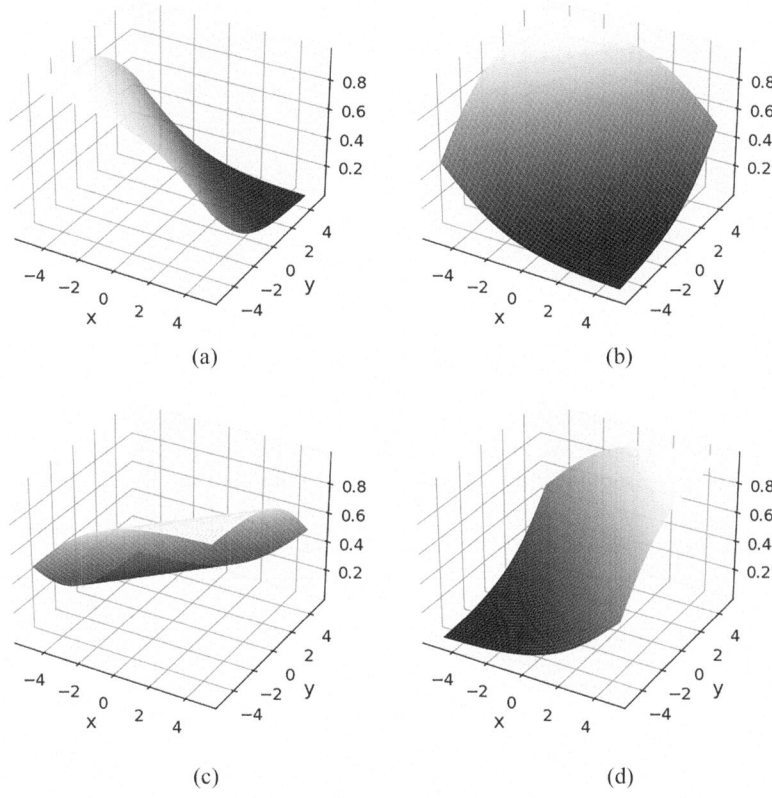

Fig. 10.2 Input-output relationship of the single neuron with different weight and bias values. (**a**) $w_1 = -0.5$, $w_2 = -0.5$, $b = 0$. (**b**) $w_1 = -0.5$, $w_2 = 0.5$, $b = 0$. (**c**) $w_1 = 0.5$, $w_2 = -0.5$, $b = 0$. (**d**) $w_1 = 0.5$, $w_2 = 0.5$, $b = 0$

Listing 10.1 Class definition for the single neuron in TensorFlow

```
class Model(tf.Module):
    def __init__(self, **kwargs):
        super().__init__(**kwargs)

    def __call__(self, x):
        return self.w * x + self.b
```

We can initialize the weight and bias terms in our object. This can be done by setting the weights under the function __init__, also known as the constructor, as in Listing 10.2. We will adjust these weights during training in Sect. 10.4. An accepted strategy for this purpose is to initialize the weight and bias terms by random numbers.

Listing 10.2 Initializing the single neuron weights in TensorFlow

```
import tensorflow as tf

class Model(tf.Module):
    def __init__(self, **kwargs):
        self.w = tf.Variable(5.0)
        self.b = tf.Variable(0.0)

    def __call__(self, x):
        return self.w * x + self.b
```

10.2.2 Obtaining Output for a Given Input

As we initialize the single neuron parameters (or train them), we can obtain neuron output for a given input. We provide a sample code for this purpose in Listing 10.3. This code uses the input-output relation formed in this section.

Listing 10.3 Input-output relation of the single neuron in TensorFlow

```
import tensorflow as tf

class Model(tf.Module):
    def __init__(self):
        self.W = tf.Variable(1.0)
        self.b = tf.Variable(1.0)

    def __call__(self, x):
        self.out= tf.keras.activations.sigmoid(self.W * x + self.b)
        return self.out

model = Model()
u=model(0.0).numpy()

print(u)
```

10.3   Forming the Single Neuron in Keras

Although TensorFlow allows forming the single neuron from basic definitions, its usage may become complex as more neurons are added to the operation. Keras simplifies neuron formation for such cases. Therefore, we will consider it in this section by defining and initializing a layer. We will also provide a way of obtaining output for a given input in Keras.

10.3.1 Layer Definition and Parameter Initialization

Keras has a layer structure which can be used to define several parallel neurons at once. We will benefit from this setup extensively in Chap. 11. Here, we will use the layer structure to define the single neuron. We provide the sample code formed for this purpose in Listing 10.4.

Listing 10.4 Single neuron definition in Keras

```
import tensorflow as tf

w0 = tf.keras.initializers.Constant(1.)
b0 = tf.keras.initializers.Constant(1.)
l0=tf.keras.layers.Dense(units=1,
                         input_shape=[1],
                         kernel_initializer=w0,
                         bias_initializer=b0,
                         activation='sigmoid')
model = tf.keras.models.Sequential([l0])

result=model.predict([0.0])

print(result)
```

As can be seen in Listing 10.4, we first define constant initializers which set the weight and bias values to 1. To do so, we should define a layer whose size is compatible with the single element. Therefore, we create the single neuron by calling the Dense class from Keras. Here, units indicates the number of neurons, and input_shape indicates the number of inputs and dimension of each input. kernel_initializer and bias_initializer are parameters to set values of the weight and bias terms, respectively. Finally, the activation parameter sets the activation function.

10.3.2 Obtaining Output for a Given Input

As we define the neuron and set its parameters under Keras, we can obtain its output for a given input by the predict function of the formed model. We provide such an example in Listing 10.4 as model.predict([0.0]). Here, we obtain output of the formed model to input 0. As the script in Listing 10.4 is executed, the result is printed as 0.7310586. Let's repeat this process by hand by using the input value, weight, and bias terms as 0, 1, and 1, respectively. Thus, output of the neuron becomes $\frac{1}{1+\exp(-(0*1+1))} = 0.7310586$. Hence, we obtain the same output by hand as well.

We had four representative input-output plots in 2D for the single neuron in Fig. 10.2. We provide the sample code formed for this purpose, using one set of w and b values, in Listing 10.5. To form the four input-output relations in Fig. 10.2, we should modify the w term with four possible combinations as

w = tf.keras.initializers.Constant([-0.5, -0.5]), w = tf.keras.initializers
.Constant([-0.5, 0.5]), w = tf.keras.initializers.Constant([0.5, -0.5]), and
w = tf.keras.initializers.Constant([0.5, 0.5]). For all these combinations, we
set b = tf.keras.initializers.Constant(0.0).

Listing 10.5 Single neuron input-output calculation on an interval

```
import tensorflow as tf
import numpy as np
import matplotlib.pyplot as plt
from matplotlib import cm

w = tf.keras.initializers.Constant([.5, -0.5])
b = tf.keras.initializers.Constant(0.)
l0=tf.keras.layers.Dense(units=1,
                         input_shape=[2],
                         kernel_initializer=w,
                         bias_initializer=b,
                         activation='sigmoid')

model = tf.keras.Sequential([l0])

xval, yval = np.meshgrid(np.arange(-5, 5, 0.1),np.arange(-5, 5,
    0.1))

Z = model.predict(np.c_[xval.ravel(), yval.ravel()])

Z = Z.reshape(xval.shape)

fig, ax = plt.subplots(subplot_kw={"projection": "3d"})
surf = ax.plot_surface(xval, yval, Z, cmap=cm.coolwarm)
plt.xlabel('x')
plt.ylabel('y')
plt.show()
```

As can be seen in Listing 10.5, we initialize the weight and bias terms similar
to Listing 10.4. However, we have two inputs this time. Hence, we create the dense
layer with the parameter input_shape=[2]. Besides, we formed the input values x
and y in the interval $[-5, 5]$. The function numpy.meshgrid produces combination of
x,y values in this interval. When the predict function of the Keras model is called,
the single neuron output is calculated for each combination. Hence, we can plot the
output as a two-dimensional surface.

10.4 Training the Single Neuron

We considered parametric and nonparametric methods for classification and regression in previous chapters. There, we either estimated parameters of a fixed model
from labeled data, or we directly used labeled data to construct the model. Therefore,
we have either an explicit or implicit model for the classifier or regressor. In neural
networks, we form the model from connected neurons. We also need labeled data.
Different from previous methods, we train the neurons by labeled data to grasp the

relationship between input and output. In other words, the neural network learns the input-output relationship from data. We will explore how this can be done on a single neuron in this section. Therefore, we will first consider the meaning of training and explain it from a broader perspective. Then, we will focus on the loss function and parameter updating by optimizers. Finally, we will explore how training can be done in TensorFlow and Keras.

10.4.1 What Is Training?

The single neuron has weight and bias terms as explained in Sect. 10.1. These terms should be set such that the neuron can represent the input-output relationship for a classifier or regressor. Therefore, we should have labeled training data such that we know the desired (expected) output for a given input on a number of samples. This will lead to adjusting the neuron weight and bias terms via training. To do so, we should define the loss function and optimization. We will focus on the loss function in Sect. 10.4.2. We will focus on the parameter updating operation via optimizers in Sect. 10.4.3.

10.4.2 The Loss Function

The loss function gives us the difference between the desired and neuron output for a given input. More specifically, we will pick an input sample from the labeled training data with its corresponding (desired) output. Then, we will feed input to the neuron (or neural network) and obtain its output. We can call this as the predicted output. The loss function will give us a measure how different the desired and predicted output values are. We will perform this operation on all training set to calculate the overall loss value.

There are several loss functions in the literature. We will focus on three most well-known ones available under Keras next. We will briefly explain each loss function and show how it can be used under Keras. For more information on all loss functions available under Keras, please see [51].

10.4.2.1 Categorical and Binary Cross-Entropy

One of the most commonly used loss functions for classification tasks is the categorical cross-entropy. This function calculates the cross-entropy between the predicted and desired output values. Binary cross-entropy is the special case of categorical cross-entropy which produces the loss if a sample belongs to a class or not. Binary cross-entropy loss value can be expressed as

$$L = - \sum_{i=1}^{2} y_i * ln(p_i) \qquad (10.2)$$

$$= -y_1 \ln(p_1) - y_2 \ln(p_2) \tag{10.3}$$

where y_i is the desired output (ground truth) value for class i. p_i is the predicted probability of class i. Since this is a binary classification task, y_i should be either 0 or 1. Hence, $y_2 = 1 - y_1$. Therefore, Eq. 10.2 can be simplified as

$$L = -y_1 \ln(p_1) - (1 - y_1) \ln(1 - p_1) \tag{10.4}$$

In other words

$$L = \begin{cases} - \ln(1 - p_1), \text{ if } y_1 = 0 \\ - \ln(p_1), \text{ if } y_1 = 1 \end{cases} \tag{10.5}$$

Neuron model output does not produce probability values directly. Therefore, probability values for classes should be inferred. This is where the activation function comes into play. We will use the sigmoid activation function introduced in Sect. 9.4 to produce probabilities while keeping the output differentiable. Therefore, binary cross-entropy is also called sigmoid loss.

In Keras, the `BinaryCrossentropy` class can be used to calculate the binary cross-entropy loss from either probabilities or model outputs. To emphasize this point, we create logits and their corresponding probabilities with the sigmoid function in Listing 10.6. Logit refers to the model output where the model produces scores depending on its weights. Logits can be converted to probabilities using the sigmoid function for the binary classification case. The binary cross-entropy loss can be calculated for both logits and probabilities. Note that we also set `from_logits = True` to produce the same result from model output. As the code in Listing 10.6 is executed, both `bce` and `bce_logits` outputs will be approximately 0.9899213. Hence, both outputs are the same except minor floating-point error.

Listing 10.6 Binary cross-entropy loss calculation

```
import tensorflow as tf

y_pred_logits = tf.constant([0.41, -0.85, -1.39, 1.39])

y_pred = tf.keras.activations.sigmoid(y_pred_logits)

y_true = [0, 1, 0, 0]

bce = tf.keras.losses.BinaryCrossentropy()
bce_logits = tf.keras.losses.BinaryCrossentropy(from_logits=True)

result_logits = bce_logits(y_true, y_pred_logits)
result= bce(y_true, y_pred)

print(result.numpy())
print(result_logits.numpy())
```

The binary cross-entropy loss can handle only binary classification tasks as its name suggests. Categorical cross-entropy can be used for multi-class classification tasks. Similar to the binary cross-entropy, categorical cross-entropy loss value is the multiplication of the ground truth and logarithmic probability values as

$$L = - \sum_{i=1}^{C} y_i * ln(p_i) \tag{10.6}$$

where C is number of classes, y_i is ground truth label for corresponding class, and p_i is the model predicted probability of the ith class. In binary classification, the sigmoid activation function is used to map model output to probability values. This yields the output to the interval [0, 1]. In multi-class classification, probability of the model output is calculated by the softmax function. This produces an output such that all class probabilities sum up to 1. The softmax function used in operation is

$$p_i = \frac{\exp(\hat{y}_i)}{\sum_{j=1}^{C} \exp(\hat{y}_j)} \tag{10.7}$$

where $\hat{y}_i$ is the predicted output value of the ith class. We divide exponentiated value of the predicted output of the ith class to the sum of all exponentiated values to find out probability of class i. Thus, softmax guarantees that resulting probabilities are in the interval [0, 1] and they sum up to 1.

In Keras, the `CategoricalCrossentropy` class can be used to calculate the categorical cross-entropy loss from either probabilities or model outputs. To emphasize this point, we create logits and their corresponding probabilities with the softmax function in Listing 10.7. Alternatively, calculation by using only arithmetic operations is also available in Listing 10.7. Similar to the binary cross-entropy case, `from_logits = True` argument is added to produce the same result from the model output. Both outputs in Listing 10.7 are the same except minor floating-point error.

Listing 10.7 Categorical cross-entropy loss calculation

```
import tensorflow as tf

y_true = tf.constant([[0, 1, 0], [0, 0, 1]])

logits = tf.constant([[-1.99, 0.95, -100], [-1.3,0.78,-1.3]])

y_pred = tf.nn.softmax(logits)

cce_logits = tf.keras.losses.CategoricalCrossentropy(from_logits=
    True)
cce = tf.keras.losses.CategoricalCrossentropy()

result_logits = cce_logits(y_true, logits)
result= cce(y_true, y_pred)

print(result.numpy())
print(result_logits.numpy())
```

10.4.2.2 Mean Absolute, Absolute Percentage, and Mean Squared Error

There are loss definitions for regression tasks besides classification loss functions introduced in the previous section. Since regression outputs are real values instead of class probabilities, we can measure the loss numerically between the model output and ground truth. To do so, we need a distance metric. Mean absolute error (MAE), mean absolute percentage error (MAPE), and mean squared error (MSE) are the common distance metrics used for this purpose. We provide the mathematical definition of MSE loss value as

$$L = \frac{1}{N} \sum_{i=1}^{N} (y_i - \hat{y}_i)^2 \tag{10.8}$$

where y_i is ground truth, $\hat{y}_i$ is predicted output by the model, and N is the number of samples.

We provide a sample usage of MSE in Listing 10.8. Here, we can easily calculate the distance between the ground truth and model output by the class MeanSquaredError under Keras. In this script, y_true and y_pred consist of two individual vectors, each consisting of two elements. MSE is calculated vector-wise, such that MSE calculation of the first vector is $((2 - 0)^2 + (1 - 1)^2)/2 = 2$. MSE calculation of the second vector is $((3-0)^2 - (0-0)^2)/2 = 4.5$. Furthermore, Keras reduces these results to obtain a scalar loss by default. We can disable this option by creating a loss object by setting it to NONE as in Listing 10.8. This will output a tensor of size two with values [2, 4.5], while MeanSquaredError object with default parameters returns 3.25 which is average of the two values.

Listing 10.8 MSE loss calculation

```
import tensorflow as tf

reduction_enum = tf.keras.losses.Reduction

y_true = [[0., 1.], [0., 0.]]
y_pred = [[2., 1.], [3., 0.]]

mse = tf.keras.losses.MeanSquaredError()
mse_reduction = tf.keras.losses.MeanSquaredError(reduction_enum.
    NONE)

result = mse(y_true, y_pred)
result_reduction = mse_reduction(y_true, y_pred)

print(result.numpy())
print(result_reduction.numpy())
```

10.4.2.3 Cosine Similarity

Another important loss function is cosine similarity. This function is generally used to compare extracted features from neural networks. A common usage of the cosine similarity loss function is in face recognition, where each face is converted to a vector called embeddings. The distance between these vectors is calculated by cosine similarity. Mathematically, the cosine similarity loss value between two vectors v_1 and v_2 can be expressed as

$$L = \frac{v_1.v_2}{\|v_1\|.\|v_2\|} \tag{10.9}$$

The cosine similarity loss function always produces a value between -1 and 1, since it is the cosine value between two vectors.

In Keras, the cosine similarity loss function can be implemented by the class `CosineSimilarity`. We provide a usage example for the cosine similarity loss function in Listing 10.9. Here, the cosine similarity between the vectors with the same and opposite directions is computed both manually and with the class `CosineSimilarity`. As the code is executed, the reader can observe that both calculations give the same result which is -1 for the opposite vectors and 1 for the vectors in the same direction. Please note that magnitude of the vectors do not affect similarity since we normalize them.

Listing 10.9 Cosine similarity loss calculation

```
import tensorflow as tf

y_true = tf.constant([[0.4, 0.3], [-0.2, 0.6]])
y_pred = tf.constant([[0.8, 0.6], [0.4,-1.2]])

true_norm = tf.norm(y_true, axis = 1)
pred_norm = tf.norm(y_pred, axis = 1)
norm_prod = true_norm * pred_norm

cos_sim = -tf.reduce_sum(y_true * y_pred, axis = 1) / norm_prod

print(cos_sim.numpy())

cosine_loss = tf.keras.losses.CosineSimilarity(reduction = tf.
    keras.losses.Reduction.NONE)

result = cosine_loss(y_true, y_pred)
print(result.numpy())
```

10.4.3 Parameter Updating by Optimizers

We can adjust parameters of the neuron (or neural network) by minimizing the loss value. Here, the reader may think that we can obtain optimal parameters by global search over all weight and bias combinations. This becomes impractical since we will have several parameters, each having a range of possible values. Let's consider a simple scenario to emphasize this issue. Assume that we should trace 100 values to find the optimal value of a parameter to minimize loss. If we have m such parameters, then our search space will consist of 100^m samples. This becomes exhaustive for a neural network with several hundred to several thousand parameters. To avoid this, we will benefit from iterative parameter updating methods. These are generally called optimizers.

There are several optimizers in the literature. We focus on the well-known ones available under Keras next. We will briefly explain each optimizer method and show how it can be used under Keras. For more information on all optimizers available under Keras, please see [52].

10.4.3.1 Stochastic Gradient Descent

The most well-known parameter optimization method is the stochastic gradient descent (SGD). We can explain it by an example. Let's consider the simplest case such that the single neuron has one input. Hence, it has one weight and bias term. Output of the neuron for a given input $x_1[n]$ is $y[n] = \varphi(w_1 x_1[n] + b_1)$. We would like to train this neuron. Assume that we have N training samples as $(x_1[n], y_d[n])$ for $n = 1, \cdots, N$ where $y_d[n]$ is the desired output for input $x_1[n]$.

We can define the error between the actual and desired output values for the neuron as $e[n] = y_d[n] - y[n]$. This value leads to the loss function definition. Assume that we use the MSE loss function. Then, the loss value becomes

$$L = \sum_{n=1}^{N} e^2[n] \tag{10.10}$$

$$= \sum_{n=1}^{N} (y_d[n] - y[n])^2 \tag{10.11}$$

$$= \sum_{n=1}^{N} (y_d[n] - \varphi(w_1 x_1[n] + b_1))^2 \tag{10.12}$$

We will use the stochastic gradient descent method to minimize the loss iteratively by updating the single neuron parameters w_1 and b_1. The aim here is finding the best possible w_1 and b_1 values to minimize the loss. In other words, the neuron will learn the characteristics of the relationship between $(x_1[n], y_d[n])$ by adjusting its parameters during training.

We can form the parameter updating formula for the given loss function and stochastic gradient descent rule as

$$w_1(i+1) = w_1(i) - \mu \frac{\partial L(w_1, b_1)}{\partial w_1} \tag{10.13}$$

$$b_1(i+1) = b_1(i) - \mu \frac{\partial L(w_1, b_1)}{\partial b_1} \tag{10.14}$$

where i is the iteration number and μ is the positive step size value. $w_1[0]$ and $b_1[0]$ values are taken random at the beginning of iteration.

Based on the loss function definition in Eq. 10.10, we can write

$$\frac{\partial L(w_1, b_1)}{\partial w_1} = \frac{2}{N} \sum_{n=1}^{N} e[n] \varphi'(w_1 x_1[n] + b_1) x_1[n] \qquad (10.15)$$

$$\frac{\partial L(w_1, b_1)}{\partial b_1} = \frac{2}{N} \sum_{n=1}^{N} e[n] \varphi'(w_1 x_1[n] + b_1) \qquad (10.16)$$

where $\varphi'(\cdot)$ is the derivative of the activation function.

We iteratively update the parameters w_1 and b_1. Each iteration is called epoch in the neural network jargon. We can stop the iteration in two different ways. First, we can define a threshold for the difference between the old and new parameter values. If this value is reached, then iteration stops. Second, we can iterate parameter updating till a given number, called epoch number, has been reached.

Parameter updating is done iteratively toward the opposite direction of gradient of that parameter. Therefore, if the gradient of the corresponding parameter is zero, then the optimizer stops learning. However, zero gradient does not always mean global minimum for the loss function. Instead, it can be a local minimum. Therefore, model training process can stuck at local minimum. Moreover, if the step size μ is chosen too large, then the gradient sign can vary in each iteration. This will result model parameters to oscillate.

The stochastic gradient descent algorithm can be implemented using the SGD class of Keras. In Listing 10.10, we create an SGD object and variable var with single element. We pick the step size as 0.1. Based on these values, our loss function becomes $L = w^2/2$. Therefore, gradient of the loss function will be $\frac{\partial L}{\partial w} = w$.

Listing 10.10 Stochastic gradient descent optimizer example

```
import tensorflow as tf

opt = tf.keras.optimizers.SGD(learning_rate=0.1)
var = tf.Variable(10.0)

loss = lambda: (var ** 2)/2.0

for i in range(10):
    opt.minimize(loss, [var])
    print(var.numpy())
```

In Listing 10.10, we start with $w_1(0) = 10$. In the next step, we update this value with its gradient, which is again w. Therefore, the weight update is proportional to the weight itself. We should update our weight as $w_1(1) = 9$ based on Eq. 10.13. Next, the weight becomes $w_1(2) = 8.1$. The iterations will continue, until either we reach a certain iteration number or the gradient becomes zero. In Listing 10.10, we can observe that the variable var is updated similar to our manual calculation.

We can illustrate working principles of the gradient descent method on an example. Assume that we have a surface $z = x^2 + y^2$. We start from an arbitrary point and seek for global minimum. To do so, we calculate gradients of z with respect to x and y as

$$\frac{\partial z}{\partial x} = 2x$$

$$\frac{\partial z}{\partial y} = 2y$$

Once the gradients are calculated, we move toward the negative direction of the corresponding gradients with the predefined step size. Let's say we started from point $x = -3$, $y = 3$. In this case, $z = -3^2 + 3^2 = 18$ and gradients are -6 and 6. In the next step, we decrease variables x and y proportional to their gradients with an arbitrary step size μ. Let's pick $\mu = 0.1$. Then, $x_{new} = -3 - 0.1 * -6 = -2.4$, $y_{new} = 3 - 0.1 * 6 = 2.4$. We repeat this process by given number of iterations. We provide the complete example in Listing 10.11.

Listing 10.11 Gradient descent steps in TensorFlow

```
import numpy as np
import tensorflow as tf

x = tf.Variable(-3.0)
y = tf.Variable(3.0)

iters = 20
step = 0.1

xval, yval = np.meshgrid(np.arange(-3, 3, 0.1),np.arange(-3, 3,
    0.1))
Z = xval * xval + yval * yval

for iter in range(iters):
    with tf.GradientTape() as tape:
        z = x * x + y * y
    x_before = tf.constant(x)
    y_before = tf.constant(y)
    dz = tape.gradient(z, [x,y])
    dx = -step * dz[0]
    dy = -step * dz[1]
    x.assign_add(dx)
    y.assign_add(dy)
    new_z = x*x + y*y
    print(f'x = {x.numpy():.2f}, y= {y.numpy():.2f}, z = {new_z.
        numpy():.2f}')
```

In Listing 10.11, GradientTape automatically calculates gradient of the variables and keeps their record. One can run the example to observe that x and y values become 0 which is global minimum for $x^2 + y^2$. Visualization of the gradient descent used in Listing 10.11 is shown in Fig. 10.3.

Fig. 10.3 Visualization of
the gradient descent method

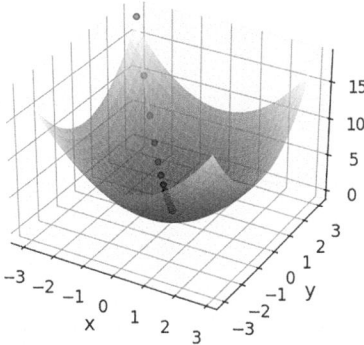

10.4.3.2 RMSProp
Despite its common use, the stochastic gradient descent method has a limitation as
the fixed learning rate for each input variable. AdaGrad, also known as adaptive
gradient, is the next method for parameter updating such that the learning rate
for each weight is adaptive based on the gradient of the corresponding weight
(partial derivatives). However, AdaGrad may suffer from very small step sizes as
it approaches to global minimum or local minima, which causes AdaGrad to miss
the optimal point. Root mean squared propagation, or RMSProp in short, is an
extension of the gradient descent method and AdaGrad which employs decaying
average of partial gradients. RMSProp overcomes the limitation of AdaGrad by
using the decaying moving average which decays early gradients and weights recent
gradients more. The update mechanism for the RMSprop algorithm is

$$s(t+1) = \rho s(t) + \left(\frac{\partial^2 L}{\partial^2 w} (1.0 - \rho) \right) \tag{10.17}$$

$$\eta(t+1) = \frac{\eta(t)}{\epsilon + \sqrt{s(t+1)}} \tag{10.18}$$

$$w(t+1) = w(t) - \eta(t+1)\frac{\partial L}{\partial w} \tag{10.19}$$

where η is the adaptive learning rate which is proportional to the square of the
gradient of the corresponding weight. Moreover, the ρ parameter sets the weight
between the current and previous gradient square values. In each iteration, we
calculate weighted sum of current gradient square and gradient squares from
previous iterations. Hence, the previous gradient effects decay over time. Note that
$s(0) = 0$, initially.

We implement the RMSProp optimizer in Listing 10.12 using the class RMSprop
in Keras. Similar to the SGD example in Listing 10.10, we pick the initial value
of the weight as 10 and initial learning rate as 0.1 to observe differences in output.
The ρ parameter is 0.9 by default in Keras implementation. Using Eq. 10.17, we can
calculate the value of the variable for next two iterations as

$$s(1) = 0.9 * 0 + 100 * 0.1 = 10$$

$$\eta(1) = \frac{0.1}{\sqrt{10}} = 0.0316$$

$$w(1) = 10 - 0.0316 * 10 = 9.68$$

$$s(2) = 0.9 * 10 + 9.68 * 0.1 = 18.68$$

$$\eta(2) = \frac{0.1}{\sqrt{18.68}} = 0.023$$

$$w(2) = 9.68 - 0.023 * 9.68 = 9.46$$

Listing 10.12 RMSProp optimizer example

```
import tensorflow as tf

opt = tf.keras.optimizers.RMSprop(learning_rate=0.1)
var = tf.Variable(10.0)

loss = lambda: (var ** 2) / 2.0

for _ in range(10):
    opt.minimize(loss, [var])
    print(var.numpy())
```

10.4.3.3 Adam

Adam optimization is an extension to RMSProp and stochastic gradient descent methods. It uses the first- and second-order moments of gradients to modify the learning rate of each weight. Similar to AdaGrad and RMSProp, the Adam optimizer adapts the learning rate for each variable. Moreover, it smooths the learning rate by using an exponentially decaying moving average of gradients. The Adam optimizer is commonly used for computer vision tasks as it generalizes better than SGD in most cases. Moment estimates used in the Adam optimization are

$$m(t) = \beta_1 m(t-1) + (1 - \beta_1)\frac{\partial L}{\partial w} \tag{10.20}$$

$$v(t) = \beta_2 v(t-1) + (1 - \beta_2)\left(\frac{\partial L}{\partial w}\right)^2 \tag{10.21}$$

where β_1 and β_2 are hyperparameters that control decay rates of gradient moment estimations. Both moments are set to 0 initially. They tend to move toward 0 as β_1 and β_2 become 1. Therefore, we compute bias-corrected moments $\hat{m}(t)$ and $\hat{v}(t)$ as

$$\hat{m}(t) = \frac{m_t}{1 - \beta_1} \tag{10.22}$$

$$\hat{v}(t) = \frac{v_t}{1 - \beta_2} \tag{10.23}$$

Eventually, the weight updating formula becomes

$$w(t + 1) = w(t) - \eta \frac{\hat{m}(t)}{\sqrt{\hat{v}(t)} + \epsilon} \tag{10.24}$$

where ϵ is the term to prevent the optimizer from dividing 0.

Keras has the class Adam for Adam optimization. We optimize the variable with this class in Listing 10.13. Here, the parameter is set as 10 initially. The objective function is $x^2/2$. For completeness, one can find the first two steps of the numerical calculation below. Default value for β_1 and β_2 is 0.9 and 0.99, respectively. Here, $\epsilon = 10^{-7}$. Based on these values, we can calculate the first two iterations as

$$m(1) = 0.9 * 0 + (1 - 0.9) * 10 = 1$$

$$v(1) = 0.999 * 0 + (1 - 0.999) * 100 = 0.1$$

$$mhat(1) = 1/(1 - 0.9) = 10$$

$$vhat(1) = 0.1/(1 - 0.999) = 100$$

$$w(1) = 10 - 0.1 * 10/10 = 9.9$$

$$m(2) = 0.9 * 1 + 0.1 * 9.9 = 1.89$$

$$v(2) = 0.999 * 0.1 + 0.001 * 9.9 * *2 = 0.198$$

$$mhat(2) = 1.89/(1 - 0.9^2) = 9.95$$

$$vhat(2) = 0.198/(1 - 0.999^2) = 99$$

$$w(2) = 9.9 - 0.1 * 9.95/sqrt(99) = 9.8$$

Listing 10.13 Adam optimizer example

```
import tensorflow as tf

opt = tf.keras.optimizers.Adam(learning_rate=0.1)
var = tf.Variable(10.0)

loss = lambda: (var ** 2) / 2.0

for _ in range(10):
    opt.minimize(loss, [var])
    print(var.numpy())
```

10.4.4 Training in TensorFlow

Since TensorFlow allows us to form basic structures, we can explain training steps of the single neuron best there. Let's consider a regression example given in [53]. We provide the simplified version of this script in Listing 10.14.

Listing 10.14   Training the single neuron in TensorFlow

```python
import tensorflow as tf
import numpy as np
import matplotlib.pyplot as plt
from numpy.random import default_rng

size=1000

rng = default_rng(0)
noise = .1*rng.standard_normal(size)

x = np.linspace(-2, 2, size)
y = 3*x + 2 + noise

class Model(tf.Module):
  def __init__(self, **kwargs):
    super().__init__(**kwargs)
    self.w = tf.Variable(5.0)
    self.b = tf.Variable(0.0)

  def __call__(self, x):
    return self.w * x + self.b

model = Model()

# Compute a single loss value for an entire batch
def loss(target_y, predicted_y):
  return tf.reduce_mean(tf.square(target_y - predicted_y))

# Given a callable model, inputs, outputs, and a learning rate...
def train(model, x, y, learning_rate):

  with tf.GradientTape() as t:
    # Trainable variables are automatically tracked by
        GradientTape
    current_loss = loss(y, model(x))

  # Use GradientTape to calculate the gradients with respect to W
      and b
  dw, db = t.gradient(current_loss, [model.w, model.b])

  # Subtract the gradient scaled by the learning rate
  model.w.assign_sub(learning_rate * dw)
  model.b.assign_sub(learning_rate * db)

model = Model()

epochs = range(10)

def training_loop(model, x, y):
  for epoch in epochs:
    # Update the model with the single giant batch
    train(model, x, y, learning_rate=0.1)
```

```
# Do the training
training_loop(model, x, y)

yp = model(x)

plt.plot(x,y,'r.')
plt.plot(x,yp,'b')
plt.show()
```

In Listing 10.14, we train a single neuron using TensorFlow. First, we define seed for reproducibility of random processes. We then create input data as 1000 equidistant numbers between -2 and 2. Our desired output (target data) is produced as a linear function of input data with added noise $y = 3x + 2 + \varepsilon$ where ε indicates the noise term. Afterward, we create the model with weight and bias values set as 5 and 0, respectively. Please note that we inherit the tf.Module class while creating the Model class. Then, we define the loss function as MSE. After all the definitions are done, the training function and training loop must be implemented. The tf.GradientTape class is used to track gradients of model parameters. Then, the loss function and variable gradients are calculated as

```
current_loss = loss(y, model(x))
dw, db = t.gradient(current_loss, [model.w, model.b])
```

Once we have the gradients, all we need to do is to multiply the gradient value with learning rate and update weights accordingly. Hence, we implement the stochastic gradient descent method step by step under TensorFlow.

10.4.5 Training in Keras

Although TensorFlow allows us to observe the training steps of a single neuron, we will not need this detail most of the times. Hence, we can benefit from Keras to train our single neuron. We can reconsider the regression example in Sect. 10.4.4. We can show how Keras can be used to train the single neuron on the same example in Listing 10.15.

Listing 10.15 Training the single neuron in Keras

```
import tensorflow as tf
import numpy as np
import matplotlib.pyplot as plt
from numpy.random import default_rng

size=1000

rng = default_rng()
noise = .1*rng.standard_normal(size)

x = np.linspace(-2, 2, size)
y = 3*x + 2 + noise
```

```
model = tf.keras.models.Sequential([
  tf.keras.layers.Dense(1, input_shape=[1]),
  ])

model.compile(loss='mean_squared_error', optimizer=tf.keras.
    optimizers.SGD(0.1))

model.fit(x, y, epochs=50, verbose=False)

yp = model.predict(x)

plt.plot(x,y,'r.')
plt.plot(x,yp,'b')
plt.show()
```

In Listing 10.15, single neuron training is implemented in Keras. Again, we create the input and desired output (target) data using the same function in Listing 10.14. Next, the single neuron is created in Keras by calling keras.layers. Dense. As the Dense layer is called with the argument num_units = 1, it will create a single neuron with one-dimensional input. Since Keras has predefined layers, we do not define each variable manually. The keras.Sequential class is used to create the Keras model. Normally, it is used to connect layers. However, we have single neuron in this case. In the final stage of model building, the model.compile is called to define the loss function and optimizer. As can be seen in Listing 10.15, our loss function is MSE and the optimizer is SGD. Finally, the model can be trained by calling the fit function of the keras.Sequential class. We provide loss values with respect to the training iteration number in Fig. 10.4. As can be seen in this figure, our loss decreases gradually. Model parameters are updated during training. We can now predict output to a new input using the function model.predict.

Fig. 10.4 Plot of loss over iteration

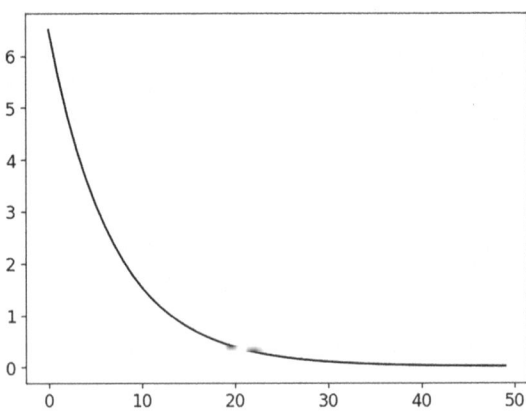

10.5   Forming a Classifier with the Single Neuron

We can feed features as inputs to the neuron. As we train it accordingly, we can use the single neuron as a classifier. Let's explain this operation on the example given Listing 10.16.

Listing 10.16 Forming a classifier with the single neuron

```
import numpy as np
from sklearn.model_selection import train_test_split
import tensorflow as tf

np.random.seed(42)   # For reproducibility

def Gaussian2D(mean,L,theta,len):
        c, s = np.cos(theta), np.sin(theta)
        R = np.array([[c, -s], [s, c]])
        cov=R@L@R.T
        return np.random.multivariate_normal(mean, cov, len).T

len=1000
mean = [-2, -2]
E = np.diag([1,10])
theta=np.radians(45)
x1, y1 =Gaussian2D(mean,E,theta,len)

cls1=np.zeros((2,len))
cls1[0,:]=x1
cls1[1,:]=y1

len=1000
mean = [2, 2]
E = np.diag([1,10])
theta=np.radians(-45)
x2, y2 = Gaussian2D(mean,E,theta,len)

cls2=np.zeros((2,len))
cls2[0,:]=x2
cls2[1,:]=y2

F=np.concatenate((cls1,cls2),axis=1)
F=F.T

X=np.append(cls1, cls2, axis=1)
labels = np.hstack((np.zeros(len, dtype=np.uint8), np.ones(len,
    dtype=np.uint8)))
train_samples, test_samples, train_labels, test_labels =
    train_test_split(F, labels, test_size=0.2)

model = tf.keras.models.Sequential([
  tf.keras.layers.Dense(1, input_shape=[2], activation='sigmoid')
  ])

model.compile(optimizer=tf.keras.optimizers.Adam(learning_rate=1e
    -3),
                loss=tf.keras.losses.BinaryCrossentropy(),
                metrics=[tf.keras.metrics.BinaryAccuracy(),
                    tf.keras.metrics.FalseNegatives()])
```

```
model.fit(train_samples, train_labels, validation_data = (
    test_samples, test_labels), epochs=50, verbose=1)
print("Finished training the model")
```

In Listing 10.16, we create two classes in two-dimensional space using the function `multivariate_normal` as in Listing 5.2. This function creates normal distributed data with given mean and covariance matrices. Here, we generate 1000 samples from class 1 with mean $[-2, -2]$ and covariance matrix:

$$\begin{bmatrix} 5.5 & -4.5 \\ -4.5 & 5.5 \end{bmatrix}$$

and class2 is generated with mean $[2, 2]$ and rotated covariance matrix as $\begin{bmatrix} 5.5 & 4.5 \\ 4.5 & 5.5 \end{bmatrix}$.

After creating data, we merge both classes and define labels to indicate classes of samples by

```
X=np.append(cls1,cls2, axis=1)
label=np.append(np.zeros(len),np.ones(len))
```

Next, we define the single neuron inside the model. Furthermore, we define our loss function as binary cross-entropy and optimizer as Adam. Binary cross-entropy calculates the entropy between model prediction and ground truth data. Also, we embed binary accuracy and false-negative metrics into the model. Therefore, the model reports binary accuracy and false negative at the end of each epoch. We use the function `fit` to train the model. The first two arguments of the function are our training samples and labels used to train the model. The parameter `validation_data` is used for feeding the model with validation data. This is a good practice to monitor the model loss and metrics on data which is not seen by the model previously. The parameter `epoch = 50` specifies that the model will iterate over the entire training dataset 50 times. The parameter `verbose = 1` monitors model progress on the terminal screen. In our case, loss, binary accuracy, and false negatives for both train and validation datasets are monitored on the terminal screen as follows.

```
Epoch 50/50
50/50 [==============================] - 0s 3ms/step -
    loss: 0.2329 - binary_accuracy: 0.9256 -
    false_negatives: 100.0000 - val_loss: 0.2119 -
    val_binary_accuracy: 0.9325 - val false negatives:
    23.0000
Finished training the model
```

We provide the decision boundary formed for this example in Fig. 10.5. As can be seen in this figure, our decision boundary now splits data into two classes since weights of the neuron are trained to minimize the binary cross-entropy between the ground truth and predicted values. As a result of training, we obtain 93% accuracy.

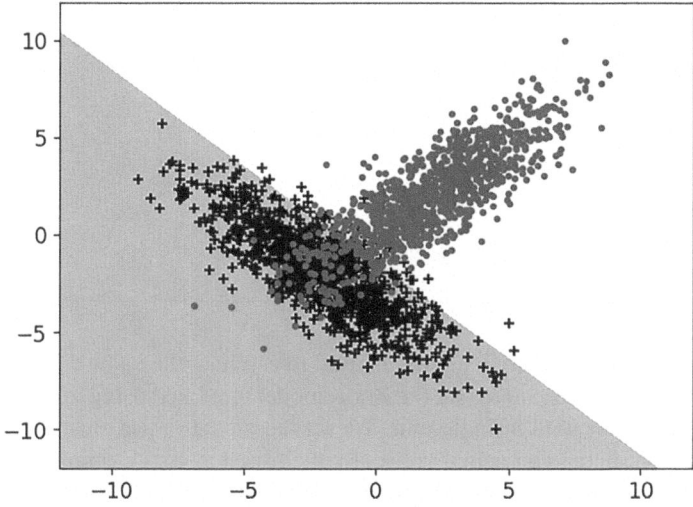

Fig. 10.5 Decision boundary formed by the single neuron

10.6 Forming a Regressor with the Single Neuron

We considered the regressor formation via single neuron in explaining the training operation in Sect. 10.4.5. Therefore, the script in Listing 10.15 can be evaluated from the regressor perspective as well. We strongly suggest the reader to consider this perspective. Combining the data we generated in Listing 5.3 with the single neuron training, we obtain the script in Listing 10.17 for the regression example.

Listing 10.17 Forming a regressor with the single neuron

```
import tensorflow as tf
import numpy as np
import matplotlib.pyplot as plt

np.random.seed(42)    # For reproducibility
size=200

x = np.array(range(size))/size*10

noise = np.random.normal(0, .5, size)
y1=3*x+4+noise

noise = np.random.normal(0, .2, size)
y2=np.sin(np.pi/2*x)+noise

model = tf.keras.models.Sequential([
  tf.keras.layers.Dense(1, input_shape=[1]),
  ])

model.compile(loss='mean_squared_error', optimizer=tf.keras.
    optimizers.SGD(0.001))
model.fit(x, y1, epochs=100)
```

```
y1_pred = model.predict(x)

model.fit(x, y2, epochs=100)
y2_pred = model.predict(x)

plt.figure(1)
plt.scatter(x, y1, c = 'k')
plt.plot(x, y1_pred, 'b')

plt.figure(2)
plt.scatter(x ,y2, c = 'k')
plt.plot(x, y2_pred, 'b')
plt.show()
```

In Listing 10.17, the single neuron is used to estimate the values of line and sine data as in Chap. 7. Please note that the Keras model is created using only one neuron, namely, Dense layer with a single unit. We set the model to use mean squared error between the predicted and actual values by the function model.compile. Hence, we set SGD as the optimizer with learning rate = 0.001. Models for both predicting the line and sine data are identical. Thus, the same model is trained for 100 epochs for both problems. We plot the predicted values to observe fitness of the model for different data in Fig. 10.6.

As can be seen in Fig. 10.6, the single neuron can only represent linear relationship in the data since it has the input-output relation as $y = wx + b$. Therefore, the model parameters converge to the actual values in Fig. 10.6a. However, the same model fails to estimate the value of sine data since there is no linear relationship there.

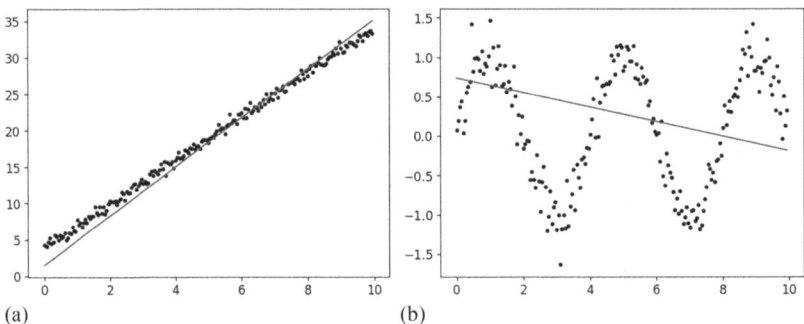

(a)   (b)

Fig. 10.6 Regression example with single neuron. (a) Noisy line data and its regression result. (b) Noisy sine data and its regression result

10.7 Application: Human Activity Recognition via Accelerometer Data

We extracted features from accelerometer data for the human activity recognition application in Sect. 5.4. We also used the Bayes classifier for the application in Sect. 6.6. Here, we use a single neuron for human activity recognition. Since the single neuron can distinguish two classes, we set the human activity recognition application classes as "walking" and "not walking" in this section.

We provide the Python script for human activity recognition with two classes in Listing 10.18. In this script, we first form the features as in Sect. 5.4. Then, we form the single neuron classifier with sigmoid activation function. We train the neuron using the Adam optimizer, binary cross-entropy loss, and binary accuracy and false negatives as metrics. As the training is done, we save the model for future use.

Listing 10.18 Training the single neuron for human activity recognition

```
import os.path as osp
from data_utils import read_data
from feature_utils import create_features
from sklearn import metrics
import tensorflow as tf
from matplotlib import pyplot as plt

DATA_PATH = osp.join("WISDM_ar_v1.1", "WISDM_ar_v1.1_raw.txt")
TIME_PERIODS = 80
STEP_DISTANCE = 40
data_df = read_data(DATA_PATH)
df_train = data_df[data_df["user"] <= 28]
df_test = data_df[data_df["user"] > 28]

train_segments_df, train_labels = create_features(df_train,
    TIME_PERIODS, STEP_DISTANCE)
test_segments_df, test_labels = create_features(df_test,
    TIME_PERIODS, STEP_DISTANCE)

model = tf.keras.models.Sequential([
    tf.keras.layers.Dense(1, input_shape = [10], activation = '
        sigmoid')
    ])

model.compile(optimizer=tf.keras.optimizers.Adam(learning_rate=1e
    -3),
              loss=tf.keras.losses.BinaryCrossentropy(),
              metrics=[tf.keras.metrics.BinaryAccuracy(),
                  tf.keras.metrics.FalseNegatives()])

train_segments_np = train_segments_df.to_numpy()
test_segments_np = test_segments_df.to_numpy()
train_labels[train_labels != "Walking"] = 1
train_labels[train_labels == "Walking"] = 0
test_labels[test_labels != "Walking"] = 1
test_labels[test_labels == "Walking"] = 0
train_labels = train_labels.astype(int)
test_labels = test_labels.astype(int)

model.fit(train_segments_np, train_labels, epochs=50, verbose=1)
perceptron_preds = model.predict(test_segments_np)
```

```
conf_matrix = metrics.confusion_matrix(test_labels,
    perceptron_preds > 0.5)
cm_display = metrics.ConfusionMatrixDisplay(confusion_matrix =
    conf_matrix, display_labels=["Walking", "Not Walking"])
cm_display.plot()
cm_display.ax_.set_title("Single Neuron Classifier Confusion
    Matrix")
plt.show()

model.save("har_neuron_model.h5")
```

10.8   Application: Keyword Spotting from Audio Signals

We extracted features from audio data for the keyword spotting application in Sect. 5.5. We also used the kNN classifier for the application in Sect. 6.7. Here, we use a single neuron for speech recognition. Since the single neuron can distinguish two classes, we set the speech recognition application classes as "zero" and "not zero" in this section.

We provide the Python script for keyword spotting with two classes in Listing 10.19. In this script, we first form the features as in Sect. 5.5. Therefore, the Python script reads audio data and creates MFCC features. Then, we form the single neuron classifier with sigmoid activation function. We train the neuron using the Adam optimizer, binary cross-entropy loss, and binary accuracy and false negatives as metrics. As the training is done, we save the model for future use.

Listing 10.19 Training the single neuron for keyword spotting from audio signals

```
import os
import scipy.signal as sig
from sklearn.metrics import confusion_matrix,
    ConfusionMatrixDisplay
from matplotlib import pyplot as plt
import tensorflow as tf
from mfcc_func import create_mfcc_features

RECORDINGS_DIR = "recordings"
recordings_list = [(RECORDINGS_DIR, recording_path) for
    recording_path in os.listdir(RECORDINGS_DIR)]

FFTSize = 1024
sample_rate = 8000
numOfMelFilters = 20
numOfDctOutputs = 13
window = sig.get_window("hamming", FFTSize)
test_list = {record for record in recordings_list if "yweweler"
    in record[1]}
train_list = set(recordings_list) - test_list
train_mfcc_features, train_labels = create_mfcc_features(
    train_list, FFTSize, sample_rate, numOfMelFilters,
    numOfDctOutputs, window)
test_mfcc_features, test_labels = create_mfcc_features(test_list,
    FFTSize, sample_rate, numOfMelFilters, numOfDctOutputs,
    window)
```

```
model = tf.keras.models.Sequential([
  tf.keras.layers.Dense(1, input_shape = [numOfDctOutputs * 2],
    activation = 'sigmoid')
])

model.compile(optimizer=tf.keras.optimizers.Adam(learning_rate=1e
  -3),
              loss=tf.keras.losses.BinaryCrossentropy(),
              metrics=[tf.keras.metrics.BinaryAccuracy(),
                  tf.keras.metrics.FalseNegatives()])

train_labels[train_labels != 0] = 1
test_labels[test_labels != 0] = 1

model.fit(train_mfcc_features, train_labels, epochs=50, verbose
  =1, class_weight = {0:10., 1:1.})
perceptron_preds = model.predict(test_mfcc_features)

conf_matrix = confusion_matrix(test_labels, perceptron_preds >
  0.5)
cm_display = ConfusionMatrixDisplay(confusion_matrix =
  conf_matrix)
cm_display.plot()
cm_display.ax_.set_title("Single Neuron Classifier Confusion
  Matrix")
plt.show()

model.save("fsdd_perceptron.h5")
```

10.9 Application: Handwritten Digit Recognition from Digital Images

We extracted features from digital images for the handwritten digit recognition application in Sect. 5.6. We also used the SVM classifier for the application in Sect. 6.8. Here, we use a single neuron for handwritten digit recognition. Since the single neuron can distinguish two classes, we set the application classes as "zero" and "not zero" in this section.

We provide the Python script for handwritten digit recognition with two classes in Listing 10.20. In this script, we first form the features as in Sect. 5.6. Therefore, the Python script reads MNIST data and computes Hu moments for each image. Then, we form the single neuron classifier with sigmoid activation function. We train the neuron using the Adam optimizer, binary cross-entropy loss, and binary accuracy and false negatives as metrics. As the training is done, we save the model for future use.

Listing 10.20 Training the single neuron for handwritten digit recognition from digital images

```
import os
import numpy as np
import cv2
```

```
from sklearn.metrics import confusion_matrix,
    ConfusionMatrixDisplay
import tensorflow as tf
from mnist import load_images, load_labels
from matplotlib import pyplot as plt

train_img_path = os.path.join("MNIST-dataset", "train-images.idx3
    -ubyte")
train_label_path = os.path.join("MNIST-dataset", "train-labels.
    idx1-ubyte")
test_img_path = os.path.join("MNIST-dataset", "t10k-images.idx3-
    ubyte")
test_label_path = os.path.join("MNIST-dataset", "t10k-labels.idx1
    -ubyte")

train_images = load_images(train_img_path)
train_labels = load_labels(train_label_path)
test_images = load_images(test_img_path)
test_labels = load_labels(test_label_path)

train_huMoments = np.empty((len(train_images),7))
test_huMoments = np.empty((len(test_images),7))

for train_idx, train_img in enumerate(train_images):
    train_moments = cv2.moments(train_img, True)
    train_huMoments[train_idx] = cv2.HuMoments(train_moments).
        reshape(7)

for test_idx, test_img in enumerate(test_images):
    test_moments = cv2.moments(test_img, True)
    test_huMoments[test_idx] = cv2.HuMoments(test_moments).
        reshape(7)

features_mean = np.mean(train_huMoments, axis = 0)
features_std = np.std(train_huMoments, axis = 0)
train_huMoments = (train_huMoments - features_mean) /
    features_std
test_huMoments = (test_huMoments - features_mean) / features_std

model = tf.keras.models.Sequential([
  tf.keras.layers.Dense(1, input_shape = [7], activation = '
      sigmoid')
  ])

model.compile(optimizer=tf.keras.optimizers.Adam(learning_rate=1e
    -3),
              loss=tf.keras.losses.BinaryCrossentropy(),
              metrics=[tf.keras.metrics.BinaryAccuracy()])

train_labels[train_labels != 0] = 1
test_labels[test_labels != 0] = 1

model.fit(train_huMoments,
          train_labels,
          batch_size = 128,
          epochs=50,
          class_weight = {0:8, 1:1},
          verbose=1,
          workers = 16,
          use_multiprocessing = True)
perceptron_preds = model.predict(test_huMoments)
```

```
conf_matrix = confusion_matrix(test_labels, perceptron_preds >
    0.5)
cm_display = ConfusionMatrixDisplay(confusion_matrix =
    conf_matrix)
cm_display.plot()
cm_display.ax_.set_title("Single Neuron Classifier Confusion
    Matrix")
plt.show()

model.save("mnist_single_neuron.h5")
```

10.10 Application: Estimating Future Temperature Values

We can estimate future temperature values using the single neuron as regressor.
Hence, we benefit from the SML2010 dataset introduced in Sect. 7.6.1. As in
Sect. 7.6, we form a Python script for this purpose. We provide the formed script
in Listing 10.21.

Listing 10.21 Python script for single neuron formation, training, and testing

```
import pandas as pd
import numpy as np
from sklearn.model_selection import train_test_split
from sklearn.metrics import mean_absolute_error
from matplotlib import pyplot as plt
import tensorflow as tf

df = pd.read_csv("temperature_dataset.csv")
y = df["Room_Temp"][::4]
prev_values_count = 5

X = pd.DataFrame()
for i in range(prev_values_count, 0, -1):
    X["t-" + str(i)] = y.shift(i)

X = X[prev_values_count:]
y = y[prev_values_count:]

X_train, X_test, y_train, y_test = train_test_split(X, y,
    test_size=0.2, random_state=0)

train_mean = X_train.mean()
train_std = X_train.std()

X_train = (X_train - train_mean) / train_std
X_test = (X_test - train_mean) / train_std

model = tf.keras.models.Sequential([tf.keras.layers.Dense(1,
    input_shape=[5])])

model.compile(
    optimizer=tf.keras.optimizers.SGD(learning_rate=5e-3),
    loss=tf.keras.losses.MeanAbsoluteError(),
)

model.fit(
```

```
    X_train,
    y_train,
    batch_size=128,
    epochs=3000,
    verbose=1,
    workers=16,
    use_multiprocessing=True,
)

y_train_predicted = model.predict(X_train)
y_test_predict = model.predict(X_test)

fig, ax = plt.subplots(1,1)
ax.plot(y_test.to_numpy(), label = "Actual values")
ax.plot(y_test_predict, label = "Predicted values")
plt.legend()
plt.show()

mae_train = np.sqrt(mean_absolute_error(y_train,
    y_train_predicted))
mae_test = np.sqrt(mean_absolute_error(y_test, y_test_predict))

print(f"Training set MAE: {mae_train}\n")
print(f"Test set MAE:{mae_test}")

model.save("temperature_prediction_perceptron.h5")
```

In Listing 10.21, we read and prepare the dataset as in Listing 7.33. We then form a single neuron with five inputs as the regressor. We train the neuron with the SGD optimizer setting learning rate to 0.005. We use the mean absolute error as loss function during training. As the training is done, we save the model for future use.

10.11 Summary of the Chapter

We started our neural network journey by the single neuron in this chapter. Although this structure is not as powerful as the multilayer neural networks to be introduced in Chap. 11, it gives valuable insight on the working principles of neural networks in general. Therefore, we started with the mathematical definition of the single neuron. Then, we considered its formation in TensorFlow and Keras. Although we will benefit from the latter option extensively, forming the neuron in TensorFlow gives valuable insight on working principles of neural networks. Therefore, we covered it in this chapter. Afterward, we focused on training the single neuron. Here, we explained the mechanism in training, loss function, and optimizers used in training. We covered training both in TensorFlow and Keras. Next, we formed a classifier and regressor by the single neuron. Finally, we considered real-life applications introduced in the previous chapters now from the single neuron perspective.

Multilayer Neural Networks

11

11.1 Background

The single neuron introduced in Chap. 10 provides a good medium to explain fundamentals of neural networks. Therefore, it served its purpose in the previous chapter. However, a single neuron may not be used in solving complex machine learning problems. Let's explain this issue by examples. The single neuron can only form a decision boundary to separate two classes. Therefore, we cannot use it to classify more than two classes. Besides, the decision boundary formed by the single neuron can only be a hyperplane in N-dimensional feature space. It becomes a straight line in two-dimensional feature space. Hence, classification performance of the single neuron will be limited. The single neuron also has limited power while forming a regressor. We will only have limited functional form in the regressor depending on the activation function used. Therefore, the single neuron cannot be used to approximate complex functions.

In order to overcome shortcomings of the single neuron explained in the previous paragraph, we should use more neurons in operation. Therefore, neural networks emerged. We will introduce them in this section by starting with their structure. Then, we will define deep and shallow neural networks. Finally, we will summarize fully connected neural networks which will be the structure to be used in the following sections.

11.1.1 Structure of the Multilayer Neural Network

We can form a layer by appending neurons in a parallel manner while using more than one neuron. Then, we can sequentially append them to form the multilayer neural network. We schematically provide one such structure in Fig. 11.1. The neural network structure in this figure has input, hidden, and output layers. In this figure, x and y stand for the input and output, respectively.

© The Author(s), under exclusive license to Springer Nature Switzerland AG 2025
C. Ünsalan et al., *Embedded Machine Learning with Microcontrollers*,
https://doi.org/10.1007/978-3-031-70912-8_11

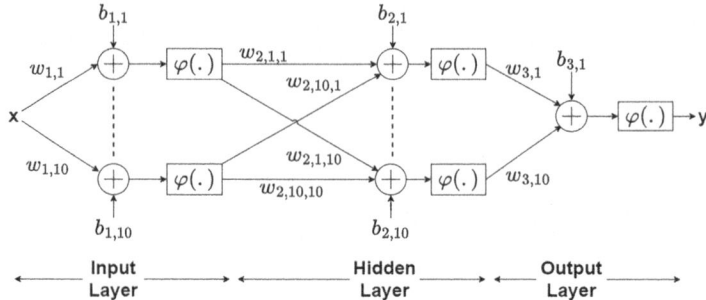

Fig. 11.1 The multilayer neural network structure

In Fig. 11.1, the first layer is called input. The number of neurons in this layer is equal to the feature space dimension to be fed to the neural network. The input layer consists of ten neurons in our case. Hence, there are ten weight terms indicated by $w_{1,k}$ for $k = 1, \cdots , 10$. Similarly, there are ten bias terms indicated by $b_{1,k}$ for $k = 1, \cdots , 10$ in the input layer. The second layer is called hidden since it cannot be directly reached by input or output. This layer consists of ten neurons in our case. Hence, there are 10×10 weight terms indicated by $w_{2,k,l}$ for $k = 1, \cdots , 10$ and $l = 1, \cdots , 10$. There are also ten bias terms indicated by $b_{2,k}$ for $k = 1, \cdots , 10$ in the hidden layer. The last layer is called output. The number of neurons in this layer is equal to the number of classes when a classifier is formed. Likewise, the number of neurons in the output layer is equal to the regressor output dimension when such a structure is formed. This layer consists of one neuron in our case. Hence, there are ten weight terms indicated by $w_{3,k}$ for $k = 1, \cdots , 10$. Finally, there is one bias term indicated by $b_{3,1}$ in the output layer. The activation function for the neurons in our neural network is indicated by $\varphi(\cdot)$. Although we set the same activation function throughout the network, this need not be the case. In other words, we can set different activation functions for different layers in the neural network. In fact, this will be the case most of the times.

11.1.2 Deep and Shallow Neural Networks

We can define deep and shallow neural networks as follows. A shallow neural network has at most one or two hidden layers. Whereas a deep neural network may have several hidden layers. Naturally, there is a difference between the decision boundary formation capability and regression power of deep and shallow neural networks. Related to this, training time and data requirements are different in deep and shallow neural networks. Throughout the book, we will use either a deep or shallow neural network for the given problem. However, we will call them as neural network in general.

11.1.3 Fully Connected Neural Networks

When there is a connection between each neuron in one layer and all neurons in the next layer, we have a specific structure called dense layer. If this is the case for all layers in the network, then we have a fully connected neural network. The formed neural network in Fig. 11.1 is fully connected since all the neurons in the input, hidden, and output layers are connected. Here, we should emphasize that a neuron can only connect to the neurons in the next layer unless otherwise stated. Therefore, all neurons in the input layer are connected to all neurons in the hidden layer in our example. Likewise, all neurons in the hidden layer are connected to the neuron in the output layer. We will use such structures while forming classifiers and regressors in the book. As a side note, we will have feedback loop formation or connection of neurons to the ones in previous layers in Chap. 14.

11.2 Implementing the Multilayer Neural Network in Keras

Keras has a simple and user-friendly interface while forming the multilayer neural network. Therefore, we will use it in this and following chapters. Next, we will introduce layer formation and sequentially appending these layers to construct the neural network. Then, we will look at the input-output relation of the formed network.

11.2.1 Forming Layers

We can form a layer under Keras by defining the number of neurons in it, the activation function used in these neurons, and other neuron properties. Let's reconsider the neural network structure in Fig. 11.1. Since this is a fully connected neural network with three layers, we use the Dense class under Keras similar to the single neuron layer case. However, this time we have more than one layer. Moreover, each layer has more than one neuron in it. These layers can be formed as follows.

```
tf.keras.layers.Dense(10, input_shape=[1], activation='sigmoid'),
tf.keras.layers.Dense(10, activation='sigmoid'),
tf.keras.layers.Dense(1, activation = 'linear')
```

In our fully connected neural network, the first layer consists of ten neurons each having a single input. The neurons in the first layer have the sigmoid activation function. The second layer consists of ten neurons again with the sigmoid activation function. The third layer consists of a single neuron with the linear activation function. These three layers are formed separately. Hence, they still do not form a network. We will form the neural network from them in the next section.

11.2.2 Sequentially Appending Layers

We can sequentially append layers to form the neural network. To do so, we should use the Sequential function under Keras. There are two ways of forming the complete neural network structure afterward. We will consider them next.

The first way of using the Sequential class is by feeding the layers to it as a parameter. We provide one such example on forming the complete neural network next.

```
model = tf.keras.models.Sequential([
tf.keras.layers.Dense(10, input_shape=[1], activation='sigmoid'),
tf.keras.layers.Dense(10, activation='sigmoid'),
tf.keras.layers.Dense(1, activation = 'linear')
])
```

Here, the input, hidden, and output layers are appended. Hence, outputs of the neurons in the first layer are fed to the input of the neurons in the second layer. Likewise, outputs of the neurons in the second layer are fed to the input of the neuron in the third layer. If we consider the overall neural network, then input to the first layer is the input of the overall structure. Output of the third layer is the output of the overall structure. This neural network can be used in classification and regression operations.

The second way of using the Sequential class is by forming an empty class. Then, we add layers to it one by one using the add function. We provide one such example next.

```
model = tf.keras.models.Sequential()
first_layer = tf.keras.layers.Dense(10, input_shape=[1],
    activation='sigmoid')
second_layer = tf.keras.layers.Dense(10, activation='sigmoid'),
third_layer = tf.keras.layers.Dense(1, activation = 'linear')

model.add(first_layer)
model.add(second_layer)
model.add(third_layer)
```

Here, the input, hidden, and output layers are formed as first_layer, second_layer, and third_layer, respectively. Then, they are appended to the neural network structure step by step. However, there is no difference on the formed structure by the first and second methods.

11.2.3 Obtaining Output for a Given Input

As in the single neuron case, we can use the function predict under Keras to obtain output of the neural network for a given input. Usage of this function will be the same as in Sect. 10.3. However, we will not set the model parameters in this section as we did there. Instead, we will provide the usage of the predict function after training the neural network in the next section.

11.3 Training the Neural Network

Training the neural network is based on the same tools introduced in Sect. 10.4. However, training requires one more step compared to the single neuron case. This is the backpropagation algorithm. Therefore, we will start with its brief definition first. Then, we will consider how training can be done under Keras.

11.3.1 The Backpropagation Algorithm

Training the neural network follows the same steps as in the single neuron case introduced in Sect. 10.4. However, there are extensions to it due to the layered structure of the neural network. We can explain the overall training operation by an example. Therefore, let's consider the simple neural network formed by two serially connected neurons. There is one input and output in the network. We provide schematic layout of the setup in Fig. 11.2.

We can define output of the first neuron as y_1 in Fig. 11.2. Hence, $y_1 = \varphi(w_1 x + b_1)$. Based on this definition, we will have $y = \varphi(w_2 y_1 + b_2)$. Now, we are ready to derive the parameter updating formulas for our neural network.

We will have the same training set as in the single neuron case considered in Sect. 10.4. Hence, we have N training samples as $(x[n], y_d[n])$ for $n = 1, \cdots, N$ where $y_d[n]$ is the desired output for input $x[n]$. Moreover, let's assume the same settings for the training. Hence, we will have the error value as $e[n] = y_d[n] - y[n]$. We can define the loss function by the mean-squared error as

$$L = \sum_{n=1}^{N} e^2[n] \tag{11.1}$$

$$= \sum_{n=1}^{N} (y_d[n] - y[n])^2 \tag{11.2}$$

As in the single neuron case, we can use the gradient descent rule to update parameters w_1, b_1, w_2, and b_2. Hence, we will have

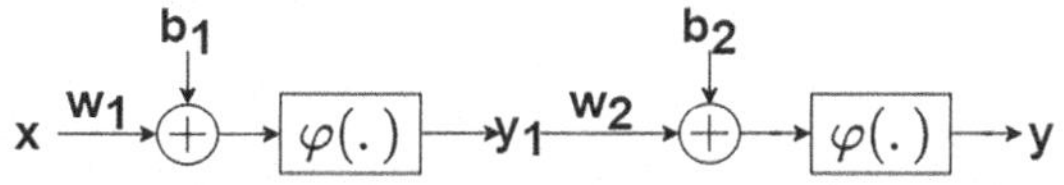

Fig. 11.2 The neural network structure for the backpropagation example

$$w_1(i+1) = w_1(i) - \mu \frac{\partial L(w_1, b_1, w_2, b_2)}{\partial w_1} \tag{11.3}$$

$$b_1(i+1) = b_1(i) - \mu \frac{\partial L(w_1, b_1, w_2, b_2)}{\partial b_1} \tag{11.4}$$

$$w_2(i+1) = w_2(i) - \mu \frac{\partial L(w_1, b_1, w_2, b_2)}{\partial w_2} \tag{11.5}$$

$$b_2(i+1) = b_2(i) - \mu \frac{\partial L(w_1, b_1, w_2, b_2)}{\partial b_2} \tag{11.6}$$

Let's first focus on updating w_2 and b_2. Hence, we will have

$$\frac{\partial L(w_1, b_1, w_2, b_2)}{\partial w_2} = \frac{2}{N} \sum_{n=1}^{N} e[n] \varphi'(w_2 y_1 + b_2) y_1 \tag{11.7}$$

$$\frac{\partial L(w_1, b_1, w_2, b_2)}{\partial b_2} = \frac{2}{N} \sum_{n=1}^{N} e[n] \varphi'(w_2 y_1 + b_2) \tag{11.8}$$

Equations 11.7 and 11.8 have the same structure as Eqs. 10.15 and 10.16, respectively. However, calculations become complicated when we want to obtain $\partial L/\partial w_1$ and $\partial L/\partial b_1$. Let's consider $\partial L/\partial w_1$ first. We will have

$$\frac{\partial L(w_1, b_1, w_2, b_2)}{\partial w_1} = \frac{2}{N} \sum_{n=1}^{N} e[n] \frac{\partial \varphi(w_2 y_1 + b_2)}{\partial w_1} \tag{11.9}$$

We should apply the chain rule to Eq. 11.9 since w_1 is not an explicit parameter in $\varphi(w_2 y_1 + b_2)$. Hence, we will have

$$\frac{\partial \varphi(w_2 y_1 + b_2)}{\partial w_1} = \frac{\partial \varphi(w_2 y_1 + b_2)}{\partial y_1} \frac{\partial y_1}{\partial w_1} \tag{11.10}$$

Applying the chain rule in Eq. 11.10 is as if backpropagating the effect of total error in the neural network to a previous layer. This way, we can calculate Eq. 11.9 in two steps as

$$\frac{\partial \varphi(w_2 y_1 + b_2)}{\partial y_1} = \varphi'(w_2 y_1 + b_2) w_2 \tag{11.11}$$

$$\frac{\partial y_1}{\partial w_1} = \varphi'(w_1 x[n] + b_1) x[n] \tag{11.12}$$

We can apply the same derivation to update the bias parameter b_1. Hence, we can form the overall parameter updating formulas for the neural network.

In general, we need the chain rule implementation in an elegant way to update the neuron parameters in hidden layer. This method is called the backpropagation algorithm [26]. We can apply it to neural networks with more than two layers besides our simple example. Then, the chain rule will be applied iteratively starting from the output to input layer. This way, it becomes possible to train neural networks with several hidden layers.

Application of the backpropagation algorithm will be the same as in our example given in this section for neural networks with more than one input and output. However, error calculations will differ based on multiple outputs and inputs in the network. For more information on the application of the backpropagation algorithm to a general neural network, please see the celebrated paper by Rumelhart et al. [26].

11.3.2 Training in Keras

Keras automatically applies the backpropagation algorithm while training the neural network. Hence, the user can directly apply the methods introduced in Sect. 10.4.5 to train the neural network having any number of hidden layers. As a result, training a single neuron or neural network is handled in the same way under Keras.

We provide an example on training a neural network under Keras in Listing 11.1. Here, the neural network is built by using the Sequential and Dense classes from Keras. It consists of three layers with the following parameters. The first layer consists of ten neurons each having the sigmoid as the activation function. The input shape of this layer is one since the neural network has one input. The second layer also has ten neurons with sigmoid activation functions. The last layer has one neuron with the linear activation function.

Listing 11.1 Training the neural network in Keras

```
import tensorflow as tf
import numpy as np
import matplotlib.pyplot as plt
from numpy.random import default_rng

size=1000

rng = default_rng()
noise = .1*rng.standard_normal(size)

x = np.linspace(0., 4*np.pi, size)
y = np.sin(x) + noise

model = tf.keras.models.Sequential([
    tf.keras.layers.Dense(10, input_shape=[1], activation='sigmoid'
        ),
    tf.keras.layers.Dense(10, activation='sigmoid'),
    tf.keras.layers.Dense(1, activation = 'linear')
    ])

model.compile(loss='mean_squared_error', optimizer=tf.keras.
    optimizers.SGD(0.1))
model.fit(x, y, epochs=500, verbose=False)
```

```
yp = model.predict(x)

plt.plot(x,y,'r.')
plt.plot(x,yp,'b')
plt.show()
```

In Listing 11.1, we start by defining the input and output samples to be used in training. To note here, we will form a regressor by the neural network here. As we form the training set, we form the model with mentioned parameters. While doing so, `model = tf.keras.models.Sequential` defines the neural network model under Keras. Next, we configure the model for training by setting the loss function and the optimizer via the function `model.compile`. In this example, we set the loss function as MSE. We set the optimizer as SGD with learning rate 0.1. We finally train the neural network by `model.fit(x, y, epochs=500)`. Here, the model is trained on the input data x and target data y for 500 epochs.

11.4   Forming a Classifier with the Multilayer Neural Network

As explained in Sect. 11.1, the multilayer neural network can form a complex decision boundary when used as a classifier. Therefore, let's expand the single neuron classifier introduced in Sect. 10.5 by using the neural network. Let's explain this operation on an example as in Listing 11.2.

Listing 11.2 Forming the classifier with the neural network

```
import os.path as osp
import numpy as np
import tensorflow as tf

DATA_DIR = "classification_data"
MODEL_DIR = "models"

train_samples = np.load(osp.join(DATA_DIR, "cls_train_samples.npy
    "))
train_labels = np.load(osp.join(DATA_DIR, "cls_train_labels.npy")
    )
test_samples = np.load(osp.join(DATA_DIR, "cls_test_samples.npy")
    )
test_labels = np.load(osp.join(DATA_DIR, "cls_test_labels.npy"))

model = tf.keras.models.Sequential([
    tf.keras.layers.Dense(20, input_shape=[2], activation='relu'),
    tf.keras.layers.Dense(1, activation = 'sigmoid')
    ])

model.compile(optimizer=tf.keras.optimizers.Adam(learning_rate=1e
    -3),
            loss=tf.keras.losses.BinaryCrossentropy(),
            metrics=[tf.keras.metrics.BinaryAccuracy(),
                    tf.keras.metrics.FalseNegatives()])
```

```
model.fit(train_samples, train_labels, validation_data = (
    test_samples, test_labels), epochs=50, verbose=1)
print("Finished training the model")

model.save(osp.join(MODEL_DIR,"nn_classification_model_tf"),
    save_format = "tf")
model.save(osp.join(MODEL_DIR, "nn_classification_model_keras.h5"
    ), save_format = "h5")
model.evaluate(test_samples, test_labels)
```

In Listing 11.2, we create two classes in two-dimensional feature space using the function `multivariate_normal` as in Listing 5.2. This function creates normal distributed data with given mean and covariance matrices. Here, we generate 1000 samples from class 1 with mean $[-2, -2]$ and covariance matrix

$$\begin{bmatrix} 5.5 & -4.5 \\ -4.5 & 5.5 \end{bmatrix}$$

and class 2 is generated with mean $[2, 2]$ and rotated covariance matrix as

$$\begin{bmatrix} 5.5 & 4.5 \\ 4.5 & 5.5 \end{bmatrix}.$$

After creating data, we merge both classes and define labels to indicate classes of samples via

```
X=np.append(cls1,cls2, axis=1)
label=np.append(np.zeros(len),np.ones(len))
```

Next, we define the multilayer neural network model with two layers. The first layer consists of 20 neurons with ReLU activation function. The second layer consists of one neuron with the sigmoid activation function. After building the model, we should decide loss function and optimizer. We pick the Adam optimizer with learning rate 0.003 such that model does not stuck in local minima. We pick the `BinaryCrossEntropy` loss since we are dealing with the two class problem. Moreover, we add two metrics, as binary accuracy and false negatives, to the **compile** function. These metrics can be monitored, and they can reveal the weak points of the model. Finally, we train the model by the function `fit`. The first two arguments of the function are our training samples and labels used to train the model. Moreover, the parameter `validation_data` is used for feeding the model with validation data. This is a good practice to monitor the model loss and metrics on data which is not seen by the model previously. Threshold for classes in binary cross entropy is set to 0.5 by default. This indicates that the input below this threshold belongs to class 1. Otherwise, it belongs to class 2. We provide the decision boundary formed for this example in Fig. 11.3. As can be seen in this figure, the decision boundary formed by the neural network can discriminate linearly nonseparable classes better.

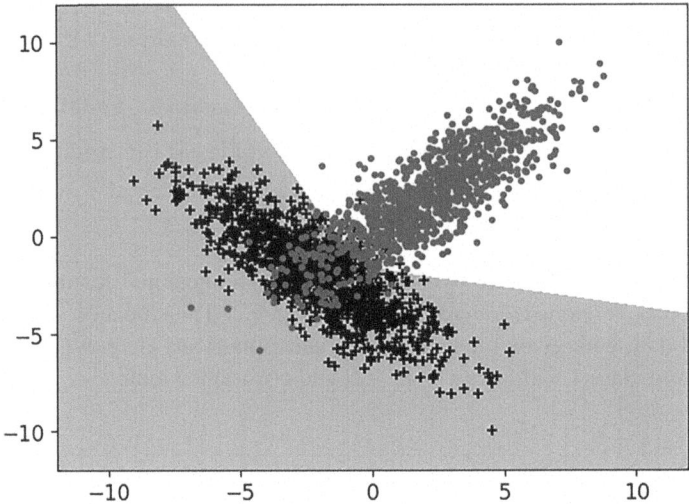

Fig. 11.3 Decision boundary formed by the neural network

11.5   Forming a Regressor with the Multilayer Neural Network

We considered regressor formation via the neural network while explaining the training operation in Sect. 11.3. Therefore, the code in Listing 11.1 can be considered from the regressor perspective as well. We strongly suggest the reader to consider evaluating this perspective. We obtain the code in Listing 11.3 for the regression example combining the data in Listing 5.3 with the multilayer neural network training.

Listing 11.3 Forming the regressor with the neural network

```
import os.path as osp
import tensorflow as tf
import numpy as np
import matplotlib.pyplot as plt

DATA_DIR = "regression_data"
MODEL_DIR = "models"

samples = np.load(osp.join(DATA_DIR, "reg_samples.npy"))
line_values = np.load(osp.join(DATA_DIR, "reg_line_values.npy"))
sine_values = np.load(osp.join(DATA_DIR, "reg_sine_values.npy"))

line_model = tf.keras.models.Sequential([
  tf.keras.layers.Dense(10, input_shape=[1], activation="relu"),
  tf.keras.layers.Dense(10, activation="relu"),
  tf.keras.layers.Dense(1)
  ])

sine_model = tf.keras.models.Sequential([
  tf.keras.layers.Dense(10, input_shape=[1], activation="relu"),
  tf.keras.layers.Dense(10, activation="relu"),
```

```
tf.keras.layers.Dense(1)
])

line_model.compile(loss='mean_squared_error', optimizer=tf.keras.
    optimizers.Adam(1e-3))
line_model.fit(samples, line_values, epochs=1000, verbose = 0)
line_pred = line_model.predict(samples)

sine_model.compile(loss='mean_squared_error', optimizer=tf.keras.
    optimizers.Adam(1e-3))
sine_model.fit(samples, sine_values, epochs=1000, verbose = 1)
sine_pred = sine_model.predict(samples)

plt.figure(1)
plt.scatter(samples, line_values, c = 'k')
plt.plot(samples, line_pred, 'b')

plt.figure(2)
plt.scatter(samples , sine_values, c = 'k')
plt.plot(samples, sine_pred, 'b')
plt.show()

line_model.save(osp.join(MODEL_DIR,"nn_line_regression_model_tf")
    , save_format = "tf")
line_model.save(osp.join(MODEL_DIR, "
    nn_line_regression_model_keras.h5"), save_format = "h5")
```

In Listing 11.3, the neural network is used to estimate the values of line and sine data as in Chap. 10. Please note that the Keras model consists of two hidden layers each having ten neurons with the ReLU activation function. The output layer in the model is a single neuron which provides the numerical estimation for the regression problem. We set the model to use MSE between the predicted and actual values by the function model.compile. Hence, we set SGD as the optimizer with learning rate 0.001. Models for both predicting the line and sine data are identical. Thus, the same model is trained for 1000 epochs for both problems. We plot the predicted values to observe fitness of the model for different data in Fig. 11.4.

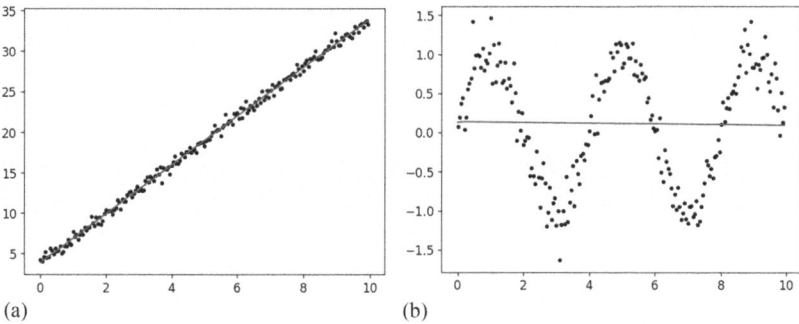

Fig. 11.4 Regression example with the neural network, first setup. (**a**) Noisy line data and its regression result. (**b**) Noisy sine data and its regression result

As can be seen in Fig. 11.4, our multilayer neural network successfully fitted a line to line data. However, it fails to estimate the sine data as in the single neuron case. As we check training loss value of the model in Listing 11.5, we can see that it is not decreasing. Hence, we can deduce that the model is not able to learn the sine data.

```
Epoch 997/1000
7/7 [==============================] - 0s 1ms/step - loss
    : 0.5363
Epoch 998/1000
7/7 [==============================] - 0s 1ms/step - loss
    : 0.5368
Epoch 999/1000
7/7 [==============================] - 0s 718us/step -
    loss: 0.5450
Epoch 1000/1000
7/7 [==============================] - 0s 1ms/step - loss
    : 0.5342
```

Therefore, we increase the number of neurons in the model from 10 to 100 and train the model again. In other words, we change the following lines

```
tf.keras.layers.Dense(10, input_shape=[1], activation='
    relu'),
tf.keras.layers.Dense(10, activation='relu'),
tf.keras.layers.Dense(1)
```

to

```
tf.keras.layers.Dense(100, input_shape=[1], activation='
    relu'),
tf.keras.layers.Dense(100, activation='relu'),
tf.keras.layers.Dense(1)
```

We obtain the regression result as in Fig. 11.5. As can be seen in this figure, the model can predict the line very well. However, increasing the number of neurons is not enough to predict the sine data, and it can only predict correct values up to $x \approx 2$.

As we check the value of the training loss function given in Listing 11.5, it is decreasing until the final epoch. This indicates that the loss value has not yet reached the minimum. In other words, number of epochs is not enough for model to learn enough.

```
Epoch 997/1000
7/7 - 0s - loss: 0.3771 - 10ms/epoch - 1ms/step
Epoch 998/1000
7/7 - 0s - loss: 0.3772 - 8ms/epoch - 1ms/step
Epoch 999/1000
```

```
7/7 - 0s - loss: 0.3771 - 9ms/epoch - 1ms/step
Epoch 1000/1000
7/7 - 0s - loss: 0.3769 - 10ms/epoch - 1ms/step
```

Hence, we increase the number of epochs while fitting the model. Therefore, we change the following line `model.fit(x, y2, epochs=1000, verbose = 2)` to `model.fit(x, y2, epochs=2000, verbose = 2)` in the code. We provide the regression result for the sine data in this setting in Fig. 11.6. As can be seen in this figure, the model can predict the sine data accurately. Hence, the error can be decreased by setting appropriate hyperparameters such as learning rate, the number of hidden layers, and number of neurons in each layer.

11.6 Application: Human Activity Recognition via Accelerometer Data

We extracted features from accelerometer data for the human activity recognition application in Sect. 5.4. We also used the single neuron to classify "walking" and "not walking" classes in Sect. 10.7. Here, we use a multilayer neural network for human activity recognition. In this application, we use all the human activity classes as "downstairs," "jogging," "sitting," "standing," "upstairs," and "walking."

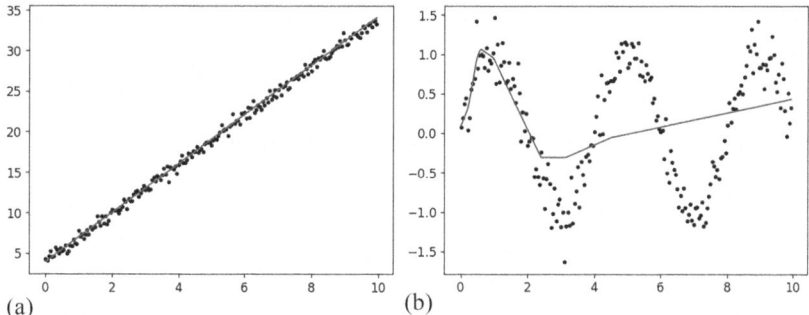

Fig. 11.5 Regression example with the neural network, second setup. (**a**) Noisy line data and its regression result. (**b**) Noisy sine data and its regression result

Fig. 11.6 Regression example with the multilayer neural network, third setup

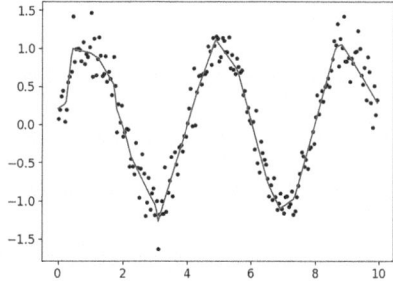

We provide the Python script for human activity recognition with six classes in Listing 11.4. In this script, we first form the features as in Sect. 5.4. Then, we form the multilayer neural network classifier with three layers. The first two layers consist of 100 neurons with the ReLU activation function. The final layer consists of six neurons to represent the total number of classes for this application. Since this is a multi-class classification application, we use the softmax activation function the final layer. We train the model using Adam optimizer and categorical cross-entropy loss functions. Categorical cross entropy expects training labels in one-hot encoded format. Hence, we convert the original labels to this form using the function OneHotEncoder from the sklearn library. As the training is done, we save the model for future use.

Listing 11.4 Training the neural network for human activity recognition

```python
import os.path as osp
import numpy as np
import tensorflow as tf
from sklearn.metrics import confusion_matrix,
    ConfusionMatrixDisplay
from matplotlib import pyplot as plt
from data_utils import read_data
from sklearn.preprocessing import OneHotEncoder
from feature_utils import create_features

DATA_PATH = osp.join("WISDM_ar_v1.1", "WISDM_ar_v1.1_raw.txt")
TIME_PERIODS = 80
STEP_DISTANCE = 40
data_df = read_data(DATA_PATH)
df_train = data_df[data_df["user"] <= 28]
df_test = data_df[data_df["user"] > 28]

train_segments_df, train_labels = create_features(df_train,
    TIME_PERIODS, STEP_DISTANCE)
test_segments_df, test_labels = create_features(df_test,
    TIME_PERIODS, STEP_DISTANCE)

model = tf.keras.models.Sequential([
    tf.keras.layers.Dense(100, input_shape=[10], activation="relu")
    ,
    tf.keras.layers.Dense(100, activation="relu"),
    tf.keras.layers.Dense(6, activation = "softmax")
    ])

train_segments_np = train_segments_df.to_numpy()
test_segments_np = test_segments_df.to_numpy()
ohe = OneHotEncoder()
train_labels_ohe = ohe.fit_transform(train_labels.reshape(-1, 1))
    .toarray()
categories, test_labels = np.unique(test_labels, return_inverse =
    True)
model.compile(loss=tf.keras.losses.CategoricalCrossentropy(),
    optimizer=tf.keras.optimizers.Adam(1e-3))
model.fit(train_segments_np, train_labels_ohe, epochs=50, verbose
    = 1)
nn_preds = model.predict(test_segments_np)
predicted_classes = np.argmax(nn_preds, axis = 1)

conf_matrix = confusion_matrix(test_labels, predicted_classes)
```

```
cm_display = ConfusionMatrixDisplay(confusion_matrix =
    conf_matrix, display_labels= categories)
cm_display.plot()
cm_display.ax_.set_title("Neural Network Confusion Matrix")
plt.show()

model.save("mlp_har_model.h5")
```

11.7 Application: Keyword Spotting from Audio Signals

We extracted features from audio data for the keyword spotting application in
Sect. 5.5. We also used the single neuron to classify "zero" and "not zero" classes
in Sect. 10.8. Here, we use a multilayer neural network for keyword spotting. In this
application, we use all spoken digit classes from zero to nine.

We provide the Python script for keyword spotting with ten classes in Listing 11.5. In this script, we first form the features as in Sect. 5.5. Therefore, the
Python script reads audio data and creates MFCC features. Then, we form the
multilayer neural network classifier with three layers. The first two layers consist
of 100 neurons with the ReLU activation function. The final layer consists of ten
neurons to represent the total number of classes for this application. Since this is
a multi-class classification application, we use the softmax activation function the
final layer. We train the model using Adam optimizer and categorical cross-entropy
loss functions. Categorical cross entropy expects training labels in one-hot encoded
format. Hence, we converted the original labels to this form using the function
OneHotEncoder from the sklearn library. As the training is done, we save the model
for future use.

Listing 11.5 Training the neural network for keyword spotting

```
import os
import numpy as np
import scipy.signal as sig
from mfcc_func import create_mfcc_features
from sklearn.metrics import confusion_matrix,
    ConfusionMatrixDisplay
import tensorflow as tf
from matplotlib import pyplot as plt
from sklearn.preprocessing import OneHotEncoder

RECORDINGS_DIR = "recordings"
recordings_list = [(RECORDINGS_DIR, recording_path) for
    recording_path in os.listdir(RECORDINGS_DIR)]

FFTSize = 1024
sample_rate = 8000
numOfMelFilters = 20
numOfDctOutputs = 13
window = sig.get_window("hamming", FFTSize)
test_list = {record for record in recordings_list if "yweweler"
    in record[1]}
train_list = set(recordings_list) - test_list
```

```
train_mfcc_features, train_labels = create_mfcc_features(
    train_list, FFTSize, sample_rate, numOfMelFilters,
    numOfDctOutputs, window)
test_mfcc_features, test_labels = create_mfcc_features(test_list,
    FFTSize, sample_rate, numOfMelFilters, numOfDctOutputs,
    window)

model = tf.keras.models.Sequential([
    tf.keras.layers.Dense(100, input_shape=[26], activation="relu")
    ,
    tf.keras.layers.Dense(100, activation="relu"),
    tf.keras.layers.Dense(10, activation = "softmax")
    ])

ohe = OneHotEncoder()
train_labels_ohe = ohe.fit_transform(train_labels.reshape(-1, 1))
    .toarray()
categories, test_labels = np.unique(test_labels, return_inverse =
    True)
model.compile(loss=tf.keras.losses.CategoricalCrossentropy(),
    optimizer=tf.keras.optimizers.Adam(1e-3))
model.fit(train_mfcc_features, train_labels_ohe, epochs=100,
    verbose = 1)
nn_preds = model.predict(test_mfcc_features)
predicted_classes = np.argmax(nn_preds, axis = 1)

conf_matrix = confusion_matrix(test_labels, predicted_classes)
cm_display = ConfusionMatrixDisplay(confusion_matrix =
    conf_matrix, display_labels=categories)
cm_display.plot()
cm_display.ax_.set_title("Neural Network Confusion Matrix")
plt.show()

model.save("mlp_fsdd_model.h5")
```

11.8 Application: Handwritten Digit Recognition from Digital Images

We extracted features from digital images for the handwritten digit recognition application in Sect. 5.6. We also used the single neuron to classify "zero" and "not zero" classes in Sect. 10.9. Here, we use a multilayer neural network for handwritten digit recognition. In this application, we use digits from zero to nine as separate classes.

We provide the Python script for handwritten digit recognition with ten classes in Listing 11.6. In this script, we first form the features as in Sect. 5.6. Therefore, the Python script accesses data from the MNIST dataset and generates Hu moments as features. Then, we form the multilayer neural network classifier with three layers. The first two layers consist of 100 neurons with the ReLU activation function. The final layer consists of ten neurons to represent the total number of classes for this application. Since this is a multi-class classification application, we use the softmax activation function the final layer. We train the model using Adam optimizer and sparse categorical cross-entropy loss functions. Sparse categorical cross entropy can handle integer labels directly. Therefore, the labels remain in their original form as digits from "0" to "9".

Listing 11.6 Training the neural network for handwritten digit recognition

```
import os
import numpy as np
import cv2
from sklearn.metrics import confusion_matrix,
    ConfusionMatrixDisplay
from mnist import load_images, load_labels
import keras
from keras.callbacks import EarlyStopping, ModelCheckpoint
from matplotlib import pyplot as plt

train_img_path = os.path.join("MNIST-dataset", "train-images.idx3
    -ubyte")
train_label_path = os.path.join("MNIST-dataset", "train-labels.
    idx1-ubyte")
test_img_path = os.path.join("MNIST-dataset", "t10k-images.idx3-
    ubyte")
test_label_path = os.path.join("MNIST-dataset", "t10k-labels.idx1
    -ubyte")

train_images = load_images(train_img_path)
train_labels = load_labels(train_label_path)
test_images = load_images(test_img_path)
test_labels = load_labels(test_label_path)

train_huMoments = np.empty((len(train_images),7))
test_huMoments = np.empty((len(test_images),7))

for train_idx, train_img in enumerate(train_images):
    train_moments = cv2.moments(train_img, True)
    train_huMoments[train_idx] = cv2.HuMoments(train_moments).
        reshape(7)

for test_idx, test_img in enumerate(test_images):
    test_moments = cv2.moments(test_img, True)
    test_huMoments[test_idx] = cv2.HuMoments(test_moments).
        reshape(7)

model = keras.models.Sequential([
  keras.layers.Dense(100, input_shape=[7], activation="relu"),
  keras.layers.Dense(100, activation="relu"),
  keras.layers.Dense(10, activation = "softmax")
  ])

categories = np.unique(test_labels)
model.compile(loss=keras.losses.SparseCategoricalCrossentropy(),
    optimizer=keras.optimizers.Adam(1e-4))
mc_callback = ModelCheckpoint("mlp_mnist_model.h5")
es_callback = EarlyStopping("loss", patience = 5)
model.fit(train_huMoments, train_labels, epochs=1000, verbose =
    1, callbacks=[mc_callback, es_callback])

nn_preds = model.predict(test_huMoments)
predicted_classes = np.argmax(nn_preds, axis = 1)

conf_matrix = confusion_matrix(test_labels, predicted_classes)
cm_display = ConfusionMatrixDisplay(confusion_matrix =
    conf_matrix, display_labels= categories)
```

```
cm_display.plot()
cm_display.ax_.set_title("Neural Network Confusion Matrix")
plt.show()
```

In Listing 11.6, the model is preserved for future use using the ModelCheckpoint function from Keras. This function enables us to preserve the model after each epoch. Only the model with the best performance is preserved, which is determined by monitoring a specific metric, such as validation loss. To prevent overfitting and conserve computational resources, EarlyStopping is also employed. This function halts the training process when a monitored metric has ceased to improve for a specified number of epochs, known as patience.

11.9 Application: Estimating Future Temperature Values

We can estimate future temperature values using the multilayer neural network as regressor. Hence, we benefit from the SML2010 dataset introduced in Sect. 7.6.1. As in Sect. 10.10, we form a Python script for this purpose. We provide the formed script in Listing 11.7.

Listing 11.7 Python script for neural network formation, training, and testing

```
import pandas as pd
import numpy as np
from sklearn.model_selection import train_test_split
import tensorflow as tf
from sklearn.metrics import mean_absolute_error
from matplotlib import pyplot as plt

df = pd.read_csv('temperature_dataset.csv')
y = df['Room_Temp'][::4]
prev_values_count = 5

X = pd.DataFrame()
for i in range(prev_values_count, 0, -1):
    X['t-' + str(i)] = y.shift(i)

X = X[prev_values_count:]
y = y[prev_values_count:]


X_train, X_test, y_train, y_test = train_test_split(X, y,
    test_size=0.2, random_state=0)
train_mean = X_train.mean()
train_std = X_train.std()

X_train = (X_train - train_mean)/ train_std
X_test = (X_test - train_mean)/ train_std

model = tf.keras.models.Sequential([tf.keras.layers.Dense(10,
    input_shape = [5]),
                                    tf.keras.layers.Dense(10),
                                    tf.keras.layers.Dense(1)])
```

```
model.compile(optimizer=tf.keras.optimizers.SGD(learning_rate=5e
    -3),
                loss=tf.keras.losses.MeanAbsoluteError())

model.fit(X_train,
            y_train,
            batch_size = 128,
            epochs=3000,
            verbose=1,
            workers = 16,
            use_multiprocessing = True)

y_train_predicted = model.predict(X_train)
y_test_predict = model.predict(X_test)

fig, ax = plt.subplots(1,1)
ax.plot(y_test.to_numpy(), label = "Actual values")
ax.plot(y_test_predict, label = "Predicted values")
plt.legend()
plt.show()

mae_train = np.sqrt(mean_absolute_error(y_train,
    y_train_predicted))
mae_test = np.sqrt(mean_absolute_error(y_test, y_test_predict))

print(f"Training set MAE: {mae_train}\n")
print(f"Test set MAE:{mae_test}")

model.save("temperature_prediction_mlp.h5")
```

In Listing 11.7, we read and prepare the dataset as in Listing 10.21. We then form a two layer neural network with five inputs as the regressor. Each layer consists of 100 neurons with the ReLU activation function. We train the model with the SGD optimizer setting the learning rate to 0.0005. We use MAE as the loss function during training. Finally, the code saves the model in Keras format as "temperature_prediction_mlp.h5."

11.10 Summary of the Chapter

We extended the single neuron case to multilayer neural networks in this chapter. The reason for this extension was that the single neuron was not sufficient enough to solve complex machine learning problems. The multilayer neural network offered a more powerful structure. To explain this structure better, we started with the background information on neural networks. We defined structure of the neural network as well as deep and shallow neural networks. We also formed the fully connected neural network in line with the background information. We then showed how to implement a neural network in Keras. We covered the layer structure and how to use it in forming the network. Then, we focused on training the neural network. Here, we briefly introduced the backpropagation algorithm. We also explored the training steps under Keras. Afterward, we formed a classifier and regressor via neural networks. Finally, we considered real-life applications introduced in previous chapters now from the neural network perspective.

Embedding the Neural Network Model to the Microcontroller

12

12.1 TensorFlow Lite

We will cover the basics of TensorFlow Lite in this section. Then, we will focus on model conversion via TensorFlow Lite. We will also show ways of interpreting the converted model on PC. Hence, we will be ready to embed the converted TensorFlow Lite model to the microcontroller.

12.1.1 What Is TensorFlow Lite?

TensorFlow models introduced in Chap. 11 are generated to work on PC. Hence, the functions and data structures used while forming these models are specialized as such. This is also the case for the convolutional and recurrent neural network models to be introduced in Chaps. 13 and 14, respectively. However, an embedded system, whether being a microprocessor or microcontroller, has limited flash and RAM size. Therefore, there is a strict restriction for them on the model size for memory usage and latency in inference.

TensorFlow has a special version, called TensorFlow Lite, to overcome the mentioned limitations in the previous paragraph. TensorFlow Lite has limited set of operators. Moreover, it has a specific structure called flatbuffer. To note here, we can divide the TensorFlow Lite usage into two parts as embedded systems as microprocessor (Android) and microcontroller based. As expected, the microcontroller-based operations are more restricted. However, they are the main focus of this book. Therefore, we will cover the microcontroller-based TensorFlow Lite usage in detail in Sect. 12.5.

TensorFlow uses protocol buffer format to store models, while the Keras H5 model uses the HDF5 format. Both are efficient ways of storing and transporting the neural network model. On the other hand, the TensorFlow Lite model is stored in a specific structure called flatbuffer. Flatbuffer is an efficient data format which

© The Author(s), under exclusive license to Springer Nature Switzerland AG 2025
C. Ünsalan et al., *Embedded Machine Learning with Microcontrollers*,
https://doi.org/10.1007/978-3-031-70912-8_12

uses cross platform serialization library. The main advantage of flatbuffer structure is that serialized data can be accessed without parsing, which means no need to deserialize the entire file. Hence, we can get information about the model without loading it entirely to memory.

Converting a TensorFlow or Keras model to TensorFlow Lite format can be done either by the Python API or in command line. The former option allows the user to add optimization properties during conversion. Therefore, we will benefit from it throughout the book. The latter option is restricted to basic model conversion. Hence, we will not pursue it further in this book.

12.1.2 Converting a Keras Model to TensorFlow Lite Format

We can convert a Keras model to TensorFlow Lite format after training it by the methods introduced in Chap. 11. To do so, we should use the TensorFlow commands listed below.

```
# The Keras "model" has been formed and trained beforehand
converter = tf.lite.TFLiteConverter.from_keras_model(model)
tflite_model = converter.convert()
```

The first line in the code snippet creates an instance of the TensorFlow Lite converter tf.lite.TFLiteConverter which takes the pretrained Keras model as a parameter. This converter is used to represent the Keras model to TensorFlow Lite format. The second line of the code calls the convert method of the converter object with its default parameters. This method takes the Keras model and converts it to the corresponding TensorFlow Lite model format. The resulting TensorFlow Lite model is stored in tflite_model. Hence, it will contain the TensorFlow Lite version of the model after executing this code snippet.

We provide an example on the usage of given commands for converting a Keras model in Listing 12.1. In this example, we pick the fully connected neural network classifier model introduced in Listing 11.2. Here, we provide the complete Python script to emphasize the overall framework. Hence, we train our model and saved it to the folder "models." The next step is converting this saved model to the TensorFlow Lite format. We use the TFLiteConverter module to convert our Keras model to TFLite format by calling the convert method. The TensorFlow Lite model is then stored in tflite_model. We save this model to a file by opening it in binary mode and write the TensorFlow Lite flatbuffer content to that file.

Listing 12.1 Converting a formed Keras model for classification to TensorFlow Lite format

```
import os.path as osp
import numpy as np
import tensorflow as tf

DATA_DIR = "classification_data"
```

```
MODEL_DIR = "models"

train_samples = np.load(osp.join(DATA_DIR, "cls_train_samples.npy
    "))
train_labels = np.load(osp.join(DATA_DIR, "cls_train_labels.npy")
    )
test_samples = np.load(osp.join(DATA_DIR, "cls_test_samples.npy")
    )
test_labels = np.load(osp.join(DATA_DIR, "cls_test_labels.npy"))

model = tf.keras.models.Sequential([
    tf.keras.layers.Dense(20, input_shape=[2], activation='relu'),
    tf.keras.layers.Dense(1, activation = 'sigmoid')
    ])

model.compile(optimizer=tf.keras.optimizers.Adam(learning_rate=1e
    -3),
                loss=tf.keras.losses.BinaryCrossentropy(),
                metrics=[tf.keras.metrics.BinaryAccuracy(),
                    tf.keras.metrics.FalseNegatives()])

model.fit(train_samples, train_labels, validation_data = (
    test_samples, test_labels), epochs=50, verbose=1)
print("Finished training the model")

model.save(osp.join(MODEL_DIR, "nn_classification_model.keras"))
model.save(osp.join(MODEL_DIR,"nn_classification_model_tf"),
    save_format = "tf")
model.save(osp.join(MODEL_DIR, "nn_classification_model_keras.h5"
    ), save_format = "h5")

# Convert the Keras model to TF Lite model
converter = tf.lite.TFLiteConverter.from_keras_model(model)
tflite_model = converter.convert()

# Save the TF Lite model
with open('models/nn_classification_model.tflite', 'wb') as f:
    f.write(tflite_model)
```

We can follow the same steps to convert the regression model considered in Chap. 11 to TensorFlow Lite format. Therefore, we provide the corresponding Python code in Listing 11.3. Here, we follow the same steps as in Listing 12.1.

Listing 12.2 Converting a formed Keras model for regression to TensorFlow Lite format

```
import os.path as osp
import tensorflow as tf
import numpy as np

DATA_DIR = "regression_data"
MODEL_DIR = "models"

samples = np.load(osp.join(DATA_DIR, "reg_samples.npy"))
line_values = np.load(osp.join(DATA_DIR, "reg_line_values.npy"))
sine_values = np.load(osp.join(DATA_DIR, "reg_sine_values.npy"))

line_model = tf.keras.models.Sequential([
    tf.keras.layers.Dense(10, input_shape=[1], activation="relu"),
```

```
    tf.keras.layers.Dense(10, activation="relu"),
    tf.keras.layers.Dense(1)
    ])

sine_model = tf.keras.models.Sequential([
    tf.keras.layers.Dense(100, input_shape=[1], activation="relu"),
    tf.keras.layers.Dense(100, activation="relu"),
    tf.keras.layers.Dense(1)
    ])

line_model.compile(loss='mean_squared_error', optimizer=tf.keras.
    optimizers.Adam(1e-3))
line_model.fit(samples, line_values, epochs=1000, verbose = 0)

sine_model.compile(loss='mean_squared_error', optimizer=tf.keras.
    optimizers.Adam(1e-3))
sine_model.fit(samples, sine_values, epochs=1000, verbose = 1)

line_model.save(osp.join(MODEL_DIR, "nn_line_regression_model.
    keras"))
sine_model.save(osp.join(MODEL_DIR, "nn_sine_regression_model.
    keras"))

# Convert the Keras line model to TF Lite model
converter = tf.lite.TFLiteConverter.from_keras_model(line_model)
tflite_model = converter.convert()

# Save the TF Lite model.
with open('models/nn_line_regression_model.tflite', 'wb') as f:
    f.write(tflite_model)

# Convert the Keras sine model to TF Lite model
converter = tf.lite.TFLiteConverter.from_keras_model(sine_model)
tflite_model = converter.convert()

# Save the TF Lite model.
with open('models/nn_sine_regression_model.tflite', 'wb') as f:
    f.write(tflite_model)
```

We can benefit from the visualization tool offered by TensorFlow Lite to observe the converted models. To do so, we should use the command python -m tensorflow .lite.tools.visualize tflite_model.tflite visualized_model.html. This command provides an html file which includes detailed information on our TensorFlow Lite model saved in the file "tflite_model.tflite." When this file is opened using a web browser, one can observe input and output nodes of the model under the "Inputs/Outputs" heading. Detailed information about the output of each layer can be observed under the "Tensors" heading. Index, name, shape, and quantization details of the tensor are located in a table here. Operations between these tensors are located under the "Ops" heading, where input-output nodes and built-in options related to corresponding operations are specified in a table. Graph visualization of our model is also given in the html file. The reader can verify nodes of the model this way. Please note that quantization variables are added as separate nodes in the graph. The "Buffer" section shows the required memory in bytes to store output nodes.

12.1.3 Converting a Saved Model to TensorFlow Lite Format

We can also convert a saved TensorFlow model to the TensorFlow Lite format. To do so, we should only change the line `converter = tf.lite. TFLiteConverter.from_keras_model(model)` in the previous section to `converter = tf.lite.TFLiteConverter.from_saved_model(saved_model_dir)` by defining the `saved_model_dir`. We provide an example to convert saved TensorFlow model in Listing 12.1 to TensorFlow Lite format in Listing 12.3.

Listing 12.3 Converting a saved TensorFlow model to TensorFlow Lite format

```
import tensorflow as tf

# Convert a saved model
saved_model_dir='models/nn_classification_model_tf'
converter = tf.lite.TFLiteConverter.from_saved_model(
    saved_model_dir)
tflite_model = converter.convert()

# Save the TF Lite model
with open('models/nn_classification_model.tflite', 'wb') as f:
    f.write(tflite_model)
```

In Listing 12.3, `saved_model_dir` is set to the directory path where the TensorFlow model is stored. The model can be saved as explained in Sect. 9.7. Here, we use the directory where we saved our model in the previous section as "models/nn_classification_model_tf." Next, we create a converter using `tf.lite .TFLiteConverter.from_saved_model(saved_model_dir)`. The `from_saved_model` method loads the saved model from the specified directory and prepares it for conversion. In Listing 12.3, `tflite_model` contains the TensorFlow Lite model after conversion. Then, we open a file for writing in binary mode with the specified path. Eventually, the `tflite_model` content is written to it.

12.1.4 Interpreting the Converted Model

We can run the TensorFlow Lite interpreter on PC to check how the converted model responds to a given input. To do so, we benefit from the TensorFlow Lite interpreter. We will not go into details of how the interpreter works. For more information on this topic, please see [54].

The TensorFlow Lite interpreter can be used in the following sections. Therefore, we picked an available example from Google and modified it for our purposes [55]. We provide the modified code in Listing 12.4. As for now, we used the code to evaluate the accuracy of the TensorFlow Lite model obtained in the previous section. However, it can accept quantized, pruned, and weight clustered models to be introduced in the following sections.

Listing 12.4 Running the TensorFlow Lite interpreter on PC to check how the converted model responds to a given input

```python
import os.path as osp
import numpy as np
import tensorflow as tf

DATA_DIR = "classification_data"
MODEL_DIR = "models"

saved_model_dir='models/nn_classification_model_tf'

train_samples = np.load(osp.join(DATA_DIR, "cls_train_samples.npy
    "))
train_labels = np.load(osp.join(DATA_DIR, "cls_train_labels.npy")
    )
test_samples = np.load(osp.join(DATA_DIR, "cls_test_samples.npy")
    )
test_labels = np.load(osp.join(DATA_DIR, "cls_test_labels.npy"))

# test the actual model on a test sample image
model = tf.keras.models.load_model(saved_model_dir)

predictions = model.predict(test_samples)
predicted_labels = np.where(predictions < 0.5, 0, 1).squeeze()

tflite_model_file= osp.join("models","nn_classification_model.
    tflite")

# Convert test_samples to float32 format
test_samples = test_samples.astype(np.float32)
accuracy = (np.sum(test_labels == predicted_labels) * 100) / len(
    test_samples)

# Helper function to run inference on a TFLite model
def tflite_predict(lite_model_path):
    # Initialize the interpreter
    interpreter = tf.lite.Interpreter(model_path=str(
        lite_model_path))
    interpreter.allocate_tensors()

    input_details = interpreter.get_input_details()[0]
    output_details = interpreter.get_output_details()[0]

    # Check if the input type is quantized, then rescale input
        data to uint8
    tflite_preds = np.empty_like(test_labels, dtype=int)
    for idx, test_sample in enumerate(test_samples):
        if input_details['dtype'] == np.uint8:
            input_scale, input_zero_point = input_details["
                quantization"]
            test_sample = test_sample / input_scale +
                input_zero_point

        interpreter.set_tensor(input_details["index"], [
            test_sample])
        interpreter.invoke()
        output = interpreter.get_tensor(output_details["index"])
            [0]
        tflite_preds[idx] = 0 if output < .5 else 1

    return tflite_preds
```

```
def evaluate_model(tflite_file):
    tflite_preds = tflite_predict(tflite_file)

    accuracy = (np.sum(test_labels== tflite_preds) * 100) / len(
        test_samples)
    return accuracy

tflite_accuracy = evaluate_model(tflite_model_file)

print(f"Keras model accuracy is {accuracy:.2f}% (Number of test
    samples={len(test_samples)})")
print(f"TFlite model accuracy is {tflite_accuracy:.2f}% (Number
    of test samples={len(test_samples)})")
```

In Listing 12.4, we load the training and test dataset as in Listing 12.1. Next, both saved Keras and TensorFlow Lite models are executed to obtain predictions for test samples. The predicted labels from the Keras model are stored in `predicted_label`. The path to the saved TensorFlow Lite model file is specified to run the TensorFlow Lite model. The `run_tflite_model` is defined to execute the TensorFlow Lite model for inference on test samples. Therefore, the interpreter is initialized, input and output details are extracted, and predictions are obtained. The TensorFlow Lite model is used to predict the label of the test samples. The predicted labels are stored in the variable `tflite_preds`. The function `evaluate_model` is defined to evaluate the accuracy of the TensorFlow Lite model on all test samples. The accuracy is calculated and printed as a percentage for both Keras and TensorFlow Lite model.

12.2   Model Optimization via Quantization

Converting a Keras or TensorFlow model to TensorFlow Lite format is not sufficient alone to decrease its size. Therefore, we should apply optimization methods on it. The first way of doing this is using quantization. We will cover it in this section. Quantization is done by using suitable data types while converting a model. Therefore, we will first cover data types and their properties in operation. Afterward, we will introduce post-training quantization and quantization aware training.

12.2.1 Data Types in Python and C Languages

Quantization is done by selecting suitable data types while converting a model to TensorFlow Lite format. Therefore, we should know data types used in Python and C languages to understand how quantization works. We summarize Python (more specifically TensorFlow) and C data types in Table 12.1 with their bit size, minimum and maximum values. Here, the first eight data types can be used for integer numbers only. The last three data types can be used to represent integer and fractional numbers. For more information on integer and fractional number representations, please see [60].

Table 12.1 Python and C data types

Python	C	Size (bits)	Minimum value	Maximum value
uint8	char	8	0	255
int8	signed char	8	−128	127
uint16	unsigned short	16	0	65,535
int16	short	16	−32,768	32,767
uint32	unsigned **int**	32	0	4,294,967,295
int32	**int**	32	−2,147,483,648	2,147,483,647
uint64	unsigned **long long**	64	0	18,446,744,073,709,551,615
int64	**long long**	64	−9,223,372,036,854,775,808	9,223,372,036,854,775,807
float16, half	–	16	−65,504	65,504
float32	**float**	32	−3.40282346e+38	3.40282346e+38
float64, double	double	64	−1.79769313e+308	1.79769313e+308

The reader can also benefit from the integer data types defined under the "stdint.h" header file in C language for microcontrollers. We can summarize them based on the number of bits assigned as follows. The 8-bit data types uint8_t and int8_t can be used instead of char and signed char, respectively. The 16-bit data types uint16_t and int16_t can be used instead of unsigned short and short, respectively. The 32-bit data types uint32_t and int32_t can be used instead of unsigned **int** and **int**, respectively. The 64-bit data types uint64_t and int64_t can be used instead of unsigned **long long** and **long long**, respectively.

Model parameters are stored in float32 format in TensorFlow. We should avoid this format in TensorFlow Lite representation whenever possible due to the excessive size of the corresponding model. Moreover, TensorFlow supports the float16 format which is not available in C language. Therefore, we will not use it in operations. TensorFlow has other data types outside the scope of this book. For more information on them, please see [48].

12.2.2 Post-training Quantization

As the name implies, post-training quantization takes the trained model and changes parameter data types for quantization. TensorFlow Lite offers four options for this approach as dynamic range, full integer, float16, and integer only: 16-bit activations with 8-bit weights quantization. Since C language does not support the float16 data type, we will not cover the float16 quantization option in this book. Besides, we will not cover the integer only: 16-bit activations with 8-bit weights quantization option since it is in experimental form as of writing this book. We will cover the remaining two post-training quantization methods next. We will provide a usage example for each method.

12.2.2.1 Dynamic Range Quantization

Model weights are converted from the float32 to int8 data type in dynamic range quantization. Moreover, activation functions are modified to work with the int8 data type. Hence, computations are done in int8 data type within the model. This reduces latency during inference. Input and output values are kept in float32 data type. Therefore, the decrease in latency is not done fully. This conversion operation is suggested as the first method to be applied since it does not need any calibration step to work.

We can apply dynamic range quantization by setting the converter. optimizations parameter to tf.lite.Optimize.DEFAULT while forming the TensorFlow Lite model. We provide a usage example in Listing 12.5. Here, converter.optimizations is set to [tf.lite.Optimize.DEFAULT] to enable default optimizations, which includes dynamic range quantization. tflite_model_quant contains the quantized TensorFlow Lite model after conversion. A file for writing is opened in binary mode with the specified path. Content of tflite_model_quant is written to it.

Listing 12.5 Dynamic range quantization while forming the TensorFlow Lite model

```
import tensorflow as tf

# Convert a saved model
saved_model_dir='models/nn_classification_model_tf'

# Dynamic range quantization
converter = tf.lite.TFLiteConverter.from_saved_model(
    saved_model_dir)
converter.optimizations = [tf.lite.Optimize.DEFAULT]
tflite_model_quant = converter.convert()

# Save the model.
with open('models/quantized_model_dyn_range.tflite', 'wb') as f:
    f.write(tflite_model_quant)
```

12.2.2.2 Full Integer Quantization

The int8 data type is used in all operations within the model in full integer quantization. There are two variants of full integer quantization as integer with float fallback and integer only. The former option uses input and output values in float32 data type. The latter option uses input and output values as int8 data type. Since the int8 data type will be used in all operations within the model, a representative dataset is needed for calibration. This dataset can be formed by a function such as **def** representative_dataset to be introduced in the following examples.

As we have the representative dataset, we can apply full integer quantization by setting the converter.optimizations parameter to tf.lite.Optimize.DEFAULT while forming the TensorFlow Lite model. Moreover, we should have extra settings for the integer with float fallback and integer only variants of full integer quantization. We provide them separately next.

```
# Integer with float fallback (using default float input/output)
converter.representative_dataset = representative_dataset

# Full integer quantization
# Integer only
converter.representative_dataset = representative_dataset

# Ensure that if any operations cannot be quantized, then the
    converter raises an error
converter.target_spec.supported_ops = [tf.lite.OpsSet.
    TFLITE_BUILTINS_INT8]

# Set the input and output tensors to int8
converter.inference_input_type = tf.uint8   # or tf.int8
converter.inference_output_type = tf.uint8   # or tf.int8
```

Here, converter.representative_dataset = representative_dataset line sets the representative_dataset as input for quantization. The representative_dataset must be a generator function or dataset that provides a representative set of input data. This helps the quantization process to determine suitable quantization range for the model input and intermediate tensors. In other words, the representative dataset helps the converter to determine whether to use integer quantization with fallback to float or not. converter.target_spec.supported_ops = [tf.lite. OpsSet.TFLITE_BUILTINS_INT8] restricts the set of supported operations to those compatible with integer quantization, specifically the operations from the set TFLITE_BUILTINS_INT8. This ensures that only operations compatible with int8 quantization are included in the model. If any operation cannot be quantized, then the converter raises an error. The lines converter.inference_input_type = tf .uint8 and converter.inference_output_type = tf.uint8 specify that both input and output tensors of the quantized model should use the unsigned 8-bit integer data type. Alternatively, one can also use the option tf.int8 if the signed 8-bit integers is preferred in operation.

We provide a usage example for full integer quantization while forming the TensorFlow Lite model in Listing 12.6. We provide both integer with float fallback and integer only options separately in this example. In Listing 12.6, two quantized versions of the model are created, one with integer and float fallback and another with integer-only quantization.

Listing 12.6 Full integer quantization while forming the TensorFlow Lite model

```
import os.path as osp
import numpy as np
import tensorflow as tf

DATA_DIR = "classification_data"

saved_model_dir='models/nn_classification_model_tf'

train_samples = np.load(osp.join(DATA_DIR, "cls_train_samples.npy
    "))

# Convert train_samples to float32 format
train_samples = train_samples.astype(np.float32)

def representative_dataset():
    for input_value in tf.data.Dataset.from_tensor_slices(
        train_samples).batch(1).take(100):
        yield [input_value]

# Full integer quantization
# Integer with float fallback (using default float input/output)
converter = tf.lite.TFLiteConverter.from_saved_model(
    saved_model_dir)
converter.optimizations = [tf.lite.Optimize.DEFAULT]
converter.representative_dataset = representative_dataset
tflite_model_quant = converter.convert()

# Save the model.
with open('models/quantized_model_int_w_float.tflite', 'wb') as f
    :
    f.write(tflite_model_quant)

# Full integer quantization
# Integer only
converter = tf.lite.TFLiteConverter.from_saved_model(
    saved_model_dir)
converter.optimizations = [tf.lite.Optimize.DEFAULT]
converter.representative_dataset = representative_dataset
# Ensure that if any ops can't be quantized, the converter throws
    an error
converter.target_spec.supported_ops = [tf.lite.OpsSet.
    TFLITE_BUILTINS_INT8]
# Set the input and output tensors to int8
converter.inference_input_type = tf.uint8   # or tf.int8
converter.inference_output_type = tf.uint8   # or tf.int8
tflite_model_quant = converter.convert()

# Save the model.
with open('models/quantized_model_full_int.tflite', 'wb') as f:
    f.write(tflite_model_quant)
```

In Listing 12.6, the training dataset is loaded as in previous examples. The function representative_dataset provides representative input data for quantization to both models. It batches and takes the first 100 samples from the training set and yields them one at a time. The first converted model uses full integer quantization with float fallback. This section configures the converter for full integer quantization with a fallback to float input/output types with default optimizations. The converter converts the model and quantized model is saved to the file "quantized_model_int_w_float.tflite." The second converted model uses full integer quantization with integer only weights, operations, and activations. Here, the converter is configured for full integer quantization without float fallback with default optimizations. `tf.lite.OpsSet.TFLITE_BUILTINS_INT8` specifies that only operations compatible with `int8` quantization are allowed. Input and output tensors are set to `tf.uint8`. The converter converts the model. Then, the quantized model is saved in the file "quantized_model_full_int.tflite." Both quantization methods aim to reduce the model size and potentially improve inference performance, especially on hardware that supports integer operations efficiently. The first version allows for float fallback if integer quantization does not work for certain operations. The second version enforces full integer quantization without fallback.

12.2.3 Quantization Aware Training

The quantization aware training approach is applied during training the model. Hence, the accuracy loss due to quantization can be minimized. Quantization-aware training simulates the effects of quantization in forward pass, while the backward propagation/gradient calculation is kept unchanged. This behavior provides observability of the quantization error which is accumulated in the model loss. Therefore, the optimizer reduces the quantization error while minimizing the loss.

We can apply quantization aware training by calling the function `quantize_model` under the `tfmot.quantization.keras` module before forming the TensorFlow Lite model. We provide a usage example in Listing 12.7. This example demonstrates the process which aims to preserve model accuracy while enabling quantization for model size reduction.

Listing 12.7 Quantization aware training while forming the TensorFlow Lite model

```
import os.path as osp
import numpy as np
import tensorflow as tf
import tensorflow_model_optimization as tfmot

DATA_DIR = "classification_data"

train_samples = np.load(osp.join(DATA_DIR, "cls_train_samples.npy
    "))
train_labels = np.load(osp.join(DATA_DIR, "cls_train_labels.npy")
    )
test_samples = np.load(osp.join(DATA_DIR, "cls_test_samples.npy")
    )
test_labels = np.load(osp.join(DATA_DIR, "cls_test_labels.npy"))

saved_model_dir='models/nn_classification_model_tf'

model = tf.keras.models.load_model(saved_model_dir)
model.summary()

# q_aware stands for for quantization aware.
```

```
q_aware_model = tfmot.quantization.keras.quantize_model(model)
# 'quantize_model' requires a recompile.
q_aware_model.compile(optimizer=tf.keras.optimizers.Adam(
    learning_rate=1e-3),
               loss=tf.keras.losses.BinaryCrossentropy(),
               metrics=[tf.keras.metrics.BinaryAccuracy(),
                     tf.keras.metrics.FalseNegatives()])

q_aware_model.summary()

train_samples_subset = train_samples[0:100] # out of 1000
train_labels_subset = train_labels[0:100]

q_aware_model.fit(train_samples_subset, train_labels_subset,
               batch_size=500, epochs=1, validation_split=0.1)
baseline_model_accuracy = model.evaluate(test_samples,
    test_labels, verbose=0)
q_aware_model_accuracy = q_aware_model.evaluate(test_samples,
    test_labels, verbose=0)
print('Baseline test accuracy:', baseline_model_accuracy)
print('Quant test accuracy:', q_aware_model_accuracy)

converter = tf.lite.TFLiteConverter.from_keras_model(
    q_aware_model)
converter.optimizations = [tf.lite.Optimize.DEFAULT]
tflite_model_quant = converter.convert()

# Save the model.
with open('models/quantized_model_w_QAT.tflite', 'wb') as f:
    f.write(tflite_model_quant)
```

In Listing 12.7, we begin by loading the MNIST dataset. Then, we load our pretrained model. The `tfmot.quantization.keras.quantize_model` method is used to create quantization aware model from the pretrained model. This prepares the model for quantization and weights are quantized during training. The quantization aware model is recompiled with the same loss, optimizer, and metrics as the original model. A subset of training data (first 1000 samples) is used for training the quantization aware model for one epoch. Both the original model `model` and the quantization aware model `q_aware_model` are evaluated on the test dataset to measure their accuracy. Next, the quantization aware model is converted to TensorFlow Lite format by calling the method `tf.lite.TFLiteConverter.from_keras_model`. The quantized TensorFlow Lite model is saved to a file. Finally, the accuracy of both original and quantization aware models on the test dataset is printed. The quantized model is then saved in the TensorFlow Lite format for deployment to the microcontroller.

12.3   Model Optimization via Pruning

Pruning is the second approach to be considered for model optimization in this book. We will start with explaining the general procedure applied in pruning. Then, we will explain how it can be implemented in TensorFlow. Finally, we will cover the joint usage of pruning and quantization in TensorFlow Lite.

12.3.1 What Is Pruning?

Pruning aims to eliminate weak connections between neurons in the model. To do so, weights in the model are inspected. The ones with magnitude less than a predefined threshold are set to zero. Hence, these connections are broken. As a result, only strong weights are kept in the model. If the weight values are stored in a matrix, then we will have its sparse version after pruning. If we can benefit from the sparsity property of the weight matrix in operation, then pruning has an effect on the model storage and inference steps. For more information on pruning, please see [56].

12.3.2 Pruning via TensorFlow

We can apply pruning on a formed TensorFlow model via available functions. We provide one such example in Listing 12.8. Here, we define pruning parameters besides data and model loading parts we introduced earlier. Therefore, the `pruning_params` dictionary is used to configure the pruning schedule for weight pruning. The value associated with the `pruning_schedule` key is an instance of the `tfmot.sparsity.keras.PolynomialDecay` class. This class is used to define a polynomial decay schedule for pruning. Provided parameters there are as follows: `initial_sparsity` specifies the initial sparsity level, which is the proportion of weights that are zero at the beginning of the pruning process. `final_sparsity` specifies the target sparsity level, which is the proportion of weights that the user wants to reach by the end of the pruning process (in our case it is 80% or 0.80). `begin_step` specifies the epoch at which the pruning process begins. `end_step` specifies the epoch at which the pruning process ends. The pruned model is created using the function `prune_low_magnitude` from the tensorflow_model_optimization library. We compile the model with the same loss and optimizer as the original one. Next, we set the callbacks for pruning including updating the pruning step and logging pruning summaries. The pruned model is trained for fine-tuning with two more epochs using the training data while applying pruning during training. The function `tfmot.sparsity.keras.strip_pruning` is used to obtain the original model with sparse weights. Finally, we save the pruned model to a file.

Listing 12.8 Pruning while forming the TensorFlow model

```
import os.path as osp
import numpy as np
import tensorflow as tf
import tensorflow_model_optimization as tfmot
import tempfile

DATA_DIR = "classification_data"
MODEL_DIR = "models"

train_samples = np.load(osp.join(DATA_DIR, "cls_train_samples.npy
    "))
train_labels = np.load(osp.join(DATA_DIR, "cls_train_labels.npy")
    )
test_samples = np.load(osp.join(DATA_DIR, "cls_test_samples.npy")
    )
test_labels = np.load(osp.join(DATA_DIR, "cls_test_labels.npy"))

saved_model_dir='models/nn_classification_model_tf'

model = tf.keras.models.load_model(saved_model_dir)
model.summary()

baseline_model_accuracy = model.evaluate(test_samples,
    test_labels, verbose=0)
prune_low_magnitude = tfmot.sparsity.keras.prune_low_magnitude

# Compute end step to finish pruning after 2 epochs.
batch_size = 128
epochs = 2
validation_split = 0.1 # 10% of training set will be used for
    validation set.

num_samples = train_samples.shape[0] * (1 - validation_split)
end_step = np.ceil(num_samples/batch_size).astype(np.int32) *
    epochs

pruning_params = {
'pruning_schedule': tfmot.sparsity.keras.PolynomialDecay(
initial_sparsity=0.50,
final_sparsity=0.80,
begin_step=0,
end_step=end_step)
}

model_for_pruning = prune_low_magnitude(model, **pruning_params)
model_for_pruning.compile(optimizer=tf.keras.optimizers.Adam(
    learning_rate=1e-3),
            loss=tf.keras.losses.BinaryCrossentropy(),
            metrics=[tf.keras.metrics.BinaryAccuracy(),
                    tf.keras.metrics.FalseNegatives()])
model_for_pruning.summary()

logdir = tempfile.mkdtemp()

callbacks = [
tfmot.sparsity.keras.UpdatePruningStep(),
tfmot.sparsity.keras.PruningSummaries(log_dir=logdir),
]

model_for_pruning.fit(train_samples,train_labels,epochs=2,
    validation_split=0.1, callbacks=callbacks)
```

```
model_for_pruning_accuracy = model_for_pruning.evaluate(
    test_samples, test_labels, verbose=0)

print('Baseline test accuracy:', baseline_model_accuracy)
print('Pruned test accuracy:', model_for_pruning_accuracy)

model_for_export = tfmot.sparsity.keras.strip_pruning(
    model_for_pruning)
pruned_keras_file = 'models/pruned_model.h5'
tf.keras.models.save_model(model_for_export, pruned_keras_file,
    include_optimizer=False)
```

12.3.3 Pruning and Quantization via TensorFlow Lite

We can apply quantization methods introduced in Sect. 12.2.2 to a pruned model. We provide one such example in Listing 12.9. Here, we convert the pruned model to TensorFlow Lite format using the function tf.lite.TFLiteConverter. from_keras_model. Then, the converted model is saved to a file. We applied dynamic range quantization using a converter by setting tf.lite.Optimize.DEFAULT. This helps in quantizing the model weights and activation functions while keeping some of them in floating-point format. The second model is saved to a file. A third model is created to obtain full integer quantization with pruning. As we did before, representative_dataset and tf.lite.Optimize.DEFAULT attributes are defined for this reason. Fully quantized model is saved to a file.

Listing 12.9 Quantization applied to the pruned TensorFlow model

```
import os.path as osp
import numpy as np
import tensorflow as tf

saved_model_dir='models/nn_classification_model_tf'

model = tf.keras.models.load_model(saved_model_dir)
model.summary()

pruned_model=tf.keras.models.load_model('models/pruned_model.h5')
pruned_model.summary()

converter = tf.lite.TFLiteConverter.from_keras_model(pruned_model
    )
pruned_tflite_model = converter.convert()

# Save the model
with open('models/pruned_model.tflite', 'wb') as f:
  f.write(pruned_tflite_model)

# Dynamic range quantization
converter = tf.lite.TFLiteConverter.from_keras_model(pruned_model
    )
converter.optimizations = [tf.lite.Optimize.DEFAULT]
tflite_model_quant = converter.convert()

# Save the model
```

```
with open('models/pruned_quant_model_dyn_range.tflite', 'wb') as
    f:
  f.write(tflite_model_quant)

DATA_DIR = "classification_data"

train_samples = np.load(osp.join(DATA_DIR, "cls_train_samples.npy
    "))

# Convert train_samples to float32 format
train_samples = train_samples.astype(np.float32)

def representative_dataset():
  for input_value in tf.data.Dataset.from_tensor_slices(
      train_samples).batch(1).take(100):
    # Model has only one input so each data point has one element
      .
    yield [input_value]

# Full integer quantization
# Integer with float fallback (using default float input/output)
converter = tf.lite.TFLiteConverter.from_keras_model(pruned_model
    )
converter.optimizations = [tf.lite.Optimize.DEFAULT]
converter.representative_dataset = representative_dataset
tflite_model_quant = converter.convert()

# Save the model.
with open('models/pruned_quant_model_int_w_float.tflite', 'wb')
    as f:
  f.write(tflite_model_quant)
```

12.4 Model Optimization via Weight Clustering

Weight clustering is the third approach to be considered for model optimization in this book. We will start with explaining the general procedure applied in it. Then, we will explain how weight clustering can be implemented in TensorFlow. Finally, we will cover the joint usage of weight clustering and quantization in TensorFlow Lite.

12.4.1 What Is Weight Clustering?

Weight clustering aims to group weight values in the neural network model. Clustering methods, such as the ones introduced in Chap. 8, are used in operation. After clustering, each group is represented by its centroid. Hence, weight values used in the model are standardized. This leads to smaller model size with a decrease in accuracy, since all of the weights lose their precision and unique clustered values are assigned to them. We will not go into details of the weight clustering method further. We refer the reader to the given reference for a detailed explanation of the method [57].

12.4.2 Weight Clustering via TensorFlow

We can apply weight clustering on a formed TensorFlow model via available functions. We provide one such example in Listing 12.10. Here, we begin with specifying the number of clusters, 16, and cluster centroids initialization method, linear in our case. In the next step, weights are clustered using the function tfmot .clustering.keras.cluster_weights which applies clustering to model weights while reducing the number of unique weight values. The clustered model is then compiled with the specified loss and optimizer. Fine-tuning is applied to the clustered model for one epoch using training data. Clustered model accuracy is evaluated against the test set. tfmot.clustering.keras.strip_clustering is utilized to recreate the original model with clustered weights, similar to pruning. Finally, the clustered model is saved to a file.

Listing 12.10 Weight clustering while forming the TensorFlow model

```python
import os.path as osp
import numpy as np
import tensorflow as tf
import tensorflow_model_optimization as tfmot

DATA_DIR = "classification_data"
MODEL_DIR = "models"

train_samples = np.load(osp.join(DATA_DIR, "cls_train_samples.npy
    "))
train_labels = np.load(osp.join(DATA_DIR, "cls_train_labels.npy")
    )
test_samples = np.load(osp.join(DATA_DIR, "cls_test_samples.npy")
    )
test_labels = np.load(osp.join(DATA_DIR, "cls_test_labels.npy"))

saved_model_dir='models/nn_classification_model_tf'

model = tf.keras.models.load_model(saved_model_dir)
model.summary()

baseline_model_accuracy = model.evaluate(test_samples,
    test_labels, verbose=0)

clustering_params = {
  'number_of_clusters': 16,
  'cluster_centroids_init': tfmot.clustering.keras.
      CentroidInitialization.LINEAR
}

# Cluster a whole model
clustered_model = tfmot.clustering.keras.cluster_weights(model,
    **clustering_params)

# Use smaller learning rate for fine-tuning clustered model
clustered_model.compile(optimizer=tf.keras.optimizers.Adam(
    learning_rate=1e-5),
                loss=tf.keras.losses.BinaryCrossentropy(),
                metrics=[tf.keras.metrics.BinaryAccuracy(),
                        tf.keras.metrics.FalseNegatives()])
```

```
clustered_model.summary()

clustered_model.fit(train_samples, train_labels, batch_size=500,
    epochs=1, validation_split=0.1)
clustered_model_accuracy = clustered_model.evaluate(test_samples,
    test_labels, verbose=0)

print('Baseline test accuracy:', baseline_model_accuracy)
print('Clustered test accuracy:', clustered_model_accuracy)

# Create 6x smaller models from clustering
model_for_export = tfmot.clustering.keras.strip_clustering(
    clustered_model)
clustered_keras_file = 'models/clustered_model.h5'
tf.keras.models.save_model(model_for_export, clustered_keras_file
    , include_optimizer=False)
```

12.4.3 Weight Clustering and Quantization via TensorFlow Lite

We can apply quantization methods introduced in Sect. 12.2.2 to a weight clustered
model. We provide one such usage example in Listing 12.11. Here, we create
three new models by quantizing the clustered model similar to pruning. First,
we convert the clustered model to TensorFlow Lite format using the function
`tf.lite.TFLiteConverter.from_keras_model`. This model is saved to a file. Next,
we apply dynamic range quantization by specifying the optimization as `tf.lite.`
`Optimize.DEFAULT` for the second model. This step quantizes the model weights and
activations, keeping some of them in floating-point format. The second model is
saved to a file. We define a representative dataset for full integer quantization for the
third model. Then, we apply full integer quantization by specifying the optimization
as `tf.lite.Optimize.DEFAULT` and setting the `representative_dataset` attribute to
the representative dataset function defined earlier. This step ensures that the model is
quantized using integer-only operations. Eventually, fully quantized model is saved
to a file.

Listing 12.11 Quantization applied to the weight clustered TensorFlow model

```
import os.path as osp
import numpy as np
import tensorflow as tf

saved_model_dir='models/nn_classification_model_tf'

model = tf.keras.models.load_model(saved_model_dir)
model.summary()

clustered_model=tf.keras.models.load_model('models/
    clustered_model.h5')
clustered_model.summary()

converter = tf.lite.TFLiteConverter.from_keras_model(
    clustered_model)
```

```
clustered_tflite_model = converter.convert()

# Save the model
with open('models/clustered_model.tflite', 'wb') as f:
  f.write(clustered_tflite_model)

# Clustering and Quantization

# Dynamic range quantization
converter = tf.lite.TFLiteConverter.from_keras_model(
    clustered_model)
converter.optimizations = [tf.lite.Optimize.DEFAULT]
tflite_model_quant = converter.convert()

# Save the model
with open('models/clustered_quant_model_dyn_range.tflite', 'wb')
    as f:
  f.write(tflite_model_quant)

DATA_DIR = "classification_data"

train_samples = np.load(osp.join(DATA_DIR, "cls_train_samples.npy
    "))

# Convert train_samples to float32 format
train_samples = train_samples.astype(np.float32)

def representative_dataset():
  for input_value in tf.data.Dataset.from_tensor_slices(
      train_samples).batch(1).take(100):
    # Model has only one input so each data point has one element

    yield [input_value]

# Full integer quantization
# Integer with float fallback (using default float input/output)
converter = tf.lite.TFLiteConverter.from_keras_model(
    clustered_model)
converter.optimizations = [tf.lite.Optimize.DEFAULT]
converter.representative_dataset = representative_dataset
tflite_model_quant = converter.convert()

# Save the model
with open('models/clustered_quant_model_int_w_float.tflite', 'wb'
    ) as f:
  f.write(tflite_model_quant)
```

12.5   Embedding the TensorFlow Lite Model to the Microcontroller

As we have the TensorFlow Lite model formed on PC, the next step is embedding it to the microcontroller. Hence, the model can be used for inference there. To do so, we will start with converting the TensorFlow Lite model to C array. Then, we will focus on methods for deploying the converted model to the microcontroller. Finally, we will test the deployed model on the microcontroller.

12.5.1 Converting the TensorFlow Lite Model to C Array

We should convert the TensorFlow Lite model to C array to embed it on the microcontroller. To note here, the TensorFlow Lite model is stored in a binary file. Therefore, we should convert its content to C array. This can be done in three ways. The first one is using the xxd utility in Linux/MacOS systems. The reader should run the command "xxd -i model.tflite > model_data.cc" in Linux/MacOS. Here, the "model.tflite" is the source file that keeps the TensorFlow Lite model. "model_data.cc" is the target file keeping the C array. The reader should form a suitable header file manually accompanying the "model_data.cc" file.

The second way for conversion is using the Python script "generate_cc_arrays.py," provided in the GitHub repository [58]. This script is platform independent. Therefore, it can be used in the Windows operating system as well as in Linux/MacOS. The reader should run the script as "python generate_cc_arrays.py outfile model.tflite" in the command prompt. Please note that the "model.tflite" should be in the same directory the script is running. The output header and source files are written to the "outfile" directory.

The third way for conversion is using the Python script "tflite2cc.py", provided in the book repository. This script is also platform independent. Therefore, it can be used in the Windows operating system as well as in Linux/MacOS. The reader should run the script as "python tflite2cc.py model.tflite > model_data.cc." The header file is also generated via this script. As in the xxd utility, the input and output file names should be fed to the script. If the path for input and output files are provided, then batch conversion can also be done.

We can also use the function convert_tflite2cc defined under "tflite2cc.py" within a Python script. We provide one such example in Listing 12.12. Here, the TensorFlow model is loaded first. Afterward, the model is converted to TensorFlow Lite format as in Listing 12.3. Then, we store the file in binary form. Afterward, we convert the model to C array using the function convert_tflite2cc(tflite_model_file, C_file). Here, tflite_model_file= ' models/nn_classification_model.tflite' is the source file. The target files are "classification.cc," "classification.cpp," and "classification.h." Please note that the "tflite2cc.py" file should be available in the working directory for this example to work.

Listing 12.12 Usage of the script for converting a TensorFlow Lite model to C array

```
import tensorflow as tf
from tflite2cc import convert_tflite2cc

# Convert a saved model
saved_model_dir='models/nn_classification_model_tf'
converter = tf.lite.TFLiteConverter.from_saved_model(
    saved_model_dir)
tflite_model = converter.convert()

tflite_model_file= 'models/nn_classification_model.tflite'
```

```
# Save the TF Lite model
with open(tflite_model_file, 'wb') as f:
  f.write(tflite_model)

C_filepath = "models/classification"

convert_tflite2cc(tflite_model, c_out= C_filepath)
```

As the Python script in Listing 12.12 is executed, it will generate the C file "classification.cc" (also "classification.cpp") and its header file "classification.h." This file contains a character array with hexadecimal entries. Length of the array in terms of bytes is also stored in the file. We provide the shortened version of the converted TensorFlow Lite model generated by the script in Listing 12.13.

Listing 12.13 Content of the converted file in shortened form

```
unsigned char converted_model_tflite[] = {0x1c,...};
unsigned int converted_model_tflite_len = 1980;
```

12.5.2 TensorFlow Lite Micro Library

TensorFlow Lite Micro library is the limited version of TensorFlow Lite. It is specifically introduced to work on microcontrollers. It can be obtained from the repository https://www.tensorflow.org/lite/microcontrollers. We should include the TensorFlow Lite Micro library to our project before deploying the converted TensorFlow Lite model to the microcontroller. We provide the library file tree containing the necessary files under the GitHub repository of this book. This library can also be obtained by following the steps given in the "new_platform_support.md" file under the path "tflite-micro/tensorflow/lite/micro/docs/" of the "tflite-micro" repository https://github.com/tensorflow/tflite-micro/tree/main.

The TensorFlow Lite Micro library consists of four folders as "examples," "signal," "tensorflow," and "third_party." TensorFlow Lite model examples can be found under the folder "examples." The remaining three folders contain source codes we should include to our STM32CubeIDE and Mbed Studio projects. We will explain these operations in Sect. 12.5.3.

We formed the "lib_model.h" and "lib_model.cpp" files to simplify the TensorFlow Lite Micro library initialization and usage. Therefore, we should first explain them before exploring how the TensorFlow Lite Micro library is imported to a project. The mentioned files contain two functions as LIB_MODEL_Init and LIB_MODEL_Run.

The function LIB_MODEL_Init requires four parameters to work. The first parameter is the address of the C array which keeps the TensorFlow Lite model data. We should define a pointer to the TfLiteTensor structure in the main code and provide it to the function as the second parameter. As the function is executed, the pointer to the TfLiteTensor structure will be initialized. Hence, it will show us the memory address of input. We should also define a C++ array of sufficient size in the main code for storing intermediate values, while the model is executed. The array and its size has to be provided as the last two parameters. We should define a pointer to the TfLiteTensor structure in the main code to store output of the model. This pointer will point to the structure which keeps data related to the model output after executing the function LIB_MODEL_Run. These two functions return 0 when successfully executed. We provide detailed descriptions of these functions in Listing 12.14.

Listing 12.14 TensorFlow Lite Micro library functions

```
int8_t LIB_MODEL_Init(const void *tfliteModel, TfLiteTensor **
    inputTensor, uint8_t *buffer, uint32_t bufferSize)
/*
Initializes the TensorFlow Lite model and library
tfliteModel: TensorFlow Lite model
inputTensor: Pointer to input TfLiteTensor structure
buffer: Pointer to buffer
bufferSize: Size of buffer
*/

int8_t LIB_MODEL_Run(TfLiteTensor **output)
/*
Runs inference
output: Pointer to output TfLiteTensor structure
*/
```

12.5.3 Deploying the Converted Model to the Microcontroller

As we convert the TensorFlow Lite model to C array, we can deploy it to the microcontroller. We handle this topic in this section. Therefore, we start with necessary settings for the STM32CubeIDE and Mbed Studio projects.

12.5.3.1 STM32CubeIDE Project Settings for Deployment

To deploy the converted TensorFlow Lite model to the microcontroller, we should first create a new C++ project under STM32CubeIDE. Then, we should copy the TensorFlow Lite Micro library folders "signal," "tensorflow," and "third_party" under the project folder.

We should add the path of copied folders and their symbol definitions to the project settings to compile the TensorFlow Lite Micro library. To do so, right click the project and select "Properties." Select the "Includes" tabs under "C/C++ General − > Paths and Symbols." Add the following include paths with "Add to all languages" checkbox checked as "<project_root>," "<project_root>/third_party/flatbuffers/include," "<project_root>/third_party/gemmlowp," "<project_root>/third_party/kissfft," and "<project_root>/third_party/ruy." Next, add the "TF_LITE_STATIC_MEMORY" symbol with the "Add to all languages" checkbox checked under the "Symbols" tab. Finally, add the folders "<project_root>/tensorflow" and "<project_root>/models" by clicking "Add Folder" under the "Source Location" tab. The project will be configured to compile the TensorFlow Lite Micro library after clicking the "Apply and Close" button.

Besides the TensorFlow Lite Micro library files, we should copy the files "lib_model.h" and "lib_model.cpp" to the "Core/Inc" and "Core/Src" folders of the STM32CubeIDE project, respectively. Then, we should copy the files "classification.h" and "classification.cc" generated in previous section to the "Core/Inc" and "Core/Src" folders under the STM32CubeIDE project. As a reminder, these file contain model parameters in C array format.

Since the TensorFlow Lite Micro library is written in C++ language, we should compile our project with a C++ compiler. To do so, we should rename the "main.c" file in the project folder as "main.cpp." Please note that when a modification is needed in STM32CubeMX after this step, it will generate a new "main.c" file. Therefore, we should copy our code to the new "main.c" file. Then, we should delete the old "main.cpp" file and rename the new "main.c" file as "main.cpp." These operations should be done after every modification in STM32CubeMX.

12.5.3.2 Mbed Studio Project Settings for Deployment

To deploy the converted TensorFlow Lite model to the microcontroller, we should first create a new blinky baremetal project under Mbed Studio. Then, we should copy the TensorFlow Lite Micro library folders "signal", "tensorflow", and "third_party" to the project folder. Afterward, we should the modify the "mbed_app.json" file to add the "TF_LITE_STATIC_MEMORY" macro definition. The modified file content will be as in Listing 12.15.

Listing 12.15 mbed_app.json file configuration

```
{
    "macros" : ["TF_LITE_STATIC_MEMORY"],
    "requires": ["bare-metal"],
    "target_overrides": {
      "*": {
        "target.c_lib": "small",
        "target.printf_lib": "minimal-printf",
        "platform.minimal-printf-enable-floating-point": false,
        "platform.stdio-minimal-console-only": true
      }
    }
}
```

Besides the TensorFlow Lite Micro library files, we should copy the files "lib_model.h" and "lib_model.cpp" to the project folder under Mbed Studio. Then, we should copy the files "classification.h" and "classification.cc" to the Mbed Studio project folder. As a reminder, these file contain model parameters in C array format.

12.5.4 Testing the Deployed Model on the Microcontroller

To check whether the deployed TensorFlow Lite model (set as classifier) works as expected, we can use the generated data on PC. Then, we can send them to the STM32 microcontroller. The classifier running on the microcontroller works on this data. Then, we can transfer the classification result back to PC and cross check with the same model working in Python on PC. Hence, we can make sure that the generated framework works as expected.

12.5.4.1 STM32CubeIDE Project Settings for Testing

At this point, we assume that the steps in Sect. 12.5.3.1 are already done. We also assume that the libraries lib_rng and lib_serial are added to the project as explained in Sects. 5.1.3.1 and 4.1.2.2, respectively. Please note that the RNG and UART peripheral units should be activated using the STM32CubeMX interface as explained in relevant sections. To use all these included files and libraries, we should include the corresponding code block to the "main.cpp" file in our project.

```
/* USER CODE BEGIN Includes */
#include "classification.h"
#include "lib_serial.h"
#include "lib_rng.h"
#include "lib_model.h"
/* USER CODE END Includes */
```

We should define macros related to the source of the input data, size of the input vector, and size of the output vector. The INPUT macro definition selects whether input data is obtained from PC or microcontroller at compile time. It can be selected as INPUT_PC or INPUT_MCU, respectively. SIZE_INPUT and SIZE_OUTPUT macros define the number of data inputs and outputs of the model, respectively. We should set them manually according to the converted TensorFlow Lite model input and output parameter size. To be more specific, these values should be the same as in the model generation step in Python on PC. In our example, we set SIZE_INPUT to be equal to two since we have a two-dimensional feature space. Likewise, we set SIZE_OUTPUT since the model output is one dimensional. Then, we should allocate a buffer on RAM for the TensorFlow Lite Micro library usage. The buffer size is defined by the expression kTensorArenaSize. In case of any error, the buffer size should be increased. Finally, we should define two TfLiteTensor pointers to handle input and output values. We will have the following settings for the example to be considered in this chapter.

```
/* USER CODE BEGIN 0 */
#define INPUT_PC 1
#define INPUT_MCU 2
#define INPUT INPUT_PC

#define SIZE_INPUT 2
#define SIZE_OUTPUT 1
```

```
constexpr int kTensorArenaSize = 136 * 1024;
alignas(16) static uint8_t tensor_arena[kTensorArenaSize];
TfLiteTensor* input = nullptr;
TfLiteTensor* output = nullptr;
/* USER CODE END 0 */
```

We provide an example on how to initialize and run the model generated in previous section in Listing 12.16. Here, we first initialize the TensorFlow Lite Micro library using the function LIB_MODEL_Init providing the required parameters. Next, we form a while loop to acquire data either from PC or microcontroller according to the INPUT macro. Then, we run the TensorFlow Lite model calling the function LIB_MODEL_Run. The only parameter of this function is a double pointer to the output tensor with data type TfLiteTensor. Finally, we transfer the classification result to PC. Please note that the Python script in Listing 12.17 should already be running on PC for executing this example successfully.

Listing 12.16 Code block for running the TensorFlow Lite model in STM32CubeIDE

```
/* USER CODE BEGIN 2 */
LIB_MODEL_Init(converted_model_tflite, &input, tensor_arena,
    kTensorArenaSize);
/* USER CODE END 2 */

/* USER CODE BEGIN WHILE */
while (1)
{
/* USER CODE END WHILE */

/* USER CODE BEGIN 3 */
#if (INPUT == INPUT_PC)
 LIB_SERIAL_Receive(input->data.f, SIZE_INPUT, TYPE_F32);
#elif (INPUT == INPUT_MCU)
 for (uint32_t i = 0; i < SIZE_INPUT; ++i)
 input->data.f[i] = (float)(LIB_RNG_GetRandomNumber() % 1000) /
    1000.0f;
LIB_SERIAL_Transmit(input->data.f, SIZE_INPUT, TYPE_F32);
#endif
LIB_MODEL_Run(&output);
LIB_SERIAL_Transmit(output->data.f, SIZE_OUTPUT, TYPE_F32);
HAL_Delay(1000);
 }
 /* USER CODE END 3 */
```

We should pick the right data type for model input. TensorFlow Lite Micro library makes use of unions to feed any data type as input to the model. In our example, we use the address defined in union input->data.f for this purpose. We pick f since the model is quantized in 32-bit floating point representation. The same situation also appears while reading the model output from output tensor. Hence, the output value is obtained by choosing the right output data type. In our example, we pick the address output->data.f to transfer the output value to PC as 32-bit floating point number. If the model was quantized in 8-bit signed integer format, then we should have used the `input->data.int8` and `output->data.int8` unions for input and output operations, respectively. Since the input and output would be in 8-bit signed integer type, we should have scaled the input and output of the model. The scale factors are stored in the `TfLiteTensor` structure for this purpose. An example for output scaling is `(output->data.int8[0]` - `output->params.zero_point)*` output `->params.scale`.

To check whether the deployed TensorFlow Lite model works as expected, we need a test setup. We provide the Python script formed for this purpose in Listing 12.17. Here, we first load the previously generated data. Then, we load our trained model. Afterward, we send the test data to the microcontroller. Next, we perform classification on the PC and microcontroller separately. We finally print both classification results on PC for comparison. Hence, we can make sure that the generated framework works as expected.

Listing 12.17 Python script for comparing TensorFlow Lite model inference results

```python
import os.path as osp
import numpy as np
import tensorflow as tf
import py_serial

py_serial.SERIAL_Init("COM4")

DATA_DIR = "classification_data"
MODEL_DIR = "models"

test_samples = np.load(osp.join(DATA_DIR, "cls_test_samples.npy")
    )
test_labels = np.load(osp.join(DATA_DIR, "cls_test_labels.npy"))

model = tf.keras.models.load_model(osp.join(MODEL_DIR, "
    nn_classification_model_keras.h5"))

i = 0
while 1:
    rqType, datalength, dataType = py_serial.
        SERIAL_PollForRequest()
    if rqType == py_serial.MCU_WRITES:
        # INPUT -> FROM MCU TO PC
        inputs = np.reshape(py_serial.SERIAL_Read(), (1,
            datalength))

    elif rqType == py_serial.MCU_READS:
        # INPUT -> FROM PC TO MCU
        inputs = test_samples[i:i+1].astype(py_serial.
            SERIAL_GetDType(dataType))
        i = i + 1
        if i >= len(test_samples):
            i = 0
        py_serial.SERIAL_Write(inputs)

    pcout = model.predict(inputs)
    rqType, datalength, dataType = py_serial.
        SERIAL_PollForRequest()
    if rqType == py_serial.MCU_WRITES:
        mcuout = py_serial.SERIAL_Read()
        print()
```

```
print("Inputs : " + str(inputs))
print("PC Output : " + str(pcout))
print("MCU Output : " + str(mcuout))
print()
```

12.5.4.2 Mbed Studio Project Settings for Testing

At this point, we assume that the steps in Sect. 12.5.3.2 are already done. We also assume that the libraries lib_rng and lib_serial are added to the Mbed Studio project. To do so, we should copy the files "lib_rng.h," "lib_rng.c," "lib_uart.h," "lib_uart.c," "lib_serial.h," and "lib_serial.c" to our Mbed Studio project folder.

We provide an example on how to initialize and run the model generated in previous section under Mbed Studio in Listing 12.18. This code is similar to the one given for the STM32CubeIDE project. Hence, the explanations there hold here as well. The only difference is that we should include the header file "lib_uart.h" and initialize the UART peripheral unit before the while loop by code in our Mbed Studio project. Please note that the Python script in Listing 12.17 should already be running on PC for running this example successfully.

Listing 12.18 TensorFlow Lite model test code for Mbed Studio

```
#include "mbed.h"
#include "classification.h"
#include "lib_uart.h"
#include "lib_serial.h"
#include "lib_rng.h"
#include "lib_model.h"

#define INPUT_PC 1
#define INPUT_MCU 2
#define INPUT INPUT_PC

#define SIZE_INPUT 2
#define SIZE_OUTPUT 1

constexpr int kTensorArenaSize = 136 * 1024;
alignas(16) static uint8_t tensor_arena[kTensorArenaSize];
TfLiteTensor* input = nullptr;
TfLiteTensor* output = nullptr;

int main()
{
LIB_MODEL_Init(converted_model_tflite, &input, tensor_arena,
    kTensorArenaSize);
LIB_UART_Init();
#if (INPUT == INPUT_MCU)
 LIB_RNG_Init();
#endif
while (true)
{
#if (INPUT == INPUT_PC)
 LIB_SERIAL_Receive(input->data.f, SIZE_INPUT, TYPE_F32);
#elif (INPUT == INPUT_MCU)
 for (uint32_t i = 0; i < SIZE_INPUT; ++i)
 input->data.f[i] = (float)(LIB_RNG_GetRandomNumber() % 1000) /
    1000.0f;
```

```
LIB_SERIAL_Transmit(input->data.f, SIZE_INPUT, TYPE_F32);
#endif
LIB_MODEL_Run(&output);
LIB_SERIAL_Transmit(output->data.f, SIZE_OUTPUT, TYPE_F32);
HAL_Delay(1000);
}
}
```

12.6 Embedding the Model to the Microcontroller via STM32Cube.AI

STMicroelectronics has a dedicated platform called STM32Cube.AI to be used by neural networks applications on their microcontrollers. This platform handles most of the operations introduced in previous sections. Hence, the user can feed the original TensorFlow model to STM32Cube.AI and obtain the final code/firmware ready to be embedded on the STM32 microcontroller. There are two versions of STM32Cube.AI as of writing this book. The first one is the add-on to be applied to STM32CubeMX called X-CUBE-AI. The second one is the web based stand-alone version called STM32Cube.AI Developer Cloud. We will consider the latter version in this book due to its ease of usage.

12.6.1 Welcome Screen and Loading the Model

STM32Cube.AI Developer Cloud can be reached from https://stm32ai-cs.st.com/home. The user should have a valid STMicroelectronics account to use it. As the user signs in to the web site, the welcome screen will be as in Fig. 12.1. Here, the user has two options. Either a custom model should be chosen from PC, or an available model from the STM32 model zoo can be chosen. If the custom model option is to be picked, then the model should be of type Keras, ONNX, or TensorFlow Lite with the extension ".h5," ".hdf5," ".keras," ".onnx," or ".tflite." If a model is to be chosen from the STM32 model zoo, then it is suitable for the STM32Cube.AI platform by default.

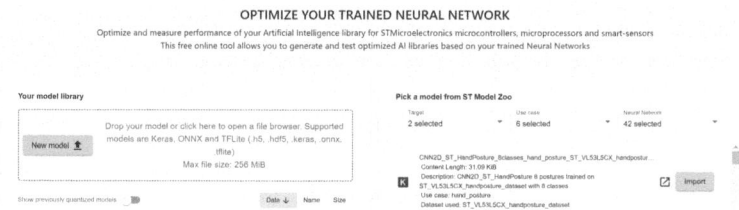

Fig. 12.1 STM32Cube.AI Developer Cloud, welcome screen

Fig. 12.2 STM32Cube.AI Developer Cloud, the quantization screen

Assume that we pick the previously formed TensorFlow model, with the file name "nn_classification_model_keras.h5," introduced in Sect. 11.4. As we drag and drop the model to the designated place on the web site, it will be shown under the "Your model library" section. Now, we can start processing this model. We can form the Neutron graph of the model by clicking on the black triangle by the model name. If the graph is OK, then we can start a new project based on our model by clicking on the "Start" button.

12.6.2 Quantization

As we click on the "Start" button after loading the model, the next screen opens up to select the "ST Edge AI Core Version" and device family to be used. We should pick the most recent version for "ST Edge AI Core Version" and "STM32 MCUs" for the device family to be used. As we do so, a new window opens as in Fig. 12.2. This screen is dedicated for the quantization operation.

As can be seen in Fig. 12.2, the selected model is summarized at the top of the screen. Its input type is 2 (32 bits). Its output type is 1 (32 bits). Its model type is STAI_Format_Float for our case. Here, post-training quantization using TensorFlow Lite converter is selected by default. Therefore, the methods introduced in Sect. 12.2.2 are applied here. The input and output data types can be selected among int8, uint8, or float32 in operation. We select int8 for the input and output for our model considered in this chapter. As we press the "Launch quantization" button in Fig. 12.2, quantization is done on the model. The result can be downloaded as the file "nn_classification_model_keras_PerChannel_quant_float32_float32_random_1.tflite."

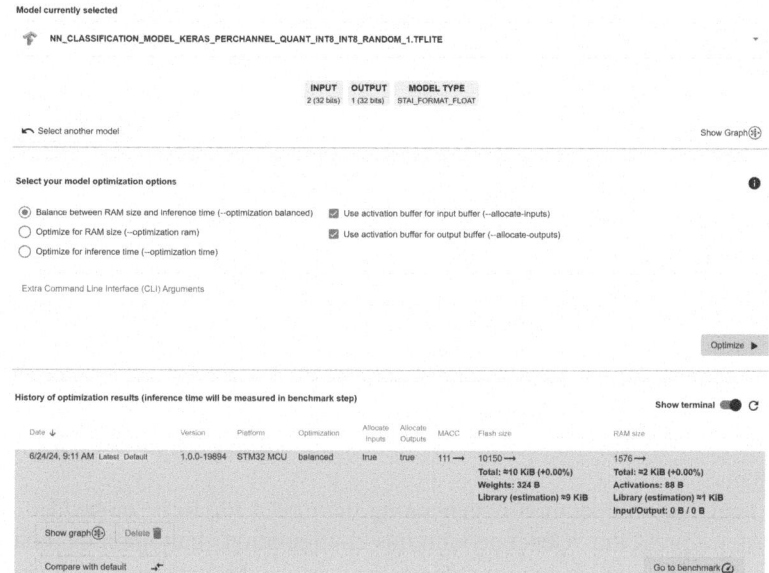

Fig. 12.3 STM32Cube.AI Developer Cloud, the quantization screen

12.6.3 Optimization Options

As the quantization is done, a new tab summarizes the quantized model. There is also a button titled "Select" next to it. As we press this button, a new window opens for optimization as in Fig. 12.3; there are also optimization options. These are "Balance between RAM size and inference time (–optimization balanced)," "Optimize for RAM size (–optimization ram)," or "Optimize for inference time (–optimization time)." The user can also select "Use activation buffer for input buffer (–allocate-inputs)" and "Use activation buffer for output buffer (–allocate-outputs)" by clicking on the corresponding boxes.

Assume that we pick the "Balance between RAM size and inference time (–optimization balanced)" with selecting "use activation buffer for input buffer (–allocate-inputs)" and "use activation buffer for output buffer (–allocate-outputs)." We can press the "Optimize" button to generate the optimized model. As the optimization ends, the result is displayed under the "History of optimization results" section in the same page as in Fig. 12.3. Then, we can press the "Go to benchmark" button.

Fig. 12.4 STM32Cube.AI Developer Cloud, benchmarking screen

12.6.4 Benchmarking and Generating the Model File for Deployment

As we press the "Go to benchmark" button in the "History of optimization results" section of the selected model, the benchmarking screen shows up. Here, the user should select the target development board. Since we are using "STM32F746G-DISCO" in this book, we can select it and start benchmarking. The result will be as in Fig. 12.4. As can be seen in this figure, the inference time for our model is 0.01358 ms.

The final step in operation is generating the model file to be embedded on the board. As we press the "Generate with this configuration" button on the bottom of the screen, a new window opens up. Here, the user should first select the board type. Afterward, the generated model can be downloaded as "C Code," "STM32CubeMX IOC file," "STM32CubeIDE Project," or "Firmware."

12.6.5 Deploying the Generated Model to the Microcontroller

As we generate the model file under STM32Cube.AI, we can deploy it to the microcontroller. We handle this topic in this section. Therefore, we start with necessary settings for the STM32CubeIDE and Mbed Studio project. Before going further, we should pay attention to some important issues. First, the STM32CubeIDE version should be recent. Second, the X-CUBE-AI package should have been added to STM32CubeMX. Third, if the reader is using the Windows operating system, then its long path property should be enabled.

12.6.5.1 STM32CubeIDE Project Settings for Deployment

We should select the download option as "C Code" in the previous section. Then, a zip file is generated with the name "nn_classification_model_keras.h5-STM32F746G-DISCO-code.zip." We will deploy the generated model to the microcontroller next. To do so, we should first create a new STM32CubeIDE project. Afterward, we should create a new folder under our project and name it as "XCUBEAI." We unzip the downloaded file and copy all the files and folders to this folder.

We should add these files to our project. To do so, right click the project and select "Properties." Select the "Includes" tabs under "C/C++ General -> Paths and Symbols." Add the following include paths "/$ProjName/XCUBEAI" and "/$ProjName/XCUBEAI/Inc." Next, select the "Libraries" tab and add ":NetworkRuntime900_CM7_GCC.a." Then, select the "Library Path" tab and add "/$ProjName/XCUBEAI/Lib." Finally, select the "Source Location" tab and add the created XCUBEAI folder there. The project will be configured to compile the X-CUBE-AI library after clicking the "Apply and Close" button.

We should include our C libraries to our project to test our deployed model as we did in previous sections. Therefore, we should add the libraries lib_rng and lib_serial to the project and initialize the UART and RNG peripheral units as explained in Sects. 5.1.3.1 and 4.1.2.2, respectively. Besides these library files, the files "lib_cube.h" and "lib_cube.c" should be copied under the "/Core/Inc" and "/Core/Src" folders of the STM32CubeIDE project, respectively.

12.6.5.2 Mbed Studio Project Settings for Deployment

We select the download option as "Download C Code" in the previous section for model deployment via Mbed Studio. Then, a zip file is generated with the name "nn_classification_model_keras.h5-STM32F746G-DISCO-code.zip." Next, we should create a new project under Mbed Studio. Please note that we should select the option "Store Mbed OS in the program folder (+1GB)." This is necessary since we will modify the compilation script of the project. After creating the project, we should create a new folder named "X-CUBE-AI." Then, we should unzip the downloaded zip file and copy all the files and folders under this folder. We should rename the file "NetworkRuntime900_CM7_GCC.a" under the "Lib" folder as "libNetworkRuntime900_CM7_GCC.a."

We should change the Mbed Studio compilation script since the X-CUBE-AI package uses the hard floating-point hardware. To do so, the reader should find the file "gcc.py" under the path "mbed-os/tools/toolchains/." We should change the code line `self.cpu.append("-mfloat-abi=softfp")` to `self.cpu.append("-mfloat-abi=hard")`. Since we are working on Arm Cortex-M7 microcontroller, we should take care of the if condition for different cores. We provide the Python script for this setup in Listing 12.19.

Listing 12.19 Mbed Studio setup script

```
elif core == "Cortex-M7F" or core.startswith("Cortex-M33F"):
    self.cpu.append("-mfpu=fpv5-sp-d16")
    self.cpu.append("-mfloat-abi=hard")
```

12.6.6 The Supplementary Library for X-CUBE-AI Usage

We formed a library to simplify the X-CUBE-AI library initialization and usage. To use our library, the model should be named as "network" within the "X-CUBE-AI Mode and Configuration" window of the STM32CubeMX interface. Our support library consists of two files as "lib_cube.h" and "lib_cube.c." These files contain two functions as LIB_CUBE_Init and LIB_CUBE_Run. The function LIB_CUBE_Init can be used to initialize the X-CUBE-AI library. The LIB_CUBE_Run function can be used to perform inference. This function requires two parameters as pointers to input and output arrays. We provide detailed description of these functions in Listing 12.20. These two functions return zero when successfully executed.

Listing 12.20 The supplementary library for X-CUBE-AI functions

```
int8_t LIB_CUBE_Init(void)
/*
Initializes the X-CUBE-AI library.
Note: Network name must be given as "network" within the
X-CUBE-AI Mode and Configuration window of STM32CubeMX interface.
*/

int8_t LIB_CUBE_Run(void * input, void * output)
/*
Runs the X-CUBE-AI model.
input: Pointer to the input data buffer
output: Pointer to the output data buffer
*/
```

12.6.7 Testing the Deployed Model on the Microcontroller

To check whether the deployed X-CUBE-AI model works as expected, we can use the test setup that we used for the TensorFlow Lite models in Sect. 12.5.4. We will again use the generated data on PC. Then, we can send them to the STM32 microcontroller. The model running on the microcontroller works on this data. Then, we can transfer the classification result back to PC and cross check with the same classifier working in Python on PC. Hence, we can make sure that the generated framework works as expected.

12.6.7.1 STM32CubeIDE Project Settings for Testing

At this point, we assume that the steps in Sect. 12.5.3.1 are already done. We also assume that the libraries lib_rng and lib_serial are added to the project as explained in Sects. 5.1.3.1 and 4.1.2.2, respectively. Please note that the RNG and UART peripheral units should be activated using the STM32CubeMX interface as explained in these sections. To use all these included files and libraries, we should include the corresponding code block to the "main.c" file in our project.

```
/* USER CODE BEGIN Includes */
#include "network.h"
#include "lib_cube.h"
#include "lib_rng.h"
#include "lib_serial.h"
/* USER CODE END Includes */
```

We should define macros related to the source of the input data, size of the input vector, and size of the output vector. The INPUT macro definition selects whether input data is obtained from PC or microcontroller at compile time. It can be selected as INPUT_PC or INPUT_MCU, respectively. SIZE_INPUT and SIZE_OUTPUT macros define the number of data inputs and outputs of the model, respectively. We should set them manually according to the converted TensorFlow Lite model input and output parameter size. Or, we can get the macro values from the X-CUBE-AI "network.h" file as shown below. The buffer allocation on RAM is done by the "lib_cube" files. We should also define two floating-point arrays to allocate required memory space for input and output of our classification model. We will have the following settings for the example considered in this chapter.

```
/* USER CODE BEGIN 0 */
#define INPUT_PC 1
#define INPUT_MCU 2
#define INPUT     INPUT_PC

#define SIZE_INPUT AI_NETWORK_IN_1_SIZE
#define SIZE_OUTPUT AI_NETWORK_OUT_1_SIZE

float input[SIZE_INPUT];
float output[SIZE_OUTPUT];
/* USER CODE END 0 */
```

We provide an example on how to initialize and run the model generated in previous section in Listing 12.21. Here, we first initialize the X-CUBE-AI library using the function LIB_CUBE_Init. Next, we form a while loop to acquire data either from PC or microcontroller according to the INPUT macro. Then, we run the X-CUBE-AI model by calling the function LIB_CUBE_Run. The first and the second parameters of this function are pointers to input and output arrays, respectively. Finally, we transfer the classification result to PC. Please note that the Python script in Listing 12.17 should already be running on PC for executing this example successfully.

Listing 12.21 Code block for running X-CUBE-AI model in STM32CubeIDE

```
/* USER CODE BEGIN 2 */
LIB_CUBE_Init();
/* USER CODE END 2 */

/* USER CODE BEGIN WHILE */
while (1)
{
/* USER CODE END WHILE */

/* USER CODE BEGIN 3 */
#if (INPUT == INPUT_PC)
  LIB_SERIAL_Receive(input, SIZE_INPUT, TYPE_F32);
#elif (INPUT == INPUT_MCU)
  for (uint32_t i = 0; i < SIZE_INPUT; ++i)
    input[i] = (float)(LIB_RNG_GetRandomNumber() % 1000) / 1000.0f;
LIB_SERIAL_Transmit(input, SIZE_INPUT, TYPE_F32);
```

```
#endif
LIB_CUBE_Run(input, output);
LIB_SERIAL_Transmit(output, SIZE_OUTPUT, TYPE_F32);
HAL_Delay(1000);
}
/* USER CODE END 3 */
```

12.6.7.2 Mbed Studio Project Settings for Testing

At this point, we assume that the steps in Sect. 12.5.3.2 are already done. We also assume that the libraries lib_rng and lib_serial are added to the Mbed Studio project. To do so, we copied the files "lib_rng.h," "lib_rng.c," "lib_uart.h," "lib_uart.c," "lib_serial.h," and "lib_serial.c" to our Mbed Studio project folder. Please note that we should change the default toolchain for Mbed Studio to GCC as explained in Sect. A.3.

We provide an example on how to initialize and run the model generated in previous section under Mbed Studio in Listing 12.22. This code is similar to the one given for the STM32CubeIDE project. Hence, the explanations there hold here as well. The only difference is that we should include the header file "lib_uart.h" and initialize the UART peripheral unit before the while loop by code in our Mbed Studio project. Please note that the Python script in Listing 12.17 should already be running on PC for running this example successfully.

Listing 12.22 X-CUBE-AI model test code for Mbed Studio

```
#include "mbed.h"
#include "network.h"
#include "lib_cube.h"
#include "lib_rng.h"
#include "lib_serial.h"
#include "lib_uart.h"

#define INPUT_PC 1
#define INPUT_MCU 2
#define INPUT INPUT_PC

#define SIZE_INPUT AI_NETWORK_IN_1_SIZE
#define SIZE_OUTPUT AI_NETWORK_OUT_1_SIZE

float input[SIZE_INPUT];
float output[SIZE_OUTPUT];

int main()
{
LIB_UART_Init();
LIB_CUBE_Init();
while (true)
{
#if (INPUT == INPUT_PC)
 LIB_SERIAL_Receive(input, SIZE_INPUT, TYPE_F32);
#elif (INPUT == INPUT_MCU)
 for (uint32_t i = 0; i < SIZE_INPUT; ++i)
 input[i] = (float)(LIB_RNG_GetRandomNumber() % 1000) / 1000.0f;
LIB_SERIAL_Transmit(input, SIZE_INPUT, TYPE_F32);
```

```
#endif
LIB_CUBE_Run(input, output);
LIB_SERIAL_Transmit(output, SIZE_OUTPUT, TYPE_F32);
HAL_Delay(1000);
}
}
```

12.7   Application: Human Activity Recognition via Accelerometer Data

We formed the fully connected neural network model for human activity recognition from accelerometer data in Sect. 11.6. After successfully training and exporting the model, we should convert it to C array to run it on the microcontroller. Therefore, we follow the steps introduced in this chapter for this purpose. As we have the C array, we can form the STM32CubeIDE or Mbed Studio project to classify human activity patterns via neural networks on the microcontroller. To do so, we should apply the necessary steps for data acquisition and scaling.

We will provide complete projects for STM32CubeIDE and Mbed Studio in the accompanying book repository. There, we will provide different settings in the implementation step. Hence, the reader will observe different options in implementation. While doing so, we will realize the complete human activity recognition system on the microcontroller from the initial accelerometer data reading to final inference step.

12.8   Application: Keyword Spotting from Audio Signals

We formed the fully connected neural network model for keyword spotting from audio signals in Sect. 11.7. After successfully training and exporting the model, we should convert it to C array to run it on the microcontroller. Therefore, we follow the steps introduced in this chapter for this purpose. As we have the C array, we can form the STM32CubeIDE or Mbed Studio project to spot keywords via neural networks on the microcontroller. To do so, we should apply the necessary steps for data acquisition and scaling.

We will provide complete projects for STM32CubeIDE and Mbed Studio in the accompanying book repository. There, we will provide different settings in the implementation step. Hence, the reader will observe different options in implementation. While doing so, we will realize the complete keyword spotting system on the microcontroller from the initial audio signal acquisition to final inference step.

12.9   Application: Handwritten Digit Recognition from Digital Images

We formed the fully connected neural network model for handwritten digit recognition from digital images in Sect. 11.8. After successfully training and exporting the model, we should convert it to C array to run it on the microcontroller. Therefore, we follow the steps introduced in this chapter for this purpose. As we have the C array, we can form the STM32CubeIDE or Mbed Studio project to recognize handwritten digits on the microcontroller. To do so, we should apply the necessary steps for data acquisition and scaling.

We will provide complete projects for STM32CubeIDE and Mbed Studio in the accompanying book repository. There, we will provide different settings in the implementation step. Hence, the reader will observe different options in implementation. While doing so, we will realize the complete handwritten digit recognition system on the microcontroller from the initial audio signal acquisition to final inference step.

12.10   Application: Estimating Future Temperature Values

We formed the fully connected neural network model for estimating future temperature values in Sect. 11.9. After successfully training and exporting the model, we should convert it to C array to run it on the microcontroller. Therefore, we follow the steps introduced in this chapter for this purpose. As we have the C array, we can form the STM32CubeIDE or Mbed Studio project to estimate future temperature values on the microcontroller. To do so, we should apply the necessary steps for data acquisition and scaling.

We will provide complete projects for STM32CubeIDE and Mbed Studio in the accompanying book repository. There, we will provide different settings in the implementation step. Hence, the reader will observe different options in implementation. While doing so, we will realize the complete temperature value estimation system on the microcontroller from the initial temperature value acquisition to final estimation step.

12.11   Summary of the Chapter

We focused on embedding the trained neural network models to the microcontroller in this chapter. To do so, we benefit from TensorFlow Lite as the specialized version of TensorFlow for embedded systems (including microcontrollers). Hence, we started with its properties. Then, we showed ways of converting TensorFlow and Keras models to TensorFlow Lite format. Model conversion is not sufficient alone to embed the model on a microcontroller. The main reason is the size of TensorFlow Lite model. In other words, the model should be optimized beforehand since microcontrollers have limited flash and RAM size. Therefore, we covered model optimization via quantization, pruning, and weight clustering. Operations performed up to this point were done on PC. The next step was embedding the final TensorFlow Lite model to the microcontroller. Hence, we considered necessary steps to be followed for this purpose. Embedding the trained neural network model to the microcontroller can also be done by the STM32Cube.AI platform. We covered it to provide a second way of embedding a TensorFlow or Keras model to the STM32 microcontroller. Throughout the chapter, we reconsidered the neural network models introduced in Chap. 11 and embed them to the microcontroller. We also applied the same procedure to the end of chapter applications.

Convolutional Neural Networks

13

13.1 The Convolution Operation

As the name implies, CNN models are based on the convolution operation.
Therefore, we will start with the mathematical definition of convolution operation
in this section. Then, we will consider implementing convolution by a single neuron
and sliding window approach. This will help us to form the link between neural
networks introduced in the previous chapters and CNN models. Finally, we will
introduce the convolution definition under Keras. Hence, we will be ready to
implement CNN models in the following section.

13.1.1 Mathematical Definition

The convolution operation can be defined on one-, two-, or three-dimensional
signals. Since our focus is on CNN models for images, we will only explore the
two-dimensional convolution operation in this book. We can define the convolution
operation on an image $I[m, n]$ and filter $F[m, n]$, both being two-dimensional
matrices, as

$$O[m, n] = \sum_k \sum_l I[m - k, n - l] F[k, l] \qquad (13.1)$$

The convolution operation gets two inputs, $I[m, n]$ and $F[m, n]$, and generates
output $O[m, n]$. Here, the filter $F[m, n]$ decides on output of the convolution
operation. In other words, we can rephrase the convolution operation as filtering the
input image by a predefined filter. Output of the operation will be a new image. Here,
filter coefficients should be selected such that we can obtain the desired output from
the convolution operation. We should mention two important and related issues

© The Author(s), under exclusive license to Springer Nature Switzerland AG 2025   319
C. Ünsalan et al., *Embedded Machine Learning with Microcontrollers*,
https://doi.org/10.1007/978-3-031-70912-8_13

before going further. First, filter coefficients will be learnt in CNN. Second, actually the correlation operation is used in CNN with the formula

$$O[m, n] = \sum_k \sum_l I[m + k, n + l]F[k, l] \tag{13.2}$$

The reader can observe that the only difference between Eqs. 13.1 and 13.2 is that one has the term $m - k, n - l$; the other has $m + k, n + l$. This means that the convolution operation requires the input image to be rotated during operation. However, since filter coefficients will be learnt during training in CNN, using the formalism in Eq. 13.2 will not cause any problem. Only location of the learnt filter coefficients will change. The representation in Eq. 13.2 is taken as the de facto form in neural networks literature. Therefore, we will use it from this point on to represent the convolution operation.

Let's explain the convolution operation by an example. Assume that we have an image $I[m, n]$ with size $M \times N$. Assume further that our filter $F[m, n]$ has size 2×2. Let the entries of these matrices be

$$I[m, n] = \begin{bmatrix} i_{0,0} & i_{0,1} & \cdots & i_{0,N-1} \\ \vdots & & & \vdots \\ i_{M-1,0} & i_{M-1,1} & \cdots & i_{M-1,N-1} \end{bmatrix} \tag{13.3}$$

and

$$F[m, n] = \begin{bmatrix} w_{0,0} & w_{0,1} \\ w_{1,0} & w_{1,1} \end{bmatrix} \tag{13.4}$$

We can apply the convolution operation to obtain its output as

$$O[m, n] = \sum_{k=0}^{1} \sum_{l=0}^{1} I[m + k, n + l]F[k, l] \tag{13.5}$$

Assume that we would like to obtain $O[1, 1]$. Then, we should calculate

$$O[1, 1] = \sum_{k=0}^{1} \sum_{l=0}^{1} I[1 + k, 1 + l]F[k, l] \tag{13.6}$$

Hence, we can write

$$O[1, 1] = i_{1,1}w_{0,0} + i_{1,2}w_{0,1} + i_{2,1}w_{1,0} + i_{2,2}w_{1,1} \tag{13.7}$$

Filter coefficients are set to emphasize or suppress some components of the input in signal processing. Therefore, the convolution operation requires these coefficients to be carefully set. This corresponds to filter design in signal processing. As an example, if we want to detect vertical edges in the image $I[m, n]$, then we should form a filter accordingly. One such option is using Sobel filters, $F[m, n]$ and $F^T[m, n]$, with coefficients

$$F[m, n] = \begin{bmatrix} 1 & 0 & -1 \\ 2 & 0 & -2 \\ 1 & 0 & -1 \end{bmatrix} \tag{13.8}$$

We will observe the output for a given test image via Sobel filters in Sect. 13.2.1. The reader will clearly see the effect of filtering there.

Although we considered filters with size 2×2 and 3×3 till now, the filter size can take any value in the convolution operation. The important point in selecting the filter size is that, it should serve its purpose. In other words, the filter should provide what we need to obtain at the output of convolution operation.

13.1.2 Implementing the Convolution Operation by a Single Neuron and Sliding Window Approach

As we check Eq. 13.7, we can observe that the convolution operation, with the filter size 2×2 in the given example, can be realized by a single neuron. The actual implementation done this way is as in Fig. 13.1. In this figure, $b = 0$ and $\varphi(\cdot)$ is the linear activation function. This figure clearly indicates that the convolution operation can be realized by a neuron. Here, weights of the neuron correspond to parameters of the applied filter.

We should emphasize that Eq. 13.7 is formed to calculate one output value, $O[1, 1]$, in Fig. 13.1. We should apply the convolution operation to all inputs such that we can form the $O[m, n]$ matrix as output of the convolution operation. This is done by feeding subparts of the input image to the single neuron successively and obtaining corresponding output values. This operation is called sliding window in signal processing literature. The mechanism of sliding window is as follows. A subwindow is taken from the original image. It is processed and the corresponding

Fig. 13.1 The convolution operation represented by a single neuron

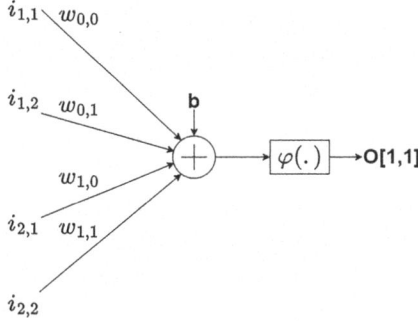

output is obtained. Next, the subwindow is shifted and the same steps are applied again. This sliding operation is continued till the overall image is covered. To emphasize again, only inputs of the neuron change by the sliding window approach in Fig. 13.1. Hence, the neuron weights (or filter coefficients) are kept fixed in the operation.

13.1.3 Convolution Definition Under Keras

Since CNN is popular in image classification applications, Keras has a dedicated convolution layer. We will explain it in detail in Sect. 13.2.1. Beforehand, we should explain the convolution definition in this layer. Based on CNN operations, the convolution should be applicable to multiple images (or tensors), in batch form, at once. Hence, the first entry of the input tensor for the convolution under Keras indicates the batch number. The convolution should process images with single channel (such as grayscale) or multichannel (such as RGB). It should also process tensors with any dimension. Moreover, there may be more than one filter in the convolution operation. Therefore, the formula used to represent the convolution operation under Keras is

$$
O[a, m, n, p] = \varphi \left(\sum_k \sum_l \sum_f I[a, m+k, n+l, q] \times F[k, l, q, p] + b \right)
$$

$$(13.9)$$

where a is the batch number. To note here, images are fed to the convolution operation in batch form. m and n values are the image matrix indices as in Eq. 13.5. p is the number of filters. As explained previously, more than one filter can be used in the convolution operation. To note here, these filters are used separately. Hence, the output tensor $O[a, m, n, p]$ will have p filtered results. q is the input channel number. Hence, filtering can be done on multiple input channels at once. Finally, b is the bias term.

Keras allows the user to select a nonlinear activation function in the convolution operation. Hence, $\varphi(\cdot)$ in Eq. 13.9 can be taken as any valid activation function available under Keras. This means that the convolution operation implemented under Keras is not exactly the same as in Sect. 13.1.1. The reader should keep these differences in mind while forming CNN models to be introduced next.

13.2   Forming a CNN Model Under Keras

We will focus on CNN model formation under Keras in this section. Therefore, we will introduce specific CNN layers as convolution, pooling, and flatten. Afterward, we will form feature extraction block of the CNN model using mentioned layers. The CNN model also has a classification block within itself. We will form it by the fully connected neural network structure introduced in Sect. 11.2. Then, we will show how to merge the feature extraction and classification blocks to reach the final CNN model.

13.2.1 The Convolution Layer

The convolution layer under Keras is defined as `tf.keras.layers.Conv2D`. We provide the parameters for this layer next.

```
tf.keras.layers.Conv2D(
    filters,
    kernel_size,
    strides=(s_x, s_y),
    padding='valid',
    activation=None,
    use_bias=True,
    kernel_initializer='glorot_uniform',
    bias_initializer='zeros',
    kernel_regularizer=None,
    bias_regularizer=None,
    activity_regularizer=None,
)
```

Now, we are in a position to explain parameters of the convolution layer. `filters` represents the number of convolution filters `num_filters`. In other words, it is the depth of convolution layer. `kernel_size` sets the size of each convolution layer. If it is set as a single number, then convolution filters become square. Alternatively, one can give a specific size with (`kernel_width`,`kernel_height`). Size of 2D convolution filters are formed as (`kernel_width`, `kernel_height`, `num_filters`) up to this point. The parameter `stride` sets the spatial shift for the convolution layer which means that the sliding window slides s_x and s_y units in x and y directions, respectively.

The image/tensor boundaries cannot be calculated as they have no neighbors around them in the convolution operation. Therefore, we either add zeros to image boundaries to keep the input shape at output or take the input as it is and formed output will have smaller size compared to input. The `padding` parameter can be set to determine this behavior. It can be set as either `same` or `valid`. `same` keeps the original input shape by adding zeros to image boundaries. `valid` produces shrunk output. The `activation` parameter is the activation function to be applied to output of the convolution layer. It can be the activation function such as `relu` or `softmax`. Alternatively, it can be any custom function or one of the functions under the `keras.activations` module. `kernel_initializer` and `bias_initializer` are the functions to initialize weights and bias values, respectively. `kernel_initializer` sets weights according to the Glorot/Xavier uniform distribution. `bias_initializer` sets all bias values to zero as default (if they are not set). The remaining parameters are regularizers, namely, `kernel_regularizer`, `bias_regularizer`, and `activity_regularizer`. They can be set to any regularizer function. Hence, trained parameters do not overfit to the problem. For more information on other parameters, please see [59].

Fig. 13.2 Grayscale image
for the Sobel filter example

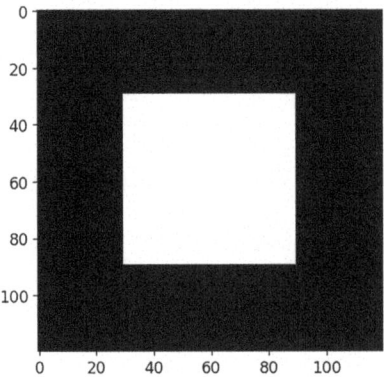

We provide an example on the usage of the `tf.keras.layers.Conv2D` layer on a grayscale image given in Fig. 13.2. We pick Sobel filters $F[m, n]$ and $F^T[m, n]$ given in Eq. 13.8. Hence, our aim is horizontal and vertical edge detection via convolution. We provide the complete script to obtain the Sobel filter output from the `tf.keras.layers.Conv2D` layer in Listing 13.1.

Listing 13.1 Sobel filtering example under Keras

```
import tensorflow as tf
from matplotlib import pyplot as plt
import numpy as np
from PIL import Image

im = Image.open('square1.jpg').convert('L')

plt.figure()
plt.imshow(im,cmap='gray')

# normalize to the range 0-1
im = np.asarray(im)
im=im/255.

img = np.expand_dims(im, 0)

sobel_x=np.array([[1, 0, -1],
                  [2, 0, -2],
                  [1, 0, -1]],dtype='float32')

sobel_y = sobel_x.T

filters=np.zeros([3,3,1,2])

filters[:,:,0,0]=sobel_x
filters[:,:,0,1]=sobel_y

init_kernel = tf.keras.initializers.constant(filters)

init_bias = np.zeros((2,))
init_bias = tf.keras.initializers.constant(init_bias)

conv_layer = tf.keras.layers.Conv2D(2,
```

```
                                            kernel_size=(3, 3),
                                            activation=None,
                                            kernel_initializer=
                                                init_kernel,
                                            bias_initializer=init_bias,
                                            padding='same',
                                            strides=[1, 1])
input_shape = (480, 640, 1)
model = tf.keras.Sequential([
        tf.keras.Input(shape=input_shape),
        conv_layer,
        ])

model.summary()

activations = model.predict(img)
print(activations.shape)

layer = model.get_layer(name="conv2d")
weights, biases = layer.get_weights()

fig, (filter_ax1, filter_ax2) = plt.subplots(1, 2)
ax = filter_ax1.imshow(weights[:, :, 0, 0], cmap = 'gray')
ax=filter_ax2.imshow(weights[:, :, 0, 1], cmap = 'gray')

fig, (sobel_ax1, sobel_ax2) = plt.subplots(1, 2)
ax=sobel_ax1.imshow(activations[0, :, :, 0], cmap = 'gray')
ax=sobel_ax2.imshow(activations[0, :, :, 1], cmap = 'gray')

plt.show()
```

In Listing 13.1, we have a white square surrounded by black pixels. We begin by loading the image using Python imaging library (PIL). By default, PIL reads the image pixels as eight-bit unsigned integer. We then display the loaded image using the Matplotlib library. Then, we convert the image to the NumPy array to perform mathematical operations on it. We also normalize it to [0, 1] range. The function np.expand_dims is used to add a new dimension to the image to make it compatible with the TensorFlow NHWC format where N is the batch, H is the height, W is the width, and C is the number of channels. Thus, the final dimension of the image becomes (1, 480, 640, 1). We continue with creating the 3×3 Sobel filters for x and y dimensions, respectively, based on Eq. 13.8. As they are both 3×3 matrices and have one channel, we create another NumPy array to stack filters together. Hence, filters variable contains sobel_x and sobel_y. A kernel initializer object is created where filters is passed to it as a parameter. Then, it is used to initialize our 2D convolution kernel. Similar to the kernel initialization, we initialize the bias terms with zeros only. Size of the bias initialization is two as we have two filters in this case. We then create the Conv2D layer using Keras by passing the number of filters, kernel shape, kernel initializer, and bias initializer, respectively. Afterward, this layer must be included to a Keras model to infer results using them. Thus, we create a Sequential model consisting of Input layer with given image shape and Conv2D layer that we have created earlier. model.summary() will print details of each layer. We can observe our model response to an input by calling the function model.predict. activations keeps the result of the applied Sobel filters on the image. To verify output shape, we

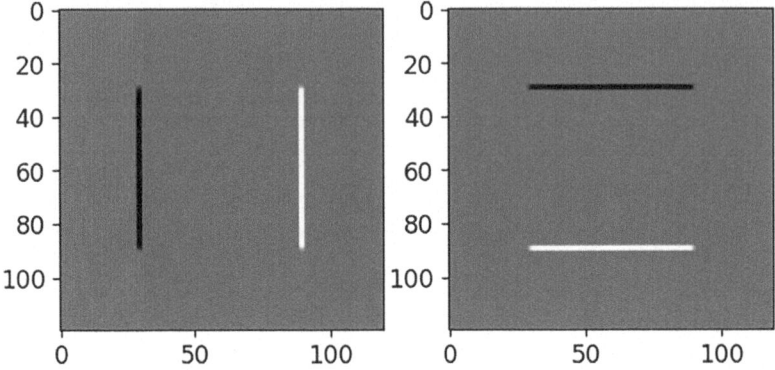

Fig. 13.3 Output of the convolution operation for the Sobel filter example

can call **print**(activation.shape). It will give us $(1, 480, 640, 2)$ in this case. Please note that height and width of the input does not change in this example since the padding = 'same' parameter is passed to the Conv2D layer. Besides, we can obtain the layers from the model using model.get_layer("conv2d"). Similarly, we can obtain weights and bias terms from that layer using layer.get_weights.

As we execute the Python script in Listing 13.1, we obtain the filter outputs as in Fig. 13.3. The first image shows output of sobel_x. The second image shows output of sobel_y. As can be seen in this figure, only horizontal edges have nonzero values for the sobel_x filter. The reason behind it is that convolution with a Sobel filter subtracts the multiplied value of right-hand neighbor pixels from the multiplied value of left-hand neighbor pixels. Therefore, output pixels become zero if there is no change horizontally. Similarly, the sobel_y filter only has response when there is pixel value change vertically. As a result, we only see horizontal and vertical edges of the square, respectively, after applying Sobel filters. Please note that darker regions in the figure represent transition from 1 to 0 (white to black) and lighter regions represent transition from 0 to 1 (black to white).

13.2.2 The Pooling Layer

The usual practice in CNN models is as follows. The obtained output tensor is downsampled in spatial domain after each convolution layer. There are two reasons for this operation. First, tensors obtained in successive layers will represent more general characteristics of the image this way. Second, the final tensor will be used as the feature vector by flattening it, as will be explained in the next section. Hence, the feature vector size obtained from the image should be decreased by downsampling.

Built-in functions to form the pooling layer under Keras can be separated into two groups as global and nonglobal. Nonglobal pooling built-in layers are MaxPooling2D and AveragePooling2D. Their working principle is similar to Conv2D. We provide the parameters used in these functions next.

```
pool_size = (pool_width, pool_height)
strides = (s_x, s_y)
padding = 'valid'
```

Let's explain the pooling layer operation based on its built-in functions. In max pooling, the maximum value in the pool sized window is taken and assigned to output. In average pooling, average value in the pool sized window is taken and assigned to output. The pooling operation is applied to the overall input image by the sliding window approach. pool_size defines size of the pooling kernel. In other words, it is the shape of the sliding window that produces either maximum or average pixel value. strides and padding parameters are the same as in the Conv2D layer parameters. strides sets the spatial shift for the sliding window s_x and s_y units in x and y directions, respectively. The padding parameter can be set as either same or valid, where the same keeps the original input shape by adding zeros to the boundaries and valid produces a shrunk output.

Global built-in pooling layers are GlobalMaxPooling2D and GlobalAveragePooling2D. They do not operate in sliding window form. Therefore, there is no pool_size, strides, and padding parameters in them. Instead, they perform maximum and averaging operations for height- and width-wise. Let's say we have a tensor with shape (1, 480, 640, 3). Then, output of the global pooling layer will have shape of (1,3). In other words, we get the max/average value of each channel. Optionally, we can define keepdims = True as parameter. Then, the output shape becomes (1,1,1,3).

13.2.3 The Flatten Layer

As explained in previous sections, each convolution layer outputs a tensor. The pooling layer downsamples this tensor in its spatial dimensions. However, the tensor structure still remains. The final layer of the convolution and pooling operations should be fed to a classifier, to be introduced in Sect. 13.2.5. Therefore, the tensor should be converted to vector form. The flatten layer, with Keras representation tf.keras.layers.Flatten, performs this operation. This layer has the optional parameter data_format which can be channels_first or channels_last as in the convolution layer. Output of the flatten layer is a one-dimensional vector.

Let's explain the flatten layer operation on an example. Assume that the final output obtained from the convolution and pooling layers is a tensor with size $10 \times 5 \times 5$. Hence, we have 10 filter outputs, each having spatial size 5×5. The flatten layer converts each filter output to vector form. Then, these vectors are concatenated. Hence, the flatten layer output will be a one-dimensional vector with 250 elements.

13.2.4 Forming the Feature Extraction Block

A CNN model is composed of two blocks as feature extraction and classification. Unlike the traditional methods introduced in Chap. 5, feature extraction is done by the convolution and pooling layers. Each layer may have different number of filters and parameters such as filter size, stride, and padding. The first layer should be formed to get input tensors. In Keras, there are two ways to create a neural network structure as `keras.Squential` and `keras.Model`. The latter is also called as the functional model. The former approach is used to create models where layers are linearly stacked. The `keras.Model` class is more flexible and supports multi-input/multi-output models. In the context of sequential model, input layer can be omitted by defining the `input_shape` parameter to the initial layer. We should define `InputLayer` or `Input` classes while creating the `Model` class. As the model is created, trainable layers including the convolution layer weight and bias terms are set by training. As output, the feature extraction block provides the features as tensors in specified shape. We will explore how to do this in Sect. 13.3.1. We will also benefit from the available feature extraction block of CNN models via transfer learning. We will cover this option in Sect. 13.4.

We provide an example on forming the feature extraction block next. To do so, we construct a CNN model for handwritten digit recognition. In Listing 13.2, we begin by loading the MNIST dataset and then form training and test images. Afterward, input images are normalized to the interval [0,1] by dividing the values to 255. As the original training dataset has shape (60,000, 28, 28), where 60,000 is the total number of images, we should add one more dimension to represent the number of channels. Applying `np.expand_dims(train_images, axis= -1)` makes the input image shape (60,000, 28, 28, 1). The function `tf.keras.utils.to_categorical` transforms labels into one-hot encoded format. In other words, it transforms labels to an array whose values are 1 for only corresponding index and 0 otherwise. For example, "Label 1" is transformed to [0, 1, 0, 0, 0, 0, 0, 0, 0, 0]. Likewise, "Label 2" is transformed to [0, 0, 1, 0, 0, 0, 0, 0, 0, 0]. Next, we build our model for feature extraction using the `Sequential` class. This model consists of an input layer, followed by a $3 \times 3 \times 32$ 2D convolution layer with the ReLU activation function, 2×2 max pooling layer, another 2D convolution layer with shape $3 \times 3 \times 64$ again with the ReLU activation function. This is followed by a 2×2 max pooling layer and final flatten layer. Therefore, we built our feature extraction block using 2D convolution, max pooling, and flatten layers. Output of the block has shape 1600, which means that we have a 1600 element feature vector.

Listing 13.2 Feature extraction block of the CNN model for handwritten digit recognition

```
import tensorflow as tf
from matplotlib import pyplot as plt
import numpy as np

# Model / data parameters
num_classes = 10
input_shape = (28, 28, 1)
```

```
# the data, split between train and test sets
(train_images, train_labels), (test_images, test_labels)  = tf.
    keras.datasets.mnist.load_data()

train_images = train_images / 255.0
test_images = test_images / 255.0

# Make sure images have shape (28, 28, 1)
train_images = np.expand_dims(train_images, -1)
test_images = np.expand_dims(test_images, -1)

print("train images shape:", train_images.shape)
print(train_images.shape[0], "train samples")
print(test_images.shape[0], "test samples")

# convert class vectors to binary class matrices
train_labels = tf.keras.utils.to_categorical(train_labels,
    num_classes)
test_labels = tf.keras.utils.to_categorical(test_labels,
    num_classes)

model = tf.keras.Sequential([
        tf.keras.Input(shape=input_shape),
        tf.keras.layers.Conv2D(32, kernel_size=(3, 3), activation
            ="relu"),
        tf.keras.layers.MaxPooling2D(pool_size=(2, 2)),
        tf.keras.layers.Conv2D(64, kernel_size=(3, 3), activation
            ="relu"),
        tf.keras.layers.MaxPooling2D(pool_size=(2, 2)),
        tf.keras.layers.Flatten(),
        ])

model.summary()
```

13.2.5 Forming the Classification Block

As explained previously, CNN models consist of classification block besides feature extraction. This block can be formed by appending a fully connected neural network to the feature extraction block. Here, the fully connected neural network is used as a classifier. To do so, the number of inputs to the fully connected neural network should be the same as length of the flatten layer. The number of outputs should be same as the class number. We can benefit from the available setup given in Sect. 11.2 for this purpose.

We can continue our handwritten digit recognition example by forming the classification block by the fully connected neural network layers. We provide them next.

```
tf.keras.layers.Dense(128, activation='relu'),
tf.keras.layers.Dense(num_classes, activation='softmax'),
```

Here, the classification block consists of two fully connected layers. The first layer has 128 neurons with the ReLU activation function. The second layer has 10 (number of classes) neurons with the softmax activation function. Therefore, the final layer produces a one-hot encoded vector as classifier output.

13.2.6 Merging the Feature Extraction and Classification Blocks

As we have the feature extraction and classification blocks, we can merge them to form the complete CNN model. We provide the merged CNN model formed for this way next.

```
model = tf.keras.Sequential([
        tf.keras.Input(shape=input_shape),
        tf.keras.layers.Conv2D(32, kernel_size=(3, 3), activation
            ="relu"),
        tf.keras.layers.MaxPooling2D(pool_size=(2, 2)),
        tf.keras.layers.Conv2D(64, kernel_size=(3, 3), activation
            ="relu"),
        tf.keras.layers.MaxPooling2D(pool_size=(2, 2)),
        tf.keras.layers.Flatten(),
        tf.keras.layers.Dense(128, activation='relu'),
        tf.keras.layers.Dense(num_classes, activation="softmax"),
        ])
model.summary()
```

As can be seen in this code snippet, we represent the feature extraction and classification layers under the same Sequential model. Therefore, these layers are sequentially connected. Flatten layer follows the convolution and pooling layers which transforms the previous tensor into one dimension. Dense layer takes the one-dimensional input and produces 128 sized vector. Final Dense layer provides output as same number as number of classes. Final output gives probabilities of each class, since softmax is used as the activation layer.

13.3   Training and Testing the CNN Model

As we form the complete CNN model by merging the feature extraction and classification blocks, the next step is its training. We will explain how to do this next. Afterward, we will pick a convolution layer and observe what the model has learnt. In other words, we will check the formed filters in the convolution layer after training. Finally, we will test the trained CNN model with different inputs.

13.3.1 Training the CNN Model

We can use the methods introduced in Sect. 11.3 to train the formed CNN model. Here, parameters in the feature extraction and classification blocks are set based on training. Hence, the training data set will be composed of input images to the model and classification labels expected as output of the model. We next train the complete CNN model formed for handwritten digit recognition in Sect. 13.2.6 this way. We provide the Keras functions used for this purpose next.

```
model.compile(loss='categorical_crossentropy', optimizer='adam',
    metrics=['accuracy'])
history=model.fit(train_images, train_labels, batch_size=128,
    epochs=5, validation_split=0.1)
```

Here, we should first compile the model and specify which loss function, optimizer, and metrics to be used while training the model. We pick the categorical cross entropy loss function since output of the model is a one-hot encoded vector and we want to minimize the error. We pick Adam as the optimizer since it is robust against local minima and adaptively updates its weights. We define the parameter `metrics = 'accuracy'` to get accuracy report of the model at the end of each epoch. The function `model.fit` takes `train_images` and `train_labels` as input and calculates the loss between output of the model and `train_labels`. Training progress scan all the images five times as we set `epochs = 5`. `batch_size = 128` means that 128 images are processed simultaneously. `validation_split=0.1` separates 10% of the training images as validation data. To note here, validation data is not used for training. Validation metrics are used to observe whether model overfits to the training data or not.

13.3.2 Visualizing the Trained Filters

We may gain insight on the working principles of the trained CNN model by checking filters (kernels) in a selected convolution layer. In other words, we can observe how the filters look like in a specific convolution layer after training. We consider this option next. To do so, we pick the trained CNN model in the previous section. We observe its filter coefficients (kernels) in one convolution layer. We benefit from the Python code for this purpose in Listing 13.3.

Listing 13.3 Visualizing the filters in the trained CNN model for handwritten digit recognition

```
import tensorflow as tf
from matplotlib import pyplot as plt
import numpy as np

model = tf.keras.models.load_model('saved_model/my_model',
    compile=False)

model.summary()

# the data, split between train and test sets
```

```
(train_images, train_labels), (test_images, test_labels)  = tf.
    keras.datasets.mnist.load_data()

train_images = train_images / 255.0
test_images = test_images / 255.0

# Make sure images have shape (28, 28, 1)
train_images = np.expand_dims(train_images, -1)
test_images = np.expand_dims(test_images, -1)

test_ind=0
im = test_images[test_ind]

plt.figure()
plt.imshow(im, cmap = "gray")

img = np.expand_dims(im, 0)
layer = model.get_layer(name="conv2d")
feature_extractor = tf.keras.Model(inputs=model.inputs, outputs=
    layer.output)
activations = feature_extractor.predict(img)
print(activations.shape)

filters, biases = layer.get_weights()

fig, ((conv_ax1, conv_ax2), (conv_ax3, conv_ax4)) = plt.subplots
    (2, 2)
ax = conv_ax1.imshow(filters[:, :, 0, 0], cmap = "gray")
ax = conv_ax2.imshow(filters[:, :, 0, 1], cmap = "gray")
ax = conv_ax3.imshow(filters[:, :, 0, 2], cmap = "gray")
ax = conv_ax4.imshow(filters[:, :, 0, 3], cmap = "gray")

fig, ((act_ax1, act_ax2), (act_ax3, act_ax4)) = plt.subplots(2,
    2)
ax = act_ax1.imshow(activations[0, :, :, 0], cmap = "gray")
ax = act_ax2.imshow(activations[0, :, :, 1], cmap = "gray")
ax = act_ax3.imshow(activations[0, :, :, 2], cmap = "gray")
ax = act_ax4.imshow(activations[0, :, :, 3], cmap = "gray")

layer = model.get_layer(name="conv2d_1")
feature_extractor = tf.keras.Model(inputs=model.inputs, outputs=
    layer.output)
activations = feature_extractor.predict(img)
print(activations.shape)

filters, biases = layer.get_weights()

fig, ((conv_ax1, conv_ax2), (conv_ax3, conv_ax4)) = plt.subplots
    (2, 2)
ax = conv_ax1.imshow(filters[:, :, 0, 0], cmap = "gray")
ax = conv_ax2.imshow(filters[:, :, 0, 1], cmap = "gray")
ax = conv_ax3.imshow(filters[:, :, 0, 2], cmap = "gray")
ax = conv_ax4.imshow(filters[:, :, 0, 3], cmap = "gray")

fig, ((act_ax1, act_ax2), (act_ax3, act_ax4)) = plt.subplots(2,
    2)
ax = act_ax1.imshow(activations[0, :, :, 0], cmap = "gray")
ax = act_ax2.imshow(activations[0, :, :, 1], cmap = "gray")
ax = act_ax3.imshow(activations[0, :, :, 2], cmap = "gray")
ax = act_ax4.imshow(activations[0, :, :, 3], cmap = "gray")

plt.show()
```

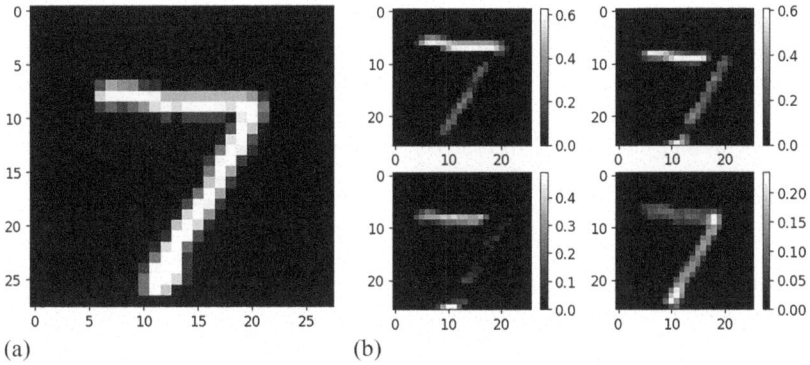

Fig. 13.4 CNN visualization example. (**a**) Input image. (**b**) Filter outputs from the first layer

In Listing 13.3, we first load the MNIST dataset. We then pick a test image with digit "7" in it. The original image is displayed using Matplotlib library. We can specify layer name to reach a layer and get its details in Keras. To do so, we use `model.get_layer(name="conv2d_1")` which gets the layer in the specified name parameter. We can also get output of this layer by defining `feature_extractor = tf.keras.Model(inputs=model.inputs, outputs=layer.output)`. In other words, we can cut the model from any layer and form another model up to that layer. By using `feature_extractor.predict(img)`, the formed feature extractor part can be used to produce activation output, which is the result of the network. Since the chosen layer is a 2D convolution layer, it has weights and bias terms which can be reached using the function `get_weights`. To note here, this is a four-channel convolution layer with 3×3 kernels. Hence, all the weights can be displayed using the function `imshow` from the Matplotlib library. The same operation can also be applied to another layer. We should follow the same procedure by only changing the name of the layer. We produce the feature extractor output as `activations` for layers. We then display them using the Matplotlib library functions.

As we execute the Python script in Listing 13.3, we obtain the filter outputs as in Fig. 13.4. As can be seen in this figure, the image we picked for testing is displayed on the left hand side. Output of the feature extractor module (activation layer) is displayed on the right hand side. Please note that specific edges of the digit have higher values, which means the feature extractor module has learnt to separate edges.

13.3.3 Testing the CNN Model

We can test the trained CNN model by feeding test images to it and observing the classification results produced by it. We provide a sample Python script for this purpose in Listing 13.4. Here, we pick an arbitrary test image after loading the MNIST dataset. Then, we add the batch dimension to match TensorFlow input dimensions. The function `model.predict(img)` predicts output of the model for the given image. We obtain the class which has maximum probability among all classes using the function `np.argmax`. Finally, we compare the actual and predicted values.

Listing 13.4 Testing the trained CNN model for handwritten digit recognition

```
test_ind=10

im = test_images[test_ind]

# Add the image to a batch where it's the only member.
img = np.expand_dims(im, 0)

# First way of predicting the label

predictions = model.predict(img)

print("Predictions")
print(predictions)

predicted_label = np.argmax(predictions)
actual_label = np.argmax(test_labels[test_ind])

plt.figure()
plt.imshow(im, cmap=plt.cm.binary)
plt.title('actual label= ' + str(actual_label) + ', predicted
    label = '+ str(predicted_label))

plt.show()
```

13.4   Transfer Learning for CNN

We can benefit from feature extraction block of an existing CNN model for our own problem. To do so, we should apply transfer learning. Hence, we will first explain the rationale behind this method. Then, we will show how to implement transfer learning under Keras.

13.4.1 What Is Transfer Learning?

Training time may be long for a CNN model due to the total number of parameters (weight and bias terms) to be adjusted. This problem can be overcome by assuming that feature extraction block of an existing CNN model may work sufficiently well for the target model. Therefore, we can benefit from it in our problem. This approach is called transfer learning. We will not go into details of the theoretical explanation of transfer learning in this book since our main focus is embedded systems. Instead, we will focus on its implementation steps.

Implementation steps of transfer learning from the application perspective are as follows. We pick pretrained feature extraction block of an existing CNN model. We freeze all or some of its weights and bias terms such that they are not trainable. Hence, weights and bias terms in the feature extraction block are kept as they are throughout the operation. In other words, we transfer the old feature extraction block to the new CNN model. Then, we append our own classification block to the new model such that weights and bias terms in this block are trainable. As a result, we form a new CNN model to be trained for the problem at hand. This way, the total number of parameters to be adjusted in the model decreases which leads to a decrease in the total training time.

There are several powerful CNN architectures available in literature. We can use them in transfer learning. However, the STM32 microcontroller has limited RAM and flash size as explained in Chap. 2. Therefore, we should pick the architectures (more precisely models formed from them) suitable to be embedded on the STM32 microcontroller. We will briefly explain them next. Besides, we will also handle the transfer learning operation under Keras based on the selected models.

13.4.2 SqueezeNet for Transfer Learning

SqueezeNet is a lightweight CNN architecture designed for efficient deep learning inference, particularly on resource-constrained devices such as embedded systems. The architecture was introduced by Iandola et al. [15]. SqueezeNet achieves a remarkably small model size, typically less than 0.5 MB while maintaining accuracy. It introduces the concept of "Fire" modules which consist of the combination of squeeze and expand layers. The squeeze layer has 1×1 convolutions to reduce the number of input channels. The expand layer uses both 1×1 and 3×3 convolutions to increase the number of channels, capturing both low- and high-level features.

The SqueezeNet model is not included in TensorFlow. Hence, we should build it by using the available layers in Keras. The reader can benefit from the SqueezeNet implementation from STMicroelectronics model zoo [42]. An open-source implementation of SqueezeNet can also be obtained from [20]. We provide the simplified version of this implementation in Listing 13.5.

Listing 13.5 Building SqueezeNet in Keras

```
import tensorflow as tf
from keras.utils import get_file

WEIGHTS_PATH_NO_TOP = "https://github.com/rcmalli/keras-
    squeezenet/releases/download/v1.0/
    squeezenet_weights_tf_dim_ordering_tf_kernels_notop.h5"

def fire(x, squeeze, expand, name):
    y = tf.keras.layers.Conv2D(
        filters=squeeze,
        kernel_size=1,
        activation="relu",
        padding="same",
        name=f"{name}_squeeze")(x)
```

```
    y1 = tf.keras.layers.Conv2D(
        filters=expand,
        kernel_size=1,
        activation="relu",
        padding="same",
        name=f"{name}_expand1x1",
    )(y)
    y3 = tf.keras.layers.Conv2D(
        filters=expand,
        kernel_size=3,
        activation="relu",
        padding="same",
        name=f"{name}_expand3x3",
    )(y)
    return tf.keras.layers.concatenate([y1, y3], name=f"{name}
        _concat")

def SqueezeNet(input_shape=(224, 224, 3), weights="imagenet",
    classes=10, dropout = None):
    model_input = tf.keras.layers.Input(shape=input_shape)
    x = tf.keras.layers.Conv2D(64, (3, 3), strides=2, padding="
        valid", activation="relu", name = "conv1")(model_input)
    x = tf.keras.layers.MaxPooling2D((3, 3), strides=2, padding="
        same")(x)
    x = fire(x, 16, 64, name="fire1")
    x = fire(x, 16, 64, name="fire2")
    x = tf.keras.layers.MaxPooling2D((3, 3), strides=2, padding="
        same")(x)
    x = fire(x, 32, 128, name="fire3")
    x = fire(x, 32, 128, name="fire4")
    x = tf.keras.layers.MaxPooling2D((3, 3), strides=2, padding="
        same")(x)
    x = fire(x, 48, 192, name="fire5")
    x = fire(x, 48, 192, name="fire6")
    x = fire(x, 64, 256, name="fire7")
    feature_extractor = fire(x, 64, 256, name="fire8")

    feature_ext_model = tf.keras.Model(inputs=[model_input],
        outputs=[feature_extractor])
    if dropout:
        x = tf.keras.layers.Dropout(dropout, name='drop9')(x)

    if weights == "imagenet":
        weights_path = get_file(
            "squeezenet_weights_tf_dim_ordering_tf_kernels_notop.
                h5",
            WEIGHTS_PATH_NO_TOP,
            cache_subdir="models",
        )

        feature_ext_model.load_weights(weights_path)

    feature_extractor_out = feature_ext_model.output
    x = tf.keras.layers.Conv2D(classes, (1, 1), name = "
        final_conv")(feature_extractor_out)
    x = tf.keras.layers.GlobalAveragePooling2D(name="
        global_avg_pool")(x)
    model_output = tf.keras.layers.Softmax()(x)
    model = tf.keras.Model(inputs = [model_input], outputs = [
        model_output])

    return model
```

In Listing 13.5, we can create a SqueezeNet model by calling the function SqueezeNet. Let us breakdown the code into pieces. The function fire (x, squeeze, expand, name) is responsible for creating the fire module, which consists of a squeeze convolution layer followed by expand convolution layers with different kernel sizes. The result of expanded convolution layers are concatenated. The fire function takes four parameters as follows. x is the input layer in which the fire model is connected. squeeze and expand are number of squeeze and expand convolution layer filters, respectively. The parameter name denotes the name of layers inside the fire module.

We can define the main function SqueezeNet that creates the SqueezeNet model after completing the definition of the fire module. This function takes parameters for input shape, weights, number of classes, and whether to use transfer learning or not. Here, the parameter input_shape defines the height, width, and depth of the input tensor. The parameter include_top = False is used for loading the model without the final fully connected layers. In other words, include_top = True case is used to classify 1000 classes in the ImageNet dataset. The weight parameter can be either None or 'imagenet'. If it is set to 'imagenet', then this parameter initializes weights to a pretrained model on the ImageNet dataset. Otherwise, weights are randomly initialized.

The SqueezeNet model consists of a convolution layer with 64 filters with kernel size (3, 3) and stride 2 followed by a max pooling layer with pool size (3, 3) and stride 2. Then, fire modules implemented in the previous step are included to the model, followed by max pooling layers. After adding all fire modules, we create an intermediate model for feature extraction. If we want to use ImageNet weights, then pretrained weights are downloaded and loaded up to the fire5_squeeze layer. Then, they are frozen for transfer learning. Feature extractor output is connected to a 1×1 convolutional layer with the specified number of classes, followed by global average pooling to reduce spatial dimensions, and a softmax activation layer to obtain class probabilities. Finally, the function returns the constructed SqueezeNet model.

We need a dataset to utilize the constructed SqueezeNet model. We use the MNIST handwritten digits dataset for this purpose. Using the script in Listing 13.6, we can download the dataset. Before feeding images to the model, the images in the dataset should be converted from 1 to 3 channels. To do so, we can use the function prepare_data, which adds an extra dimension as last axis and duplicate the images using the tf.image function. Next, we convert our labels to categorical labels. Then, we create a SqueezeNet object as implemented in Listing 13.5 with the specified input shape (32,32,3) and freezing one third of the layers by setting layer.trainable = False for transfer learning. To do so, we should rename the Python script in Listing 13.5 as "squeezeNet.py" and keep it in the same folder as with Listing 13.6. model.summary prints summary of the model architecture, providing a concise overview of its layers, parameters, and connections. Then, the model is compiled with the categorical cross-entropy loss function, Adam optimizer, and accuracy as the evaluation metric. A callback function is set for model checkpointing, which saves the model only when the validation loss decreases, and the file is named "squeezenet_tl_mnist.h5." The model is trained using the provided training images train_images and labels train_labels. Training is conducted with a batch size of 128, over 10 epochs, and includes a validation split of 10%. The model is monitored by the specified callback during training and training history is stored in the variable history.

Listing 13.6 Transfer learning application with the SqueezeNet model for handwritten digit recognition

```
import tensorflow as tf
from squeezeNet import SqueezeNet

num_classes = 10
(train_images, train_labels), (test_images, test_labels)  = tf.
    keras.datasets.mnist.load_data()
data_shape = (32, 32, 3)

def prepare_tensor(images, out_shape):
    images = tf.expand_dims(images, axis=-1)
    images = tf.repeat(images, 3, axis=-1)
    images = tf.image.resize(images, out_shape[:2])
    images = images / 255.0
    return images

train_images = prepare_tensor(train_images, data_shape)
test_images = prepare_tensor(test_images, data_shape)

# convert class vectors to binary class matrices
train_labels = tf.keras.utils.to_categorical(train_labels,
    num_classes)
test_labels = tf.keras.utils.to_categorical(test_labels,
    num_classes)
model = SqueezeNet(input_shape=data_shape, dropout=0.2)
num_layers_to_train = len(model.layers)//3
for layer in model.layers[:num_layers_to_train]:
    layer.trainable = False

model.summary()
model.compile(loss='categorical_crossentropy', optimizer='adam',
    metrics=['accuracy'])
callbacks = [tf.keras.callbacks.ModelCheckpoint("models/
    squeezenet_tl_mnist.h5", monitor = "val_loss", save_best_only
    = True, mode = "min", verbose = 1)]
history=model.fit(train_images, train_labels, batch_size=128,
    epochs=10, validation_split=0.1, callbacks=callbacks)
```

13.4.3 ResNet for Transfer Learning

ResNet, short for residual convolutional networks, is a deep learning architecture for image classification and computer vision tasks. It was introduced by He et al. [9]. In their paper, the authors introduced residual blocks where layers have skip connections unlike normal CNN models. These skip connections help feature reuse and facilitate gradient flow during training. Moreover, bottleneck layers are utilized in residual blocks. They reduce the number of parameters and provide a computationally efficient network. Transition layers, consisting of convolution and pooling layers, are utilized between dense blocks. They keep spatial dimensions while increasing the number of feature maps gradually. Hence, computational complexity of the model is kept low. As a result, the ResNet architecture provides feature reuse and parameter efficiency in addition to classical CNN models.

We can use the pretrained ResNet model provided by STMicroelectronics under their model zoo for transfer learning [42]. The exact link for the "resnetv1.py" Python script formed there is [43]. In this script, there is the ResNet function defined as `ResNet`. Here, the parameter `input_shape` defines the height, width, and depth of the input tensor. The parameter `num_classes` is used for defining the number of output classes. The `depth` parameter specifies the number of layers to be used in the ResNet model. The dropout parameter sets the fraction of fully connected layers to drop.

We should load pretrained weights for transfer learning. To do so, we first download the pretrained ResNet model file from the STMicroelectronics model zoo and load its weights. Then, one third of the layers are frozen. We provide the code snippet formed for this purpose next.

```
pretrained_model_path = get_file(origin = "https://github.com/
    STMicroelectronics/stm32ai-modelzoo/raw/main/
    image_classification/pretrained_models/resnetv1/
    ST_pretrainedmodel_public_dataset/cifar10/resnet_v1_8_32_tfs/
    resnet_v1_8_32_tfs.h5",
cache_subdir= "models")
model.load_weights(pretrained_model_path)
num_layers_to_train = len(model.layers)//3
for layer in model.layers[:num_layers_to_train]:
layer.trainable = False
```

In Listing 13.7, we provide the transfer learning application with the formed ResNet model. Data loading and preparation parts of the code are similar to previous models. Then, we create the ResNet model by calling `ResNet` after importing the ResNet function. To do so, we should keep the downloaded "resnetv1.py" as "resnet.py" and keep it in the same folder as with Listing 13.7. Afterward, we load model weights by the code line `model.load_weights(pretrained_model_path)`. Then, we define which layers to be frozen and continue with model training.

Listing 13.7 Transfer learning application with the ResNet model for handwritten digit recognition

```
import tensorflow as tf
from resnet import ResNet
from keras.utils import get_file

num_classes = 10
(train_images, train_labels), (
    test_images,
    test_labels,
) = tf.keras.datasets.mnist.load_data()
data_shape = (32, 32, 3)
```

```
def prepare_tensor(images, out_shape):
    images = tf.expand_dims(images, axis=-1)
    images = tf.repeat(images, 3, axis=-1)
    images = tf.image.resize(images, out_shape[:2])
    images = images / 255.0
    return images

train_images = prepare_tensor(train_images, data_shape)
test_images = prepare_tensor(test_images, data_shape)

# convert class vectors to binary class matrices
train_labels = tf.keras.utils.to_categorical(train_labels,
    num_classes)
test_labels = tf.keras.utils.to_categorical(test_labels,
    num_classes)

model = ResNet(num_classes, data_shape, 8, 0.15)
pretrained_model_path = get_file(origin = "https://github.com/
    STMicroelectronics/stm32ai-modelzoo/raw/main/
    image_classification/pretrained_models/resnetv1/
    ST_pretrainedmodel_public_dataset/cifar10/resnet_v1_8_32_tfs/
    resnet_v1_8_32_tfs.h5",
                                    cache_subdir= "models")

model.load_weights(pretrained_model_path)
num_layers_to_train = len(model.layers)//3
for layer in model.layers[:num_layers_to_train]:
    layer.trainable = False

model.compile(loss="categorical_crossentropy", optimizer="adam",
    metrics=["accuracy"])
callbacks = [
    tf.keras.callbacks.ModelCheckpoint(
        "models/resnet_tl_mnist.h5",
        monitor="val_loss",
        save_best_only=True,
        mode="min",
        verbose=1,
    )
]
history = model.fit(
    train_images,
    train_labels,
    batch_size=128,
    epochs=10,
    validation_data=(test_images, test_labels),
    callbacks=callbacks,
)
```

13.4.4 EfficientNet for Transfer Learning

EfficientNet covers a family of neural network architectures designed to achieve high performance in computer vision tasks while maintaining computational efficiency. The architecture was introduced by Tan and Le [45]. EfficientNet introduces a novel approach called compound scaling, which scales neural networks uniformly in three dimensions as the number of layers, number of filters, and input image size. Compound scaling balances model capacity and computational resources. This coefficient allows users to choose from different model sizes (B0–B7), with larger coefficients indicating larger and more powerful models.

EfficientNet uses a basic building block called the mobile inverted residual bottleneck convolution block, besides squeeze-and-excitation blocks. This block is efficient and leverages depthwise separable convolutions to reduce computation while maintaining representational power. Moreover, EfficientNet models are often trained with a technique called noisy student training which involves adding noise to training data. This approach aims the model to become more robust to variations in input.

Keras has eight variants of EfficientNet from `tf.keras.applications.` `EfficientNetB0` to `tf.keras.applications.EfficientNetB7` where B7 has the highest number of parameters. Unfortunately, these models are too large to fit in a microcontroller. Therefore, we benefit from the lightweight EfficientNet implementation provided by STMicroelectronics under their model zoo. The exact link for the "st_efficientnet_lc_v1.py" Python script formed there is [44]. In this script, there is the EfficientNet function defined as `EfficientNet`.

In Listing 13.8, we provide the transfer learning application with the EfficientNet model. Data loading and preparation parts of the code are similar to previous models. Then, we create the EfficientNet model by calling `EfficientNet` after importing the EfficientNet function. To do so, we should rename the Python script "st_efficientnet_lc_v1.py" as "efficientnet.py" and keep it in the same folder as with Listing 13.8. Afterward, we load model weights by the code line `model.` `load_weights(pretrained_model_path)`. Then, we define which layers to be frozen and continue with model training.

Listing 13.8 Transfer learning application with the EfficientNet model for handwritten digit recognition

```
import tensorflow as tf
from keras.utils import get_file, to_categorical
from efficientnet import EfficientNet

num_classes = 10
(train_images, train_labels), (
    test_images,
    test_labels,
) = tf.keras.datasets.mnist.load_data()
data_shape = (32, 32, 3)

def prepare_tensor(images, out_shape):
    images = tf.expand_dims(images, axis=-1)
    images = tf.repeat(images, 3, axis=-1)
    images = tf.image.resize(images, out_shape[:2])
    images = images / 255.0
    return images

train_images = prepare_tensor(train_images, data_shape)
test_images = prepare_tensor(test_images, data_shape)

# convert class vectors to binary class matrices
train_labels = to_categorical(train_labels, num_classes)
test_labels = to_categorical(test_labels, num_classes)
```

```
model = EfficientNet(data_shape, classes=num_classes)
pretrained_model_path = get_file(origin = "https://github.com/
    STMicroelectronics/stm32ai-modelzoo/raw/main/
    image_classification/pretrained_models/efficientnet/
    ST_pretrainedmodel_public_dataset/flowers/
    st_efficientnet_lc_v1_128_tfs/st_efficientnet_lc_v1_128_tfs.
    h5",
                                        cache_subdir= "models")

model.load_weights(pretrained_model_path, by_name= True,
    skip_mismatch= True)
num_layers_to_train = len(model.layers)//3
for layer in model.layers[:num_layers_to_train]:
    layer.trainable = False

model.compile(loss="categorical_crossentropy", optimizer="adam",
    metrics=["accuracy"])
callbacks = [
    tf.keras.callbacks.ModelCheckpoint(
        "models/efficientnet_tl_mnist.h5",
        monitor="val_loss",
        save_best_only=True,
        mode="min",
        verbose=1,
    )
]
history = model.fit(
    train_images,
    train_labels,
    batch_size=128,
    epochs=10,
    validation_data=(test_images, test_labels),
    callbacks=callbacks,
)
```

13.4.5 MobileNet for Transfer Learning

MobileNet is a family of lightweight convolutional neural network architecture designed for efficient and low-latency image classification and computer vision tasks, particularly on mobile and embedded systems. MobileNetV1 was introduced by Howard et al. [12]. MobileNet family is extended as V2 in [27]. Furthermore, MobileNetV3 is introduced in [13].

MobileNetV1 heavily depends on separable convolutions, which split the standard convolution operation into two separate steps as depthwise convolution (applying a single filter per input channel) and pointwise convolution (mixing the depthwise outputs to generate final feature maps). This reduces the computational cost significantly while maintaining good accuracy. Moreover, width and resolution multipliers are proposed by authors. Here, the width multiplier controls the number of filters in each layer. It allows the user to balance model size and accuracy. The resolution multiplier scales down the input image size, further reducing computational requirements. MobileNetV2 improved the performance of its predecessor by refining depthwise separable convolution layers with inverted residuals to improve accuracy and efficiency. It also introduced skip connections (residual connections) between layers to facilitate gradient flow and improve training stability. MobileNetV3 added further enhancements to its predecessor by introducing activation functions like h-swish and h-sigmoid. Moreover, it optimized the architecture for different hardware accelerators and introduced various features like squeeze-and-excitation blocks. They adaptively recalibrate channel-wise feature maps. In summary, accuracy and efficiency of MobileNet family has evolved over time. The choice of which version to use depends on specific requirements, such as model size, inference speed, and hardware constraints.

We can use the pretrained MobileNet feature extraction block by available functions under Keras. In fact, Keras has four variants of MobileNet as `tf.keras.applications.MobileNet`, `tf.keras.applications.MobileNetV2`, `tf.keras.applications.MobileNetV3Small`, and `tf.keras.applications.MobileNetV3Large`. These models are pretrained by the Imagenet dataset and have similar parameters. Here, the `alpha` parameter is the width multiplier hyperparameter which adjusts the fractional number of filters used in each layer. We provide the sample code snippet formed for this purpose next.

```
feature_extractor = tf.keras.applications.MobileNet(
    input_shape=(28,28,1),
    alpha=0.25,
    include_top=False,
)
```

The reader can benefit from the MobileNet implementation provided by STMicroelectronics under their model zoo [42]. Different from the previous two models, we use Keras implementation of MobileNet. In Listing 13.9, we create the MobileNetV2 object with parameter `include_top = False`. Therefore, we only obtain the feature extractor part of the MobileNet model. `MobileNetV2` is trained on ImageNet dataset by default. Therefore, we do not need to load weights separately. Then, classifier part must be added to end of the model. This part consists of a convolution layer, batch normalization layer, ReLu6, global average pooling, and fully connected layer. Dropout layer is optional for this case. As number of classes passed as an argument to the function formed, MobileNet already has the final layer, and it is ready to be trained on the MNIST dataset. We provide the Python script to train the MobilenetV2 model on MNIST dataset in Listing 13.10.

Listing 13.9 Creating the MobileNetV2 model for handwritten digit recognition

```
from keras.applications import MobileNetV2
from keras import layers
from keras import Model

def make_divisible(v, divisor, min_value=None):
    if min_value is None:
        min_value = divisor
```

```
new_v = max(min_value, int(v + divisor / 2) // divisor *
    divisor)

if new_v < 0.9 * v:
    new_v += divisor
return new_v


def mobileNetV2(input_shape, num_classes, alpha=0.35, dropout=
    None):

    feature_extractor = MobileNetV2(
        input_shape= input_shape,
        alpha=alpha,
        include_top=False,
    )

    x = feature_extractor.output
    last_block_filters = make_divisible(1280 * alpha, 8)
    x = layers.Conv2D(last_block_filters, kernel_size=1, padding=
        'same', use_bias=False)(x)
    x = layers.BatchNormalization()(x)
    x = layers.ReLU(6.)(x)

    x = layers.GlobalAveragePooling2D()(x)
    if dropout:
        x = layers.Dropout(rate=dropout, name="dropout")(x)
    outputs = layers.Dense(num_classes, activation="softmax")(x)

    return Model(inputs = feature_extractor.input, outputs=
        outputs)
```

Listing 13.10 Transfer learning application with the MobileNetV2 model for handwritten digit recognition

```
import tensorflow as tf
from mobilenet import mobileNetV2

num_classes = 10
(train_images, train_labels), (
    test_images,
    test_labels,
) = tf.keras.datasets.mnist.load_data()
data_shape = (32, 32, 3)


def prepare_tensor(images, out_shape):
    images = tf.expand_dims(images, axis=-1)
    images = tf.repeat(images, 3, axis=-1)
    images = tf.image.resize(images, out_shape[:2])
    images = images / 255.0
    return images


train_images = prepare_tensor(train_images, data_shape)
test_images = prepare_tensor(test_images, data_shape)

# convert class vectors to binary class matrices
train_labels = tf.keras.utils.to_categorical(train_labels,
    num_classes)
```

```
test_labels = tf.keras.utils.to_categorical(test_labels,
    num_classes)

model = mobileNetV2(data_shape, 10, dropout = 0.2)
num_layers_to_train = len(model.layers)//3
for layer in model.layers[:num_layers_to_train]:
    layer.trainable = False

model.summary()
model.compile(loss="categorical_crossentropy", optimizer="adam",
    metrics=["accuracy"])
callbacks = [
    tf.keras.callbacks.ModelCheckpoint(
        "models/mobilenet_tl_mnist.h5",
        monitor="val_loss",
        save_best_only=True,
        mode="min",
        verbose=1,
    )
]
history = model.fit(
    train_images,
    train_labels,
    batch_size=128,
    epochs=10,
    validation_split=0.1,
    callbacks=callbacks,
)
```

13.4.6 ShuffleNet for Transfer Learning

ShuffleNet is a neural network architecture designed for efficient and lightweight implementation on resource-constrained devices, particularly mobile phones and embedded systems. The architecture was proposed by Zhang et al. [63]. ShuffleNet introduces channel shuffling, which allows information exchange between different feature map groups. This enables efficient learning of cross-group features and improves network performance. Moreover, group convolution is employed in ShuffleNet, where channels of the input feature map are divided into groups. Then, convolution is applied separately to each group. This reduces the computational cost compared to standard convolution. ShuffleNet uses bottleneck building blocks with 1×1 pointwise convolutions for dimension reduction. This is followed by group convolution and another 1×1 pointwise convolution for dimension expansion.

ShuffleNet implementation is not included in Keras library. Hence, we should implement its layers similar to SqueezeNet. The ShuffleNetV2 model for Keras can be obtained from [28]. We ask the reader to download the corresponding Python script and name it as "ShuffleNet.py." In this Python script, convolutional layers are repeated accordingly. Then, channels are shuffled for each block in the ShuffleNet model. Finally, we can create the ShuffleNetV2 model by calling the function ShuffleNet.

We can train the ShuffleNet model on the MNIST dataset as in Listing 13.11. In this script, we first create a ShuffleNet instance. To do so, we should keep the downloaded "ShuffleNet.py" script it in the same folder as with Listing 13.11. We use the ShuffleNet model with parameters groups=3, input_shape = (32,32,3), classes = 10. The parameter scale_factor is set to 0.25 to keep the model lightweight. Once the model ist trained, it is ready to be deployed to the microcontroller.

Listing 13.11 Training ShuffleNet model for handwritten digit recognition

```python
import tensorflow as tf
from matplotlib import pyplot as plt
from ShuffleNet import ShuffleNet

num_classes = 10
(train_images, train_labels), (
    test_images,
    test_labels,
) = tf.keras.datasets.mnist.load_data()

data_shape = (32, 32, 3)

def prepare_tensor(images, out_shape):
    images = tf.expand_dims(images, axis=-1)
    images = tf.repeat(images, 3, axis=-1)
    images = tf.image.resize(images, out_shape[:2])
    images = images / 255.0
    return images

train_images = prepare_tensor(train_images, data_shape)
test_images = prepare_tensor(test_images, data_shape)

# convert class vectors to binary class matrices
train_labels = tf.keras.utils.to_categorical(train_labels,
    num_classes)
test_labels = tf.keras.utils.to_categorical(test_labels,
    num_classes)
model = ShuffleNet(
    scale_factor = 0.25,
    groups = 3,
    input_shape=data_shape,
    classes = 10
)

model.summary()
model.compile(loss="categorical_crossentropy", optimizer="adam",
    metrics=["accuracy"])
callbacks = [
    tf.keras.callbacks.ModelCheckpoint(
        "models/shufflenet_tl_mnist.h5",
        monitor="val_loss",
        save_best_only=True,
        mode="min",
        verbose=1,
    )
]
history = model.fit(
    train_images,
    train_labels,
    batch_size=128,
    epochs=10,
```

```
        validation_split=0.1,
        callbacks=callbacks,
)
```

13.5   Embedding the CNN Model to the Microcontroller

This section will be fairly short. The main reason is that we direct the reader to the methods introduced in Chap. 12. Hence, the CNN model, formed by either custom feature extraction and classification blocks or transfer learning, can be converted to the TensorFlow Lite format. Then, the C array corresponding to this model can be obtained. This array can be used for inference on the microcontroller. Since we formed the methods in Chap. 12 for general usage, the CNN model can also be processed by them. This also applies to the STM32Cube.AI usage.

The reader should only pay attention to the input of the model specific for this chapter. If we obtain the image to be processed by the model from the camera, then the input data type will be unsigned integer. Therefore, the model input should be set as such. To note here, the input of the model can also be set as float data type. Then, the image should be converted to this form before fed to the model.

13.6   Application: Keyword Spotting from Audio Signals

In this application, we form a CNN model for keyword spotting from audio signals. Different from the related application in Sect. 12.8, there is no need to extract features from audio signals. Instead, we feed the spectrogram to the CNN model directly. Since spectrogram is basically the stacked short-time Fourier transforms of the audio signal, it can be processed as an image. In Listing 13.12, we provide the Python script to transform audio signal into spectrograms and form the dataset. Here, the function get_spectrogram takes a path to a wav audio file and tuple of audio parameters as input. It starts by extracting the file specifications from the file path. Then, it reads the audio file and decodes it into a tensor. The tensor is then squeezed to remove dimensions of size one from its shape. A condition is checked to see whether the shape of the sample is greater than the fixed size. If this is the case, then the sample is sliced to the fixed size. Otherwise, the sample is padded to reach the fixed size. Short-time Fourier transform of the sample is then computed, which is a common way to represent audio data. Absolute value of the spectrogram is computed to ensure all values are positive and an extra dimension is added to the spectrogram tensor.

Listing 13.12 Creating spectrogram dataset from audio files

```
import os
import tensorflow as tf

def split_fn(wav_path, train):
```

```
    file_specs = tf.strings.split(wav_path, ".")[0]
    file_specs = tf.strings.split(file_specs, os.path.sep)[1]
    person = tf.strings.split(file_specs, "_")[1]
    if train:
        return person != b"yweweler"
    else:
        return person == b"yweweler"

def get_spectrogram(wav_path, audio_params):
    file_specs = tf.strings.split(wav_path, ".")[0]
    file_specs = tf.strings.split(file_specs, os.path.sep)[1]
    digit = tf.strings.to_number(tf.strings.split(file_specs, "_"
        )[0], out_type = tf.int32)
    wavfile = tf.io.read_file(wav_path)
    sample, _ = tf.audio.decode_wav(wavfile)
    sample = tf.squeeze(sample, axis=-1)
    fixed_size = tf.constant([audio_params[0]])
    sample_shape = tf.shape(sample)
    def pad_func():
        pad_size = fixed_size - sample_shape
        padding = tf.concat([tf.zeros_like(sample_shape),
            pad_size], axis=0)
        padding = tf.reshape(padding,(-1,2))
        return tf.pad(sample, padding)
    def slice_func():
        return tf.slice(sample, tf.constant([0]), fixed_size)

    sample = tf.cond(sample_shape > fixed_size, slice_func,
        pad_func)

    spectrogram = tf.signal.stft(sample, frame_length=
        audio_params[1], frame_step=audio_params[2])
    spectrogram = tf.abs(spectrogram)
    spectrogram = spectrogram[..., tf.newaxis]
    return spectrogram, digit

def create_datasets(sample_length, fft_size, step_size,
    batch_size):
    RECORDINGS_DIR = "recordings/*.wav"
    ds = tf.data.Dataset.list_files(RECORDINGS_DIR)
    ds_size = tf.data.experimental.cardinality(ds).numpy()
    ds = ds.map(lambda x: get_spectrogram(x, (sample_length,
        fft_size, step_size))).shuffle(ds_size)
    train_ds = ds.take(int(0.8 * ds_size))
    val_ds = ds.skip(int(0.8 * ds_size))
    test_ds = val_ds.shard(num_shards=2, index=0)
    val_ds = val_ds.shard(num_shards=2, index=1)

    train_ds = train_ds.batch(batch_size).cache().prefetch(tf.
        data.AUTOTUNE)
    val_ds = val_ds.batch(batch_size).cache().prefetch(tf.data.
        AUTOTUNE)
    test_ds = test_ds.batch(batch_size).cache().prefetch(tf.data.
        AUTOTUNE)
    spec_shape = (sample_length // step_size -1, step_size + 1,
        1)
    return train_ds, val_ds, test_ds, spec_shape
```

In Listing 13.12, the function `create_datasets` is used to create training, validation, and test datasets from the spectrograms. It starts by creating a dataset of all files matching a certain pattern. The number of elements in the dataset is then determined. The `get_spectrogram` function is applied to each element in the dataset, and the dataset is shuffled. The dataset is then split into training and validation datasets. The validation dataset is further split into validation and test datasets. The datasets are then batched, cached, and prefetched for efficient training. Finally, the shape of the spectrogram is calculated.

Next, we build a model to utilize this data. Therefore, Listing 13.13 defines the CNN model for this purpose. Here, the `create_model` function is where the CNN model is defined. It takes three parameters as the shape of the input data, number of labels, and normalization layer. The model begins with an input layer that specifies shape of the input data. This is followed by a resizing layer that resizes the input to a fixed size of 32×32. The normalization layer is then applied to normalize the input data for better generalization and quicker convergence. The model then has two `Conv2D` layers, each with a kernel size of 3 and `ReLU` activation function. The first convolutional layer has 32 filters. The second one has 64 filters. These layers are used to extract features from the input data. The `MaxPooling2D` layer follows the convolutional layers to reduce the spatial dimensions of the output from the preceding layer. The `Dropout` layer is then applied with a rate of 0.25 to prevent overfitting by randomly setting a fraction of the input units to 0 during training. The output is then flattened using the `Flatten` layer to convert the 2D spatial data into a vector. This is followed by a `Dense` layer with 64 units and `ReLU` activation function for classification. Another `Dropout` layer is applied with a rate of 0.5, followed by the final `Dense` layer with the number of units equal to the number of labels. This layer outputs the logits for each class.

Listing 13.13 Forming a CNN model for keyword spotting

```
from keras import layers, models, losses,optimizers

def normalize_input(input_ds):
    norm_layer = layers.Normalization()
    norm_layer.adapt(input_ds.map(map_func=lambda spec, label:
        spec))
    return norm_layer

def create_model(input_shape, num_labels, norm_layer):
    cnn_model = models.Sequential([
        layers.Input(shape=input_shape),
        layers.Resizing(32,32),
        norm_layer,
        layers.Conv2D(32, 3, activation='relu'),
        layers.Conv2D(64, 3, activation='relu'),
        layers.MaxPooling2D(),
        layers.Dropout(0.25),
        layers.Flatten(),
        layers.Dense(64, activation='relu'),
        layers.Dropout(0.5),
        layers.Dense(num_labels),
    ])
```

```
cnn_model.compile(optimizer = optimizers.Adam(0.0005,
    weight_decay = 1e-6),
                        loss = losses.SparseCategoricalCrossentropy
                            (from_logits=True),
                        metrics="accuracy")

return cnn_model
```

In Listing 13.13, the model is compiled with the Adam optimizer with a learning rate of 0.0005 and a weight decay of 1e-6. The loss function is set to SparseCategoricalCrossentropy with logits. The metric for evaluation is accuracy. The create_model function then returns the compiled CNN model ready for training with spectrogram data. The model is then trained using the fit method for 50 epochs, with the training and validation datasets specified. The training process outputs loss and accuracy to monitor and control the training process. Upon completion, the script results in a trained keyword spotting model using a CNN, ready for deployment in speech recognition tasks. The next step is converting the trained model to TensorFlow Lite model and C array.

Listing 13.14 Training the formed CNN model for keyword spotting

```
from data_loader import create_datasets
from keras import callbacks
from keras.models import load_model

train_ds, val_ds, test_ds, input_shape = create_datasets(8000,
    512, 256, 32)
kws_cnn_model = load_model("resnet_tl_mnist.h5")
model_cp_callback = callbacks.ModelCheckpoint("kws_cnn_model.h5",
    save_best_only=True)
es_callback = callbacks.EarlyStopping(verbose=1, patience=5)
kws_cnn_model.fit(train_ds,
            epochs=50,
            validation_data= val_ds,
            verbose=1,
            callbacks = [model_cp_callback, es_callback])
```

The Python script in Listing 13.15 is used for converting the trained model to TensorFlow Lite form. Here, the converter is configured to use default optimizations and support the operations that are built into TensorFlow Lite form. The TensorFlow Lite model is then saved to a file at the path specified by TFLITE_MODEL_PATH. Finally, the convert_tflite2cc function is called to convert the TensorFlow Lite model to a C++ source file. This file will be deployed to the microcontroller for inference.

Listing 13.15 Converting the trained CNN model to TensorFlow Lite model and C++ array

```
import tensorflow as tf
import keras
from tflite2cc import convert_tflite2cc

TFLITE_MODEL_PATH = "kws_cnn_model.tflite"
kws_cnn_model = keras.models.load_model("kws_cnn_model.h5")
```

```
converter = tf.lite.TFLiteConverter.from_keras_model(
    kws_cnn_model)
converter.optimizations = [tf.lite.Optimize.DEFAULT]
kws_cnn_lite = converter.convert()

with open(TFLITE_MODEL_PATH, "wb") as tflite_file:
    tflite_file.write(kws_cnn_lite)

convert_tflite2cc(TFLITE_MODEL_PATH, "kws_cnn_model.cc")
```

We will provide complete projects for STM32CubeIDE and Mbed Studio in the accompanying book repository. There, we will provide different settings in the implementation step. Hence, the reader will observe different options in implementation. While doing so, we will realize the complete keyword spotting system on the microcontroller from the initial audio signal acquisition to final inference step.

13.7 Application: Handwritten Digit Recognition from Digital Images

In this application, we use the previously trained ResNet model on the MNIST dataset for handwritten digit recognition. In Listing 13.16, data and pretrained model is loaded. Then, all layers of the model are trained to create new fine-tuned model for the MNIST data.

Listing 13.16 Training the ResNet model for handwritten digit recognition

```
import tensorflow as tf
import numpy as np
import keras
from sklearn.model_selection import train_test_split

num_classes = 10
(train_images, train_labels), (val_images, val_labels)  = tf.
    keras.datasets.mnist.load_data()
data_shape = (32, 32, 3)

def prepare_tensor(images, out_shape):
    images = tf.expand_dims(images, axis=-1)
    images = tf.repeat(images, 3, axis=-1)
    images = tf.image.resize(images, out_shape[:2])
    images = images / 255.0
    return images

train_images = prepare_tensor(train_images, data_shape)
val_images = prepare_tensor(val_images, data_shape)

# convert class vectors to binary class matrices
train_labels = tf.keras.utils.to_categorical(train_labels,
    num_classes)
val_labels = tf.keras.utils.to_categorical(val_labels,
    num_classes)
```

```
mnist_cnn_model = keras.models.load_model("resnet_tl_mnist.h5")
model_cp_callback = keras.callbacks.ModelCheckpoint("
    mnist_cnn_model.h5", save_best_only=True)
es_callback = keras.callbacks.EarlyStopping(verbose=1, patience
    =5)
mnist_cnn_model.fit(x = train_images,
                    y = train_labels,
                    epochs=50,
                    validation_data= (val_images, val_labels),
                    verbose=1,
                    callbacks = [model_cp_callback, es_callback])
```

The Python script in Listing 13.17 is used for converting the trained model to
TensorFlow Lite form. Here, the converter is configured to use default optimizations
and support the operations that are built into TensorFlow Lite form. The TensorFlow
Lite model is then saved to a file at the path specified by TFLITE_MODEL_PATH. Finally,
the convert_tflite2cc function is called to convert the TensorFlow Lite model to a
C++ source file. This file will be deployed to the microcontroller for inference.

Listing 13.17 Converting the trained ResNet model to TensorFlow Lite model and C++ array

```
import tensorflow as tf
import keras
from tflite2cc import convert_tflite2cc

TFLITE_MODEL_PATH = "mnist_cnn_model.tflite"
mnist_cnn_model = keras.models.load_model("mnist_cnn_model.h5")
converter = tf.lite.TFLiteConverter.from_keras_model(
    mnist_cnn_model)
converter.optimizations = [tf.lite.Optimize.DEFAULT]
mnist_cnn_lite = converter.convert()

with open(TFLITE_MODEL_PATH, "wb") as tflite_file:
    tflite_file.write(mnist_cnn_lite)

convert_tflite2cc(TFLITE_MODEL_PATH, "mnist_cnn_model.cc")
```

We will provide complete projects for STM32CubeIDE and Mbed Studio
in the accompanying book repository. There, we will provide different settings
in the implementation step. Hence, the reader will observe different options in
implementation. While doing so, we will realize the complete handwritten digit
recognition system on the microcontroller from the initial audio signal acquisition
to final inference step.

13.8   Summary of the Chapter

Convolutional neural networks are extensively used in image classification applications. Therefore, we considered them in this chapter. We introduced mathematical
definition of the convolution operation and its implementation by a single neuron
first. Afterward, we introduced the convolution definition under Keras. Then, we

explored how to form a complete CNN model under Keras. Here, we focused on feature extraction and classification blocks forming the CNN model. Afterward, we considered training and testing steps of the formed model. We then explored transfer learning to benefit from existing CNN models to be modified for our own problem. We next considered implementation steps for embedding the trained and tested CNN model on the STM32 microcontroller. Finally, we provided examples on the usage of CNN models to solve real-life problems.

Recurrence in Neural Networks

14

14.1 About Recurrence and Memory

Neural network models considered thus far do not have memory. In other words, output is calculated without paying attention to previous input/output values when an input is fed to the neural network. Adding memory to the model may improve its performance on certain applications. Therefore, we will start with forming a memory element by neurons. To do so, we will add feedback loop between neurons. To explain the concept better, we will initially assume that input to the neuron (or neurons) can only take values zero or one. Besides, we will use the hard limiter, introduced in Sect. 9.4, as the nonlinear activation function in our neuron(s). As we cover the basics of memory formation via these limitations, we will generalize the structure to other inputs and activation functions in the following sections.

14.1.1 Forming a Memory Element by Single Neuron

Let's consider the single neuron in Fig. 10.1. We can feed its output back to its input to form feedback. We provide the modified neuron structure this way in Fig. 14.1.

We can obtain/input output relation of the formed structure in Fig. 14.1. We should emphasize one important property before going further. Since we have feedback in Fig. 14.1, we should take the time parameter into account while forming the input output relation. Hence, we will have

$$y^{<t>} = \varphi(w_1 x^{<t>} + w_2 y^{<t-1>} + b) \tag{14.1}$$

where $y^{<t>}$ and $x^{<t>}$ indicate the output and input at the current time step t, respectively. $y^{<t-1>}$ is the output obtained in the previous time step.

The difference of Eq. 14.1 from the single neuron case in Eq. 10.1 is the term $y^{<t-1>}$ in the activation function. This is the part where the output calculated in the

© The Author(s), under exclusive license to Springer Nature Switzerland AG 2025 355
C. Ünsalan et al., *Embedded Machine Learning with Microcontrollers*,
https://doi.org/10.1007/978-3-031-70912-8_14

previous time step is fed back to input. Now, let's observe how the newly formed neuron structure with input and output relation given in Eq. 14.1 behaves. Let $w_1 = w_2 = 1$ and $b = -0.5$. Let's pick the activation function, $\varphi(\cdot)$, as hard limiter to emphasize the input and output relation better.

Let the input fed to the neuron with feedback in Eq. 14.1 be as $x^{<0>} = 1$ and $x^{<t>} = 0$ for $t = \{1, 2, \cdots\}$. Hence, we feed one as input once. Then, we feed zero as input afterward. As we apply these inputs to Eq. 14.1 sequentially, we will get $y^{<t>} = 1$ for $t = \{0, 1, 2, \cdots\}$. Based on this result, we can deduce that the neuron in Fig. 14.1 memorized what has been fed to it. In other words, although we feed the value one once at the beginning, the neuron has output one from that point on. This is due to the feedback loop (recurrence) in Fig. 14.1. Hence, we formed memory which can keep the input value fed to it. To note here, the neuron can only keep its output at one due to the hard limiter activation function used in operation.

14.1.2 Forming a Memory Element by Two Neurons

The alert reader can realize that output cannot be set to zero once it is set to one in Fig. 14.1. To overcome this limitation, we can extend the single neuron structure with feedback to two neurons. Hence, we can form a new memory element as in Fig. 14.2. Here, there are two feedback loops connecting the output of one neuron to input of the other neuron different from the single neuron structure in Fig. 14.1.

Let's set the weight and bias values in Fig. 14.2 as $w_1 = w_2 = w_3 = w_4 = -1$ and $b_1 = b_2 = 0.5$. Again, we use the hard limiter activation function in operation. Let's consider working principles of this structure. Hence, let's feed input values

Fig. 14.1 The single neuron with feedback

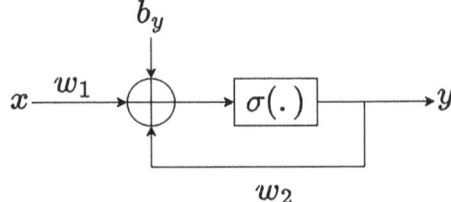

Fig. 14.2 Two neurons with feedback

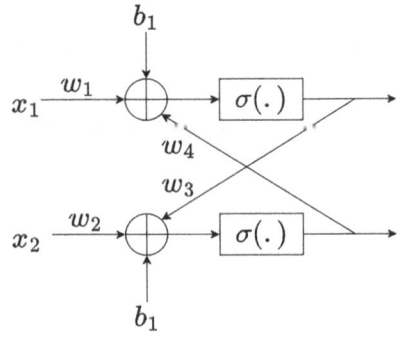

$(x_1, x_2) = \{(1, 0), (0, 0), (0, 1), (0, 1), (0, 0), (0, 0), (1, 0), (0, 0) \cdots \}$ to it. The corresponding output from the first neuron will be $\{0, 0, 0, 1, 1, 1, 1, 0, 0, \cdots \}$. Hence, we can deduce that whenever $(x_1, x_2) = (1, 0)$, the output is set to zero. Whenever $(x_1, x_2) = (0, 1)$, the output is set to one. As a side note, we should feed this input value twice to get the output. Whenever $(x_1, x_2) = (0, 0)$, the previous output value is kept as it is.

We can observe that the structure formed in Fig. 14.2 is actually a variant of SR latch based on the provided input output values. SR latch is a well-known memory element used in digital systems. We can set the SR latch by applying S=1, R=0 to it. Hence, it keeps the value one at output. We can reset the SR latch by applying S=0, R=1 to it. Hence, it keeps the value zero at output. When both S and R values are zero, then the previous output value is kept. Therefore, we can say that the inputs x_1 and x_2 in Fig. 14.2 correspond to R and S in the SR latch, respectively.

Although the structures given in this section are restricted by their input and feedback weight values, they emphasize one important property. Recurrence leads to memory. We will generalize these structures further in the following sections to form RNN, GRU, and LSTM.

14.2 Recurrent Neural Networks

Recurrent neural networks (RNN) is the first model with feedback to be considered in this chapter. Being the first model, we will cover its properties and training in detail. Hence, explanations in this section will form basis for the following sections.

14.2.1 General Structure

We can expand the structure in Fig. 14.2 by adding self-feedback loops and extra output layer. Hence, we will have the generalized model as in Fig. 14.3. This is the RNN model formed by two inputs and three neurons. Here, input values are not limited by zero or one. The activation function used in operation is also not limited by the hard limiter.

We will use the mentioned RNN model in Fig. 14.3 on a sequence of inputs in time. Weight values used in operation will be fixed in this form. However, we will feed input values sequentially. Therefore, we can represent the RNN model in modified form as in Fig. 14.4 where $h_1^{<t>}$ and $h_2^{<t>}$ are hidden states.

We can represent the RNN model in Fig. 14.4 in compact form by using the below matrix definitions. Hence, we can represent the model as in Fig. 14.5. To note here, bias values are not shown in the figure. They are implicit in definitions.

$$\mathbf{x}^{<t>} = \begin{bmatrix} x_1^{<t>} \\ x_2^{<t>} \end{bmatrix} \qquad \mathbf{h}^{<t>} = \begin{bmatrix} h_1^{<t>} \\ h_2^{<t>} \end{bmatrix} \tag{14.2}$$

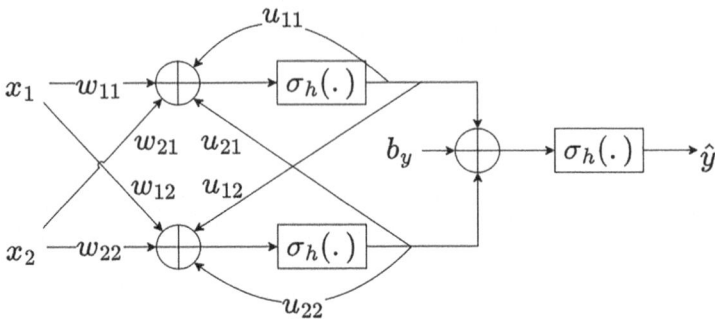

Fig. 14.3 The RNN model by two neurons

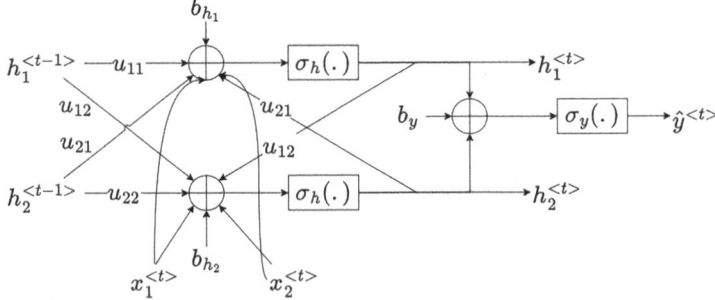

Fig. 14.4 The generalized RNN model by two neurons

Fig. 14.5 General RNN
model

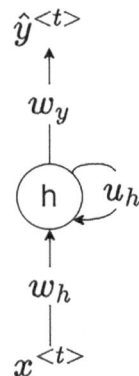

$$\mathbf{W_h} = \begin{bmatrix} w_{11} & w_{12} \\ w_{21} & w_{22} \end{bmatrix} \qquad \mathbf{U_h} = \begin{bmatrix} u_{11} & u_{12} \\ u_{21} & u_{22} \end{bmatrix} \qquad (14.3)$$

$$\mathbf{b_h} = \begin{bmatrix} b_{h1} \\ b_{h2} \end{bmatrix} \qquad \mathbf{W_y} = \begin{bmatrix} w_{y1} \\ w_{y2} \end{bmatrix} \qquad (14.4)$$

The structure in Fig. 14.5 is the general RNN form for any number of inputs and hidden states. Here, $x^{<t>}$ is the input vector at time t. $\hat{y}^{<t>}$ is the output at time t. $\mathbf{W_h}$ is the weight matrix between the input and hidden states. $\mathbf{U_h}$ is the weight matrix forming feedback between hidden states. $\mathbf{W_y}$ is the weight matrix between hidden states and output.

We can represent the relationship between the input and output for the RNN model in Fig. 14.5 as

$$\mathbf{h}^{<t>} = \sigma_h(\mathbf{W_h}\mathbf{x}^{<t>} + \mathbf{U_h}\mathbf{h}^{<t-1>} + b_h) \tag{14.5}$$

$$\hat{y}^{<t>} = \sigma_y(\mathbf{W_y}\mathbf{h}^{<t>} + b_y) \tag{14.6}$$

In Eqs. 14.5 and 14.6, $\sigma_h(\cdot)$ and $\sigma_y(\cdot)$ are nonlinear activation functions for the hidden state and output, respectively. $\sigma_h(\cdot)$ is most of the times taken as the tanh or sigmoid activation function.

14.2.2 Unfolding the RNN

We can unfold (unroll) the RNN model in Fig. 14.5. This helps us to display time dependence in the model better. Moreover, we will use unfolding in Sect. 14.7 to convert a trained RNN model to be embedded on the microcontroller. The general model in Fig. 14.5 will unfold in time as in Fig. 14.6.

When we unfold the RNN model as in Fig. 14.6, we can observe that it resembles the feedforward neural network introduced in Sect. 11.1.1. The main difference here is that weight values in RNN will not change between different time steps. This will allow us to use backpropagation while training the model. We will cover this topic in Sect. 14.2.4.

14.2.3 RNN Formation in Keras

We can represent the RNN model in Fig. 14.5 under Keras. Therefore, we feed weight values by hand first to understand the working principles of RNN. To do so, we form the Python script in Listing 14.1. In fact, this script corresponds to the memory element formation by single neuron in Sect. 14.1.1.

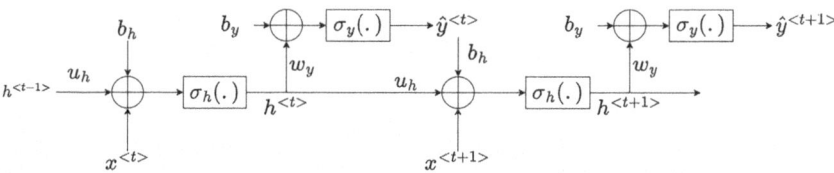

Fig. 14.6 General RNN model in unfolded form

Listing 14.1 Manual RNN formation, first case

```
import tensorflow as tf
import numpy as np

n_input = 1
n_units = 1

batch_size = 8;
x=np.zeros((1, batch_size, n_input))
x[0,0,:]=1

# RNN layer definition
Wh = 1
init_Wh = tf.keras.initializers.constant(Wh)

Uh = 1
init_Uh = tf.keras.initializers.constant(Uh)

bh = -0.5
init_bh = tf.keras.initializers.constant(bh)

def hard_limiter(x):
    return tf.math.maximum(tf.math.sign(x),0)

rnn_layer=tf.keras.layers.SimpleRNN(
    units=n_units,
    bias_initializer=init_bh,
    kernel_initializer=init_Wh,
    recurrent_initializer=init_Uh,
    activation=hard_limiter,
    return_sequences=True,
    return_state=True
    )

whole_sequence_output, final_state = rnn_layer(x)

np.set_printoptions(formatter={'float': '{: 0.2f}'.format})
out_conc=np.concatenate((x, whole_sequence_output.numpy()), axis
    =-1)
print("   Input, Output")
print(out_conc)
```

In Listing 14.1, we set the batch size as 8. Hence, the RNN model processes eight inputs at once. Then, we form input to be fed to the model. Afterward, we define weight values $\mathbf{W_h}$, $\mathbf{U_h}$ and bias term b_h in line with Eq. 14.5. Then, we form the RNN layer, rnn_layer using the function tf.keras.layers.SimpleRNN by setting model parameters. Finally, we obtain the RNN layer output by feeding input to it. As we execute the script, we obtain the same results given in Sect. 14.1.1.

We can also represent the memory element formation by two neurons introduced in Sect. 14.1.2. We provide the corresponding Python script in Listing 14.2. This script is similar to the one in Listing 14.1. Different from it, we form the output layer by defining $\mathbf{W_y}$ and b_y in Eq. 14.6. Therefore, we form the dense_layer using the function tf.keras.layers.Dense by setting model parameters. Then, we form the overall RNN model by merging both layers using the function tf.keras.Sequential.

Finally, we obtain the RNN model output by feeding input to it. As we execute the script, we obtain the same results given in Sect. 14.1.2.

Listing 14.2 Manual RNN formation, second case

```
import tensorflow as tf
import numpy as np

n_input = 2
n_units = 2

batch_size = 10;
x=np.zeros((1, batch_size, n_input))

x[0,0,:]=[1., 0.]
x[0,2,:]=[0., 1.]
x[0,3,:]=[0., 1.]
x[0,6,:]=[1., 0.]
x[0,7,:]=[1., 0.]

# RNN layer definition
Wh= np.array([[-1., 0.],
              [0., -1.]])
init_Wh = tf.keras.initializers.constant(Wh)

Uh= np.array([[0., -1.],
              [-1., 0.]])
init_Uh = tf.keras.initializers.constant(Uh)

bh = np.array([.5,.5])
init_bh = tf.keras.initializers.constant(bh)

def hard_limiter(x):
    return tf.math.maximum(tf.math.sign(x),0)

rnn_layer=tf.keras.layers.SimpleRNN(
    units=n_units,
    bias_initializer=init_bh,
    kernel_initializer=init_Wh,
    recurrent_initializer=init_Uh,
    activation=hard_limiter,
    return_sequences=True,
    )

# Dense layer definition

Wy= np.array([[1],
              [0]],)
init_Wy = tf.keras.initializers.constant(Wy)

by = -0.5
init_by = tf.keras.initializers.constant(by)

dense_layer=tf.keras.layers.Dense(
    units=1,
    kernel_initializer=init_Wy,
    bias_initializer=init_by,
    activation=hard_limiter,
    )

model = tf.keras.Sequential([
    rnn_layer,
```

```
    dense_layer
    ])

y = model(x)

model.summary()

np.set_printoptions(formatter={'float': '{: 0.2f}'.format})
out_conc=np.concatenate((x, y.numpy()), axis=-1)
print("   Input,         Output")
print(out_conc)
```

We can unfold the RNN layer in Keras by adding the property unroll=True to its definition. We will use unfolding while embedding the RNN model to the microcontroller. We will explain this operation in detail in Sect. 14.7.

14.2.4 Training the RNN

We can train the RNN model by adjusting its weight values represented by matrices $\mathbf{W_h}$, $\mathbf{U_h}$, $\mathbf{W_y}$ and bias values b_h, b_y in Eqs. 14.5 and 14.6. To do so, we can benefit from the backpropagation algorithm. The exact name for this method is backpropagation in time [62]. This method is similar to the backpropagation method introduced in Sect. 11.3. The only difference here is that we form loss values using actual outputs in time, $y^{<t>}$, and their predicted values, $\hat{y}^{<t>}$.

We can train the RNN model in Keras as in fully connected neural network and CNN models. We will provide such training examples in Sects. 14.5 and 14.6 for regression and image classification, respectively. Moreover, we can extract model parameters after training the RNN model. We provide a way of doing this in Listing 14.3. We should append this code block to the end of Listing 14.2. Hence, we can observe the model parameters we have set beforehand. In actual implementation, we will append the code block to the end of our RNN model trained by data. Then, we will be able to observe the learnt parameters after training.

Listing 14.3 Extracting RNN model parameters

```
print("\n RNN layer parameters")
layer = model.get_layer(name="simple_rnn")
matrices = layer.get_weights()

print("Wh:", matrices[0])
print("Uh:", matrices[1])
print("bh:", matrices[2])

print("\n Dense layer parameters")
layer = model.get_layer(name="dense")
matrices = layer.get_weights()

print("Wy:", matrices[0])
print("by:", matrices[1])
```

14.3   Gated Recurrent Unit

The RNN model may not work well while processing long data sequences. Researchers proposed LSTM to overcome this problem. We will consider it in detail in Sect. 14.4. Researchers also proposed a simpler architecture inspired by LSTM. This architecture is called gated recurrent unit (GRU) [5]. Being simpler, we will cover its properties in this section before the LSTM.

14.3.1 General Structure

GRU modifies the hidden state update formula, given in Eq. 14.5, by adding two gates and candidate hidden state. These modifications aim to overcome the problems in RNN. Before explaining working principles of GRU, let's provide formulas for the two gates, update and reset, and candidate hidden state.

The update gate is defined as

$$\mathbf{\Gamma}_z^{<t>} = \sigma_g(\mathbf{W_z}\mathbf{x}^{<t>} + \mathbf{U_z}\mathbf{h}^{<t-1>} + b_z) \tag{14.7}$$

where $\sigma_g(\cdot)$ is generally taken as the sigmoid activation function. Therefore, the update gate can take values between zero and one. The parameters $\mathbf{W_z}$, $\mathbf{U_z}$, and b_z are learnt during training the GRU. To note here, the update gate value changes in time depending on the present input and previous hidden state values.

The reset gate is defined as

$$\mathbf{\Gamma}_r^{<t>} = \sigma_g(\mathbf{W_r}\mathbf{x}^{<t>} + \mathbf{U_r}\mathbf{h}^{<t-1>} + b_r) \tag{14.8}$$

where $\sigma_g(\cdot)$ is the same activation function used in Eq. 14.7. Therefore, the reset gate can also take values between zero and one. The parameters $\mathbf{W_r}$, $\mathbf{U_r}$, and b_r are learnt during training the GRU. To note here, the reset gate value changes in time depending on the present input and previous hidden state values.

The candidate hidden state is defined as

$$\tilde{\mathbf{h}}^{<t>} = \sigma_h(\mathbf{W_h}\mathbf{x}^{<t>} + \mathbf{U_h}(\mathbf{\Gamma}_r^{<t>} \odot \mathbf{h}^{<t-1>}) + b_h) \tag{14.9}$$

where $\sigma_h(\cdot)$ is generally taken as the hyperbolic tangent activation function. Therefore, the candidate hidden state values can be between -1 and 1. The parameters $\mathbf{W_h}$, $\mathbf{U_h}$, and b_h are learnt during training the GRU. In Eq. 14.9, we have the element-wise multiplication operator $\odot$ such that gating is applied to each hidden state separately.

Based on the gate and candidate hidden state formulas in Eqs. 14.7, 14.8, and 14.9, the hidden state update formula in GRU becomes

$$\mathbf{h}^{<t>} = \mathbf{\Gamma}_z^{<t>} \odot \mathbf{h}^{<t-1>} + (1 - \mathbf{\Gamma}_z^{<t>}) \odot \tilde{\mathbf{h}}^{<t>} \tag{14.10}$$

Here, we used the notation suggested by Cho et al. [5]. This formulation is also used in Keras implementation of the GRU layer. Therefore, there is consistency in both ends.

Now, let's explain working principles of the GRU, hence the hidden state update formula in Eq. 14.10. As can be seen here, the current hidden state value is the weighted sum of previous and candidate hidden state values. Weight values are taken from the update gate vector $\Gamma_z^{<t>}$. Hence, $\Gamma_z^{<t>}$ controls "how much information from the previous hidden state will carry over to the current hidden state." In other words, it decides on "how much past should matter now." Moreover, since we have $\Gamma_z^{<t>}$ and $(1 - \Gamma_z^{<t>})$ as element-wise multipliers of $\mathbf{h}^{<t-1>}$ and $\tilde{\mathbf{h}}^{<t>}$, respectively; incrementing the effect of one block will decrement the effect of other block.

We should also take a close look at the candidate hidden state formula in Eq. 14.9. As can be seen here, this formula is similar to the one in Eq. 14.5. The difference here is the gating applied to the previous hidden state value as $\Gamma_r^{<t>} \odot \mathbf{h}^{<t-1>}$. Therefore, $\Gamma_r^{<t>}$ decides on how much information to carry from the previous hidden state to the candidate hidden state value.

As we have $\mathbf{h}^{<t>}$, we can calculate output of the GRU at time t as

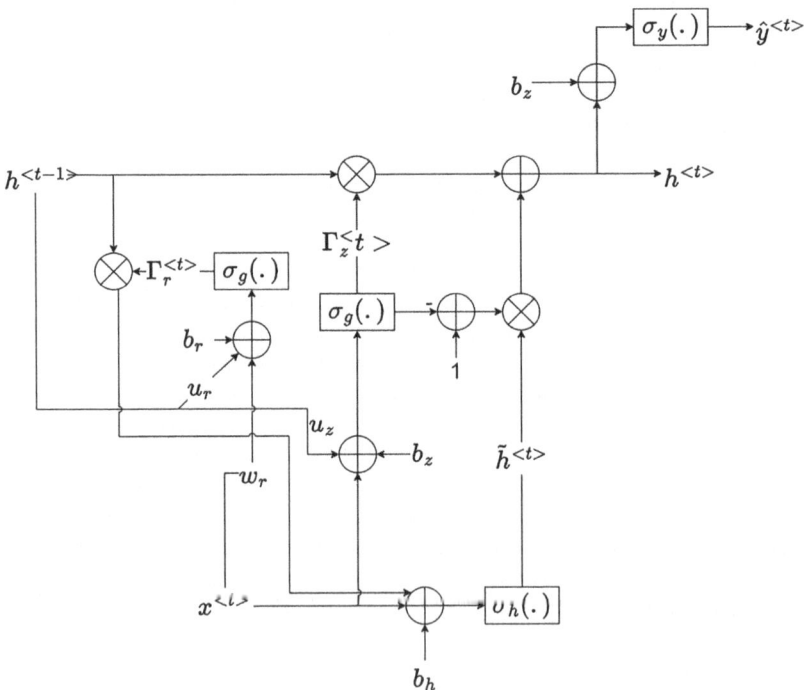

Fig. 14.7 Schematic representation of the GRU

$$\hat{y}^{<t>} = \sigma_y(\mathbf{W_y}\mathbf{h}^{<t>} + b_y) \tag{14.11}$$

where $\mathbf{W_y}$ is the weight matrix for the hidden states and b_y is the bias term as in RNN. We can schematically represent the GRU input output and hidden state update formulas as in Fig. 14.7.

14.3.2 GRU Formation and Training in Keras

We can represent the GRU model in Fig. 14.7 under Keras. Therefore, we feed weight values by hand first to understand working principles of GRU. To do so, we form the Python script in Listing 14.4.

Listing 14.4 Manual GRU formation

```
import tensorflow as tf
import numpy as np

n_input = 2
n_units = 2

batch_size = 10;
x=np.zeros((1, batch_size, n_input))

x[0,0,:]=[1., 0.]
x[0,2,:]=[0., 1.]
x[0,3,:]=[0., 1.]
x[0,6,:]=[1., 0.]
x[0,7,:]=[1., 0.]

# GRU layer definition

Wz= np.array([[0., 0.],
              [0., 0.]])

Wr= np.array([[0., 0.],
              [0., 0.]])

Wh= np.array([[-1., 0.],
              [0., -1.]])

kernel_weights = np.concatenate((Wz, Wr, Wh), axis=1)
init_kw = tf.keras.initializers.constant(kernel_weights)

Uz= np.array([[0., 0.],
              [0., 0.]])

Ur= np.array([[0., 0.],
              [0., 0.]])

Uh= np.array([[0., -1.],
              [-1., 0.]])

recurrent_weights = np.concatenate((Uz, Ur, Uh), axis=1)
init_rw = tf.keras.initializers.constant(recurrent_weights)

bz=np.array([-.1, -.1])
```

```
br=np.array([.5,  .5])

bh=np.array([.5,  .5])

bias= np.concatenate((bz,  br,  bh),  axis=0)
bias = tf.keras.initializers.constant(bias)

def hard_limiter(x):
    return tf.math.maximum(tf.math.sign(x),0)

GRU_layer=tf.keras.layers.GRU(
    units=n_units,
    use_bias=True,
    bias_initializer=bias,
    kernel_initializer=init_kw,
    recurrent_initializer=init_rw,
    activation=hard_limiter,
    recurrent_activation=hard_limiter,
    reset_after=False,
    return_sequences=True,
    )

# Dense layer definition

Wy= np.array([[1.],
              [0.]],)
init_Wy = tf.keras.initializers.constant(Wy)

by = -0.5
init_by = tf.keras.initializers.constant(by)

dense_layer=tf.keras.layers.Dense(
    units=1,
    kernel_initializer=init_Wy,
    bias_initializer=init_by,
    activation=hard_limiter,
    )

model = tf.keras.Sequential([
    GRU_layer,
    dense_layer
    ])

y = model(x)

model.summary()

np.set_printoptions(formatter={'float':  '{:  0.2f}'.format})
out_conc=np.concatenate((x,  y.numpy()),  axis=-1)
print("   Input,        Output")
print(out_conc)
```

The Python script in Listing 14.4 performs the same operations as in Listing 14.2. Hence, we form a memory element introduced in Sect. 14.1.2 by GRU. To do so, we set the model parameters to specific, but not unique, values. Hence, we obtain the same results as in Listing 14.2.

We can unfold the GRU layer in Keras by adding the property `unroll=True` to its definition. We will use unfolding while embedding the GRU model to the microcontroller. We will explain this operation in detail in Sect. 14.7.

We can train the GRU model by adjusting its weight and bias values. Training steps will be the same for the RNN case as in Sect. 14.2.4. Moreover, we can extract the model parameters after training the GRU model. We provide a way of doing this in Listing 14.5. We should append this code block to the end of Listing 14.4. Therefore, we can observe the model parameters we have set beforehand. In actual implementation, we will append the code block to the end of our GRU model trained by data. Hence, we will be able to observe the learnt parameters after training.

Listing 14.5 Extracting GRU model parameters

```
print("\n GRU layer parameters")
layer = model.get_layer(name="gru")
matrices = layer.get_weights()

print("Kernel Weights:", matrices[0].shape)
print("Recurrent Weights:", matrices[1].shape)
print("Biases:", matrices[2].shape)

W = matrices[0]
U = matrices[1]
b = matrices[2]

Wz = W[:, :n_units]
Wr = W[:, n_units: n_units * 2]
Wh = W[:, n_units * 2: n_units * 3]

print("Wz:", Wz)
print("Wr:", Wr)
print("Wh:", Wh)

Uz = U[:, :n_units]
Ur = U[:, n_units: n_units * 2]
Uh = U[:, n_units * 2: n_units * 3]

print("Uz:", Uz)
print("Ur:", Ur)
print("Uh:", Uh)

bz = b[:n_units]
br = b[n_units: n_units * 2]
bh = b[n_units * 2: n_units * 3]

print("bz:", bz)
print("br:", br)
print("bh:", bh)

print("\n Dense layer parameters")
layer = model.get_layer(name="dense")
matrices = layer.get_weights()

print("Wy:", matrices[0])
print("by:", matrices[1])
```

14.4   Long Short-Term Memory

The long short-term memory (LSTM) is the more advanced version of GRU [10]. Moreover, it has been introduced earlier than GRU. We postponed its coverage to this section since LSTM is more complex.

14.4.1 General Structure

As in GRU, LSTM modifies the hidden state update formula for RNN given in Eq. 14.5. This is done by adding three gates as input, forget, and output to the operation. Besides, two new variables are added to the hidden state update formula as cell and candidate cell values. Before explaining working principles of LSTM, let's provide formulas to all these new additions.

The input gate is defined as

$$\Gamma_i^{<t>} = \sigma_g(\mathbf{W_i}\mathbf{x}^{<t>} + \mathbf{U_i}\mathbf{h}^{<t-1>} + b_i) \qquad (14.12)$$

where $\sigma_g(\cdot)$ is generally taken as the sigmoid activation function. Therefore, the gate value can be between zero and one. The parameters $\mathbf{W_i}$, $\mathbf{U_i}$, and b_i are learnt during training the LSTM model. To note here, the input gate value changes in time depending on the present input and previous hidden state values.

The forget gate is defined as

$$\Gamma_f^{<t>} = \sigma_g(\mathbf{W_f}\mathbf{x}^{<t>} + \mathbf{U_f}\mathbf{h}^{<t-1>} + b_f) \qquad (14.13)$$

As in Eq. 14.12, $\sigma_g(\cdot)$ is generally taken as the sigmoid activation function. Therefore, the gate value can be between zero and one. The parameters $\mathbf{W_f}$, $\mathbf{U_f}$, and b_f are learnt during training the LSTM model. The forget gate value changes in time depending on the present input and previous hidden state values as in the input gate.

The output gate is defined as

$$\Gamma_o^{<t>} = \sigma_g(\mathbf{W_o}\mathbf{x}^{<t>} + \mathbf{U_o}\mathbf{h}^{<t-1>} + b_o) \qquad (14.14)$$

Here, $\sigma_g(\cdot)$ is also generally taken as the sigmoid activation function. Therefore, the gate value can be between zero and one. The parameters $\mathbf{W_o}$, $\mathbf{U_o}$, and b_o are learnt during training the LSTM model. The output gate value changes in time depending on the present input and previous hidden state values as in the input gate.

The candidate cell value is defined as

$$\tilde{\mathbf{c}}^{<t>} = \sigma_c(\mathbf{W_c}\mathbf{x}^{<t>} + \mathbf{U_c}\mathbf{h}^{<t-1>} + b_c) \qquad (14.15)$$

where $\sigma_c(\cdot)$ is generally taken as the hyperbolic tangent activation function. Therefore, the candidate cell value can be between -1 and 1. The parameters $\mathbf{W_c}$, $\mathbf{U_c}$, and b_c are learnt during training the LSTM model.

Based on the above definitions, we can form the cell value as

$$\mathbf{c}^{<t>} = \mathbf{\Gamma}_f^{<t>} \odot \mathbf{c}^{<t-1>} + \mathbf{\Gamma}_i^{<t>} \odot \tilde{\mathbf{c}}^{<t>} \tag{14.16}$$

where the cell value is calculated by the gated sum of $\tilde{\mathbf{c}}^{<t>}$ and $\mathbf{c}^{<t-1>}$.

As we have the cell value, we can define the hidden state update formula for LSTM as

$$\mathbf{h}^{<t>} = \mathbf{\Gamma}_o^{<t>} \odot \sigma_h(\mathbf{c}^{<t>}) \tag{14.17}$$

where $\sigma_h(\cdot)$ is generally taken as the hyperbolic tangent or linear activation function.

At this point, we can form an analogy between the GRU and LSTM. As mentioned previously, the GRU has two gates as $\mathbf{\Gamma}_z^{<t>}$ and $\mathbf{\Gamma}_r^{<t>}$ given in Eqs. 14.7 and 14.8, respectively. LSTM has three gates as $\mathbf{\Gamma}_i^{<t>}$, $\mathbf{\Gamma}_f^{<t>}$, and $\mathbf{\Gamma}_o^{<t>}$ given in Eqs. 14.12, 14.13, and 14.14, respectively. All these five gates have the same structure with their own parameters. The cell value in LSTM has a similar form as the hidden state update formula in the GRU. Both are formed by the sum of gated values. However, LSTM has more freedom in the gating operation since it has two different gates for this purpose. The hidden state update formula in LSTM is also different from the one in GRU in two different ways. First, there is an output gate in calculation. This gate controls "how much to reveal of the cell value." Second, the cell value in LSTM is passed through the nonlinear activation function $\sigma_h(\cdot)$ to calculate the hidden state value.

As we have $\mathbf{h}^{<t>}$, we can calculate output of the LSTM at time t as

$$\hat{y}^{<t>} = \sigma_y(\mathbf{W_y}\mathbf{h}^{<t>} + b_y) \tag{14.18}$$

where $\mathbf{W_y}$ is the weight matrix for the hidden states and b_y is the bias term as in RNN. We can schematically represent the LSTM input/output and hidden state update formulas as in Fig. 14.8.

14.4.2 LSTM Formation and Training in Keras

We can represent the LSTM structure in Fig. 14.8 under Keras. Therefore, we first feed weight values by hand to understand working principles of LSTM. To do so, we form the Python script in Listing 14.6.

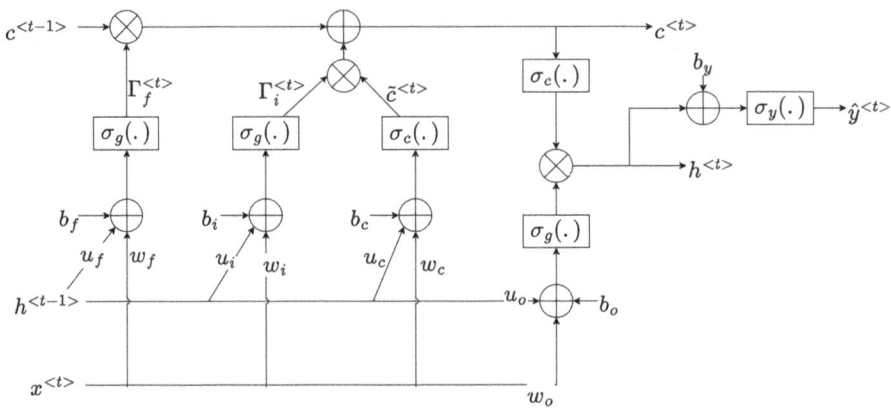

Fig. 14.8 Schematic representation of LSTM unit

Listing 14.6 Manual LSTM formation

```
import tensorflow as tf
import numpy as np

n_input = 2
n_units = 2

batch_size = 10;
x=np.zeros((1, batch_size, n_input))

x[0,0,:]=[1.,  0.]
x[0,2,:]=[0.,  1.]
x[0,3,:]=[0.,  1.]
x[0,6,:]=[1.,  0.]
x[0,7,:]=[1.,  0.]

# LSTM layer definition

Wi= np.array([[-1.,  0.],
              [0.,  -1.]])

Wf= np.array([[0.,  0.],
              [0.,  0.]])

Wc= np.array([[0.,  0.],
              [0.,  0.]])

Wo= np.array([[0.,  0.],
              [0.,  0.]])

kernel_weights = np.concatenate((Wi, Wf, Wc, Wo), axis=1)
init_kw = tf.keras.initializers.constant(kernel_weights)

Ui= np.array([[0.,  -1.],
              [-1.,  0.]])

Uf= np.array([[0.,  0.],
              [0.,  0.]])

Uc= np.array([[0.,  0.],
```

```
                        [0., 0.]])
Uo= np.array([[0., 0.],
              [0., 0.]])

recurrent_weights = np.concatenate((Ui, Uf, Uc, Uo), axis=1)
init_rw = tf.keras.initializers.constant(recurrent_weights)

bi=np.array([.5, .5])
bf=np.array([-.1, -.1])
bc=np.array([.5, .5])
bo=np.array([.1, .1])

bias= np.concatenate((bi, bf, bc, bo), axis=0)
bias = tf.keras.initializers.constant(bias)

def hard_limiter(x):
    return tf.math.maximum(tf.math.sign(x),0)

LSTM_layer=tf.keras.layers.LSTM(
    units=n_units,
    use_bias=True,
    bias_initializer=bias,
    unit_forget_bias=False,
    kernel_initializer=init_kw,
    recurrent_initializer=init_rw,
    activation=hard_limiter,
    recurrent_activation=hard_limiter,
    return_sequences=True,
    )

# Dense layer definition

Wy= np.array([[1.],
              [0.]],)
init_Wy = tf.keras.initializers.constant(Wy)

by = -0.5
init_by = tf.keras.initializers.constant(by)

dense_layer=tf.keras.layers.Dense(
    units=1,
    kernel_initializer=init_Wy,
    bias_initializer=init_by,
    activation=hard_limiter,
    )

model = tf.keras.Sequential([
    LSTM_layer,
    dense_layer
    ])

y = model(x)

model.summary()

np.set_printoptions(formatter={'float': '{: 0.2f}'.format})
out_conc=np.concatenate((x, y.numpy()), axis=-1)
print("    Input,        Output")
print(out_conc)
```

The Python script in Listing 14.6 performs the same operations as in Listing 14.2. Hence, we form a memory element introduced in Sect. 14.1.2 by LSTM. To do so, we set the model parameters to specific, but not unique, values. Hence, we obtain the same results as in Listing 14.2.

We can unfold the LSTM layer in Keras by adding the property unroll=True to its definition. We will use unfolding while embedding the LSTM model to the microcontroller. We will explain this operation in detail in Sect. 14.7.

We can train the LSTM model by adjusting its weight and bias values. Training steps will be the same as the RNN case in Sect. 14.2.4. Moreover, we can extract the model parameters after training the LSTM model. We provide a way of doing this in Listing 14.7. We should append this code block to the end of Listing 14.6. Therefore, we can observe the model parameters we have set beforehand. In actual implementation, we will append the code block to the end of our LSTM model trained by data. Hence, we will be able to observe the learnt parameters after training.

Listing 14.7 Extracting LSTM model parameters

```
print("\n LSTM layer parameters")
layer = model.get_layer(name="lstm")
matrices = layer.get_weights()

print("Kernel Weights:", matrices[0].shape)
print("Recurrent Weights:", matrices[1].shape)
print("Biases:", matrices[2].shape)

W = matrices[0]
U = matrices[1]
b = matrices[2]

Wi = W[:, :n_units]
Wf = W[:, n_units: n_units * 2]
Wc = W[:, n_units * 2: n_units * 3]
Wo = W[:, n_units * 3:]

print("Wi:", Wi)
print("Wf:", Wf)
print("Wc:", Wc)
print("Wo:", Wo)

Ui = U[:, :n_units]
Uf = U[:, n_units: n_units * 2]
Uc = U[:, n_units * 2: n_units * 3]
Uo = U[:, n_units * 3:]

print("Ui:", Ui)
print("Uf:", Uf)
print("Uc:", Uc)
print("Uo:", Uo)

bi = b[:n_units]
bf = b[n_units: n_units * 2]
bc = b[n_units * 2: n_units * 3]
bo = b[n_units * 3:]

print("bi:", bi)
```

```
print("bf:", bf)
print("bc:", bc)
print("bo:", bo)

print("\n Dense layer parameters")
layer = model.get_layer(name="dense")
matrices = layer.get_weights()

print("Wy:", matrices[0])
print("by:", matrices[1])
```

14.5 Regression via Recurrence Models

We can form a regressor via recurrence models. Before doing so, we should first adjust the data to be used in operation. Hence, it becomes suitable to be processed by the recurrence model of our choice. We provide a sample code formed for this purpose in Listing 14.8.

Listing 14.8 Preparing dataset for regression via recurrence models

```
import tensorflow as tf
import os.path as osp
import numpy as np

samples = np.load(osp.join("regression_data", "reg_samples.npy"))
line_values = np.load(osp.join("regression_data", "
    reg_line_values.npy"))
sine_values = np.load(osp.join("regression_data", "
    reg_sine_values.npy"))

series=sine_values

n_input = 1
time_length=9

ls=len(samples)
dataset=tf.keras.utils.timeseries_dataset_from_array(
    samples[:-1],
    np.roll(series, -n_input)[:-1],
    sequence_length=n_input,
    batch_size=time_length,
    end_index=ls-(ls-ls//time_length*time_length),
)

x_train=[]; y_train=[]
for inputs, targets in dataset:
  x_train.append(inputs.numpy())
  y_train.append(targets.numpy())

x_train = np.array(x_train)
y_train = np.array(y_train)

print(x_train.shape, y_train.shape)
```

In Listing 14.8, we first load data via the function np.load. Hence, we have three arrays as samples, line_values, and sine_values. To show the operations performed on these arrays, we pick the samples and sine_values, the latter being copied to the variable series. Then, we use the TensorFlow function timeseries_dataset_from_array to generate time series data from the given arrays. For our case, inputs to the function are samples and series. The parameters used in operation are sequence_length and batch_size. We set them as n_input =1 and time_length =3, respectively. We will explain these parameters in detail in the next example. Let's summarize the output obtained from the function timeseries_dataset_from_array. In our example, we obtain dataset as output. Next, we extract the **input** and output values from dataset and assign these values to variables, x_train and y_train, respectively. Based on the given parameters, shape of these variables become (22, 9, 1, 1) and (22, 9, 1), respectively. This means that the x_train variable holds a tensor with 22 elements each having size (9, 1, 1). Likewise, the y_train variable holds a tensor with 22 elements each having size (9, 1). We will explain the importance of these structures in the regression via recurrence example next.

As we modify the data to be suitable for regression via recurrence models, the next phase is forming the model for this purpose. Since RNN, GRU, and LSTM share similar properties, we provide one Python script for regression for all models in Listing 14.9. Initially only the RNN layer is enabled in the script. The GRU and LSTM are disabled. The user can enable the corresponding code line to use GRU or LSTM.

Listing 14.9 Regression example via RNN, GRU, and LSTM

```
import tensorflow as tf
import os.path as osp
import numpy as np
from matplotlib import pyplot as plt

samples = np.load(osp.join("regression_data", "reg_samples.npy"))
line_values = np.load(osp.join("regression_data", "
    reg_line_values.npy"))
sine_values = np.load(osp.join("regression_data", "
    reg_sine_values.npy"))

series=sine_values

print(samples.shape)

plt.figure()
plt.plot(samples,series)

n_input = 1
time_length=9

ls=len(samples)

dataset=tf.keras.utils.timeseries_dataset_from_array(
    samples[:-1],
    np.roll(series, -n_input)[:-1],
    sequence_length=n_input,
```

```
      batch_size=time_length,
      end_index=ls-(ls-ls//time_length*time_length),
)

x_train=[]; y_train=[]
for inputs, targets in dataset:
  x_train.append(inputs.numpy())
  y_train.append(targets.numpy())

x_train = np.array(x_train)
y_train = np.array(y_train)

print(x_train.shape, y_train.shape)

units=10

model = tf.keras.Sequential([
  tf.keras.layers.Input(shape=(time_length,n_input)),
#  tf.keras.layers.SimpleRNN(units, unroll=True),
#  tf.keras.layers.GRU(units,unroll=True),
#  tf.keras.layers.LSTM(units,unroll=True),
  tf.keras.layers.LSTM(units),
  tf.keras.layers.Dense(1)
])

model.compile(
  loss=tf.keras.losses.MeanSquaredError(),
  optimizer=tf.keras.optimizers.Adam(1e-3))

history=model.fit(x_train,y_train, epochs=500, verbose=False)

model.summary()

plt.figure()
plt.xlabel('Epoch Number')
plt.ylabel("Loss")
plt.plot(history.history['loss'])

MODEL_NAME = 'LSTM'
MODEL_DIR = 'models'

model.save(osp.join(MODEL_DIR, MODEL_NAME + '_model_tf'),
    save_format = "tf")
model.save(osp.join(MODEL_DIR, MODEL_NAME + 'model_keras.h5'),
    save_format = "h5")

y_pred = model.predict(x_train)

x_t=x_train[:,0,:,:]
x_t = x_t.reshape(-1)

plt.figure()
plt.plot(samples,series)
plt.plot(x_t,y_pred,'r')
plt.show()
```

in Listing 14.9, the most important part while forming the model is setting the input shape for the recurrence layer. Therefore, we set the input shape to the model as shape=(time_length, n_input). Here, the time_length value indicates the number of time steps to be handled in recurrence. The n_input value indicates size of the input to each recurrence state.

14.6   Image Classification via Recurrence Models

We can perform image classification via recurrence models. As explained in the previous section, we should set the input shape to the model as shape=(time_length , n_input). Here, the time_length value indicates the number of time steps to be handled in recurrence. The n_input value indicates size of the input to each recurrence state. Therefore, we should treat the image at hand as a sequence of vectors. As an example, if we have an MNIST dataset image with size 28×28 pixels, then we can represent it as 28 sequential rows each having 28 pixels. Therefore, we can feed 28 sequential features, each having 28 values, to the recurrence model. In other words, we feed one row of the image, with 28 pixels, at each time step. Thus, the image is fed to the model through 28 time steps.

We provide a sample code to classify MNIST images through recurrence models in Listing 14.10. Since RNN, GRU, and LSTM share similar properties, we provide one Python script for all recurrence models. Initially the RNN layer is enabled in the code. The GRU and LSTM units are disabled. The user can enable the corresponding code line to use the GRU or LSTM.

Listing 14.10   Image classification example via RNN, GRU, and LSTM

```
import tensorflow as tf
from matplotlib import pyplot as plt
import numpy as np
import os.path as osp

num_classes = 10

# the data, split between train and test sets
(train_images, train_labels), (test_images, test_labels)  = tf.
    keras.datasets.mnist.load_data()

train_images = train_images / 255.0
test_images = test_images / 255.0

# Make sure images have shape (28, 28, 1)
train_images = np.expand_dims(train_images, -1)
test_images = np.expand_dims(test_images, -1)

print("train images shape:", train_images.shape)
print(train_images.shape[0], "train samples")
print(test_images.shape[0], "test samples")

units = 16

model = tf.keras.models.Sequential([
    tf.keras.layers.Input(shape=(28, 28)),
#     tf.keras.layers.SimpleRNN(units, unroll=True),
#     tf.keras.layers.GRU(units,unroll=True),
    tf.keras.layers.LSTM(units,unroll=True),
    tf.keras.layers.Dense(num_classes, activation="softmax"),
```

```
])

model.summary()

model.compile(optimizer='adam', loss='
    sparse_categorical_crossentropy', metrics=['accuracy'])

model.fit(train_images, train_labels, batch_size=64, epochs=5,
    validation_split=0.1)
MODEL_NAME = 'LSTM'
MODEL_DIR = 'models'

model.save(osp.join(MODEL_DIR, MODEL_NAME + '_model_tf'),
    save_format = "tf")
model.save(osp.join(MODEL_DIR, MODEL_NAME + 'model_keras.h5'),
    save_format = "h5")
```

14.7   Embedding the Recurrence Model to the Microcontroller

As in Sect. 13.5, this section will also be fairly short. The main reason is that we direct the reader to the methods introduced in Chap. 12. Hence, the recurrence model can be converted to the TensorFlow Lite format. Since we formed the methods in Chap. 12 for general usage, the recurrence model can also be processed by them. This also applies to the STM32Cube.AI usage.

The reader should only pay attention to the unrolling operation for the recurrence models. To emphasize again, the unrolled recurrence model in fact has the same characteristics as with feedforward neural networks. The only difference will be the repetitive usage of the same weight values in the unrolled model. Therefore, as we unroll the recurrence model, we can convert it to the corresponding TensorFlow Lite format. Then, the C array corresponding to this model can be obtained. This array can be used for inference on the microcontroller.

14.8   Application: Keyword Spotting from Audio Signals

In this application, we use an LSTM model for keyword spotting from audio signals. Unlike the applications in Chaps. 6 and 13, this model learns temporal relationship between Mel frames. We provide the Python script to transform audio signals to spectrograms and forming a dataset in Listing 14.11. Here, the function mel_spectrogram takes a path to a wav audio file and parameters as input. The parameters include the sample length, FFT length, step size, and number of Mel bins. The function reads the audio file and decodes and processes it to create a Mel spectrogram.

Listing 14.11 Forming a data loader to generate MFCC features for keyword spotting from audio signals

```
import os
import tensorflow as tf

def mel_spectrogram(wav_path, params):
    sample_len = params[0]
    fft_len = params[1]
    step = params[2]
    mel_bins = params[3]
    sample_rate = 8000

    file_specs = tf.strings.split(wav_path, ".")[0]
    file_specs = tf.strings.split(file_specs, os.path.sep)[1]
    digit = tf.strings.to_number(tf.strings.split(file_specs, "_"
        )[0], out_type = tf.int32)
    wavfile = tf.io.read_file(wav_path)
    sample, _ = tf.audio.decode_wav(wavfile)
    sample = tf.squeeze(sample, axis=-1)
    fixed_size = tf.constant([sample_len])
    sample_shape = tf.shape(sample)
    def pad_func():
        pad_size = fixed_size - sample_shape
        padding = tf.concat([tf.zeros_like(sample_shape),
            pad_size], axis=0)
        padding = tf.reshape(padding,(-1,2))
        return tf.pad(sample, padding)
    def slice_func():
        return tf.slice(sample, tf.constant([0]), fixed_size)

    sample = tf.cond(sample_shape > fixed_size, slice_func,
        pad_func)

    spectrogram = tf.signal.stft(sample, frame_length = fft_len,
        frame_step= step)
    spectrogram = tf.abs(spectrogram)

    num_spectrogram_bins = fft_len // 2 + 1   # spectrogram.shape
        [-1]
    lower_edge_hertz, upper_edge_hertz = 40.0, sample_rate / 2
    linear_to_mel_weight_matrix = tf.signal.
        linear_to_mel_weight_matrix(
      mel_bins, num_spectrogram_bins, sample_rate,
         lower_edge_hertz,
      upper_edge_hertz)
    mel_spectrograms = tf.tensordot(spectrogram,
        linear_to_mel_weight_matrix, 1)
    mel_spectrograms.set_shape(spectrogram.shape[:-1].concatenate
        (linear_to_mel_weight_matrix.shape[-1:]))

    # Compute a stabilized log to get log-magnitude mel-scale
        spectrograms.
    log_mel_spectrograms = tf.math.log(mel_spectrograms + 1e-6)
    avg = tf.math.reduce_mean(log_mel_spectrograms)
    std = tf.math.reduce_std(log_mel_spectrograms)
    normalized_mel = (log_mel_spectrograms - avg) / std
    normalized_mel = normalized_mel[..., tf.newaxis]

    return normalized_mel, digit
```

```
def create_datasets(sample_length, fft_size, step_size,
    batch_size, mel_bins):
    RECORDINGS_DIR = "recordings/*.wav"
    ds = tf.data.Dataset.list_files(RECORDINGS_DIR)
    ds_size = tf.data.experimental.cardinality(ds).numpy()
    ds = ds.map(lambda x: mel_spectrogram(x, (sample_length,
        fft_size, step_size, mel_bins))).shuffle(ds_size)
    train_ds = ds.take(int(0.8 * ds_size))
    val_ds = ds.skip(int(0.8 * ds_size))
    test_ds = val_ds.shard(num_shards=2, index=0)
    val_ds = val_ds.shard(num_shards=2, index=1)

    train_ds = train_ds.batch(batch_size).cache().prefetch(tf.
        data.AUTOTUNE)
    val_ds = val_ds.batch(batch_size).cache().prefetch(tf.data.
        AUTOTUNE)
    test_ds = test_ds.batch(batch_size).cache().prefetch(tf.data.
        AUTOTUNE)
    spec_shape = (sample_length // step_size -1, mel_bins, 1)
    return train_ds, val_ds, test_ds, spec_shape
```

The Python script in Listing 14.11 is typically used to train a machine learning model on audio data, where the model is expected to learn from Mel spectrogram representations. The digit extracted from the filename can be used as a label for supervised learning. The create_datasets function makes it easy to create datasets for training, validation, and testing. Use of caching and prefetching can significantly speed up the training process. The returned spectrogram shape can be useful when defining the input shape of the model.

In Listing 14.12, the function LSTMKWSModel is defined to build the LSTM model for keyword spotting using Keras. The model is designed to classify input audio data into digits. The function takes two arguments. The first argument nCategories specifies the number of output categories. The second argument input_shape defines the shape of input data. Two convolutional layers are applied to input data. The first convolution layer has 10 filters with a (5, 1) kernel size. The second layer has one filter with a (5, 1) kernel size. Both layers use the ReLU activation function. After each convolution layer, a batch normalization layer is applied to normalize the activations of previous layer. Output of the second batch normalization layer is then reshaped to match the first two dimensions of the input shape using layers. Reshape(input_shape[0:2]). The reshaped data is then passed through two LSTM layers. These layers, defined by layers.LSTM(128, return_sequences=True, unroll = True), apply LSTM units in a forward direction to data.

Listing 14.12 Forming the LSTM model for keyword spotting from audio signals

```
from keras import layers
from keras import Model

def LSTMKWSModel(nCategories, input_shape):
    inputs = layers.Input(input_shape, name='input')
    x = layers.Conv2D(10, (5, 1), activation='relu', padding='
        same')(inputs)
    x = layers.BatchNormalization()(x)
```

```
x = layers.Conv2D(1, (5, 1), activation='relu', padding='same
    ')(x)
x = layers.BatchNormalization()(x)
x = layers.Reshape(input_shape[0:2])(x)
x = layers.LSTM(128, return_sequences=True, unroll = True)(x)
x = layers.LSTM(128, return_sequences=True, unroll = True)(x)
xFirst = x[:, -1]
query = layers.Dense(128)(xFirst)

scores = layers.Dot(axes=[1, 2])([query, x])
scores = layers.Softmax(name='softmax')(scores)
attVector = layers.Dot(axes=[1, 1])([scores, x])

x = layers.Dense(64, activation='relu')(attVector)
x = layers.Dense(32)(x)

output = layers.Dense(nCategories, activation='softmax', name
    ='output')(x)

model = Model(inputs=[inputs], outputs=[output])
model.compile(optimizer='adam', loss=['
    sparse_categorical_crossentropy'], metrics=['
    sparse_categorical_accuracy'])

return model
```

Output of the last LSTM layer is then used to compute a vector consists of states and sequence itself in Listing 14.12. This is done by first extracting the last time step of the LSTM output x[:, -1] and then transforming it to a query vector using a dense layer layers.Dense(128)(xFirst). The query vector is then used to compute scores over the LSTM output using a dot product attScores = layers.Dot(axes=[1, 2])([query, x]), followed by a softmax operation to obtain a probability value layers.Softmax()(attScores). These scores are then used to compute a weighted sum of the LSTM output, resulting in a vector layers.Dot(axes=[1, 1])([attScores , x]). The vector is then passed through two dense layers, the first one with 64 units and the second one with 32 units. Finally, the output layer is a dense layer with nCategories units and softmax activation function, which will output a probability value over nCategories categories. The model is then compiled with the Adam optimizer, sparse categorical cross-entropy loss, and sparse categorical accuracy metric. The function returns the compiled model. This model can then be trained on data for keyword spotting. In Listing 14.13, all previously implemented data and model functions are combined to train the model.

Listing 14.13 Training the formed LSTM model for keyword spotting from audio signals

```
from data_loader import create_datasets
from model import LSTMKWSModel
from keras import callbacks

SAMPLE_RATE = 8000
FFT_LEN = 512
STEP_SIZE = 256
BATCH_SIZE = 32
NUM_MEL_BINS = 80
```

```
train_ds, val_ds, test_ds, input_shape = create_datasets(
    SAMPLE_RATE, FFT_LEN, STEP_SIZE, BATCH_SIZE, NUM_MEL_BINS)

kws_rnn_model = LSTMKWSModel(10, input_shape)

model_cp_callback = callbacks.ModelCheckpoint("kws_LSTM_model.h5"
    , save_best_only=True)
es_callback = callbacks.EarlyStopping(verbose=1, patience=5)

kws_rnn_model.summary()

kws_rnn_model.fit(train_ds,
            epochs=50,
            validation_data= val_ds,
            verbose=1,
            callbacks = [model_cp_callback, es_callback])
```

The Python script in Listing 14.14 is used for converting the trained Keras model to TensorFlow Lite form. Here, the converter is configured to use default optimizations and to support only the operations that are built into TensorFlow Lite format. The TensorFlow Lite model is then saved to a file at the path specified by TFLITE_MODEL_PATH. Finally, the convert_tflite2cc function is called to convert the TensorFlow Lite model to a C++ source file. This file will be deployed to the microcontroller for inference.

Listing 14.14 Converting the trained LSTM model to TensorFlow Lite format and C++ array

```
import tensorflow as tf
import keras
from tflite2cc import convert_tflite2cc

TFLITE_MODEL_PATH = "kws_LSTM_model.tflite"
kws_LSTM_model = keras.models.load_model("kws_LSTM_model.h5")
converter = tf.lite.TFLiteConverter.from_keras_model(
    kws_LSTM_model)
converter.optimizations = [tf.lite.Optimize.DEFAULT]
converter.target_spec.supported_ops = [tf.lite.OpsSet.
    TFLITE_BUILTINS]
kws_LSTM_lite = converter.convert()

with open(TFLITE_MODEL_PATH, "wb") as tflite_file:
    tflite_file.write(kws_LSTM_lite)

convert_tflite2cc(TFLITE_MODEL_PATH, "kws_LSTM_model.cc")
```

We will provide complete projects for STM32CubeIDE and Mbed Studio in the accompanying book repository. There, we will provide different settings in the implementation step. Hence, the reader will observe different options in implementation. While doing so, we will realize the complete keyword spotting system on the microcontroller from the initial audio signal acquisition to final inference step.

14.9   Application: Handwritten Digit Recognition from Digital Images

In this application, we use an LSTM model to classify handwritten digits from digital images. We formed the Python script in Listing 14.15 for this purpose. Here, create_model is defined to build a simple recurrence model using Keras. The model is designed to classify input data into num_classes different categories. Model construction begins with an input layer defined by Input(input_shape). This layer simply takes the input data and passes it to the next layer. Next, an LSTM layer is applied to input data. This layer has 16 units and returns only computed states. The unroll=True argument is necessary for deploying model as the TensorFlow Lite model. To note here, this operation can speed up computations, although it tends to use more memory. Output of the LSTM layer is the final state. This state is then passed through a dense layer. The model is then instantiated using Model(inputs =[inp], outputs=[output]), with the input and output layers passed as arguments. An optimizer is then defined using Adam(learning_rate=1e-4). The model is then compiled with the defined optimizer, sparse categorical cross-entropy loss, and sparse categorical accuracy metric using model.compile. Finally, the function returns the compiled model. This model can then be trained on data for handwritten digit recognition task.

Listing 14.15 The LSTM model for handwritten digit recognition

```
from keras import Input
from keras import layers
from keras import Model
from keras.optimizers import Adam

def LSTMMNISTModel(input_shape, num_classes):
    inp = Input(input_shape)
    state = layers.LSTM(16, unroll=True)(inp)
    output = layers.Dense(num_classes, activation="softmax")(
        state)
    model = Model(inputs=[inp], outputs=[output])
    opt = Adam(learning_rate=1e-3)
    model.compile(
        optimizer=opt,
        loss=["sparse_categorical_crossentropy"],
        metrics=["sparse_categorical_accuracy"],
    )
    return model
```

In Listing 14.16, the LSTM model training and validation steps are shown for the handwritten digit recognition task. In this script, the MNIST dataset is loaded and modified as in previous applications. The model is then created using the function create_model. Shape of the input data and number of output categories are passed as arguments. Two callback functions are then defined to save the best model to a file and stop training when a monitored quantity has stopped improving. Finally, the model is trained using the fit method.

Listing 14.16 Training the LSTM model on the MNIST dataset

```
import keras
from sklearn.model_selection import train_test_split
from model import LSTMMNISTModel

(train_imgs, train_labels), (test_imgs, test_labels) = keras.
    datasets.mnist.load_data()
train_imgs = train_imgs / 255.0
test_imgs = test_imgs / 255.0
num_samples, *input_shape = train_imgs.shape
train_imgs, val_imgs, train_labels, val_labels = train_test_split
    (
        train_imgs, train_labels, test_size=0.1
)

MNIST_LSTM_model = LSTMMNISTModel(input_shape, 10)
model_cp_callback = keras.callbacks.ModelCheckpoint(
    "MNIST_LSTM_model.h5", save_best_only=True
)
es_callback = keras.callbacks.EarlyStopping(verbose=1, patience
    =5)
MNIST_LSTM_model.fit(
    x=train_imgs,
    y=train_labels,
    epochs=50,
    validation_data=(val_imgs, val_labels),
    verbose=1,
    callbacks=[model_cp_callback, es_callback],
)
```

The Python script in Listing 14.17 is used for converting the trained Keras model to TensorFlow Lite format. Here, the converter is configured to use default optimizations and to support only the operations that are built into the TensorFlow Lite format. The TensorFlow Lite model is then saved to a file at the path specified by TFLITE_MODEL_PATH. Finally, the convert_tflite2cc function is called to convert the TensorFlow Lite model to a C++ source file. This file will be deployed to the microcontroller for inference.

Listing 14.17 Converting the trained LSTM model to TensorFlow Lite model and C++ array

```
import tensorflow as tf
import keras
from tflite2cc import convert_tflite2cc

TFLITE_MODEL_PATH = "MNIST_LSTM_model.tflite"
MNIST_LSTM_model = keras.models.load_model("MNIST_LSTM_model.h5")
converter = tf.lite.TFLiteConverter.from_keras_model(
    MNIST_LSTM_model)
converter.optimizations = [tf.lite.Optimize.DEFAULT]
converter.target_spec.supported_ops = [tf.lite.OpsSet.
    TFLITE_BUILTINS]
MNIST_LSTM_lite = converter.convert()

with open(TFLITE_MODEL_PATH, "wb") as tflite_file:
```

```
tflite_file.write(MNIST_LSTM_lite)

convert_tflite2cc(TFLITE_MODEL_PATH, "MNIST_LSTM_model.cc")
```

We will provide complete projects for STM32CubeIDE and Mbed Studio in the accompanying book repository. There, we will provide different settings in the implementation step. Hence, the reader will observe different options in implementation. While doing so, we will realize the complete handwritten digit recognition system on the microcontroller from the initial audio signal acquisition to final inference step.

14.10 Application: Estimating Future Temperature Values

In this application, recurrent neural networks are employed to estimate future temperature values based on the previous readings. Similar to Listing 11.7, we start by loading and preprocessing the temperature readings. Preprocessing steps include forming five previous readings as features and standard normalization. After preprocessing step, we need to create a model using recurrent neural networks. This time GRU is used as data does not contain complicated time relation. In Listing 14.18, model function is shown.

Listing 14.18 Forming the GRU model in Keras

```
from keras.models import Sequential
from keras.layers import Dense, GRU, Dropout
from keras.optimizers import Adam
from keras.losses import MeanAbsoluteError

def gru_temperature_model(input_shape):
    model = Sequential([GRU(units = 50, return_sequences = True,
        input_shape = (input_shape,1), unroll = True),
                        Dropout(0.2),
                        GRU(units = 50, unroll= True),
                        Dense(units = 1)])

    model.compile(optimizer=Adam(learning_rate=5e-4),
                loss=MeanAbsoluteError())
    return model
```

Once we have the model function, we can simply train it under Listing 14.19. Note that the np.expand_dims function is used for both train and test samples as GRU expects two-dimensional input. The trained model is saved as H5 file.

Listing 14.19 Training the GRU model for future temperature value estimation

```
import numpy as np
import pandas as pd
from matplotlib import pyplot as plt
```

```
from sklearn.metrics import mean_absolute_error
from sklearn.model_selection import train_test_split
from model import gru_temperature_model

df = pd.read_csv('temperature_dataset.csv')
y = df['Room_Temp']
prev_values_count = 5

X = pd.DataFrame()
for i in range(prev_values_count, 0, -1):
    X['t-' + str(i)] = y.shift(i)

X = X[prev_values_count:]
y = y[prev_values_count:]

X_train, X_test, y_train, y_test = train_test_split(X, y,
    test_size=0.2)
train_mean = X_train.mean()
train_std = X_train.std()

X_train = (X_train - train_mean)/ train_std
X_test = (X_test - train_mean)/ train_std
temperature_model = gru_temperature_model(prev_values_count)
X_train = np.expand_dims(X_train, 2)
temperature_model.fit(X_train, y_train, epochs = 250, batch_size
    = 128)

X_test = np.expand_dims(X_test, 2)
y_test_predict = temperature_model.predict(X_test)

fig, ax = plt.subplots(1,1)
ax.plot(y_test.to_numpy(), label = "Actual values")
ax.plot(y_test_predict, label = "Predicted values")
plt.legend()
plt.show()

mae_test = np.sqrt(mean_absolute_error(y_test, y_test_predict))
print(f"Test set MAE:{mae_test}")

temperature_model.save("mlp_temperature_prediction.h5")
```

The final step to deploy the temperature estimation model is converting it to TensorFlow Lite model. This conversion can be completed as shown in Listing 14.20.

Listing 14.20 Converting future temperature value estimation model to TensorFlow Lite format

```
import tensorflow as tf
import keras
from tflite2cc import convert_tflite2cc

TFLITE_MODEL_PATH = "mlp_temperature_prediction.tflite"
mnist_rnn_model = keras.models.load_model("
    mlp_temperature_prediction.h5")
converter = tf.lite.TFLiteConverter.from_keras_model(
    mnist_rnn_model)
converter.optimizations = [tf.lite.Optimize.DEFAULT]
converter.target_spec.supported_ops = [tf.lite.OpsSet.
    TFLITE_BUILTINS]
kws_rnn_lite = converter.convert()

with open(TFLITE_MODEL_PATH, "wb") as tflite_file:
```

```
tflite_file.write(kws_rnn_lite)
```
```
convert_tflite2cc(TFLITE_MODEL_PATH, "mnist_rnn_model.cc")
```

The `create_datasets` function creates training, validation, and test datasets from .wav files in a specified directory. It lists all the .wav files in the directory, computes their Mel spectrograms, and shuffles them. It then splits the shuffled dataset into training, validation, and test datasets. The split is 80% for training and 20% for validation and testing. The validation and test datasets are further split into two equal parts. The datasets are batched, cached, and prefetched for efficient training. The function returns the three datasets and the shape of the spectrograms.

We will provide complete projects for STM32CubeIDE and Mbed Studio in the accompanying book repository. There, we will provide different settings in the implementation step. Hence, the reader will observe different options in implementation. While doing so, we will realize the complete temperature value estimation system on the microcontroller from the initial temperature value acquisition to final estimation step.

14.11 Summary of the Chapter

We had only forward connections in neural network models up to this point. In this chapter, we added feedback connection between neurons to form recurrence. This helped us in forming neural network structures with memory capability. Hence, they became more suitable for sequential data processing. To explain these concepts better, we first associated recurrence with memory in this chapter. Afterward, we introduced three popular neural network models based on recurrence as RNN, GRU, and LSTM. While introducing each model, we first provided general information about it. Then, we focused on its formation and training in Keras. Then, we provided the usage examples of recurrence models on PC. Afterward, we considered implementing recurrence models on the STM32 microcontroller. Finally, we provided examples on the usage of recurrence based models to solve real-life problems.

Appendix A

A.1 Review of Probability Theory

We will benefit from probability theory extensively while explaining traditional machine learning methods. Here, the aim will be modeling uncertainty while making decisions. Probability theory can also be used in explaining random number generation considered in Sect. 5.1. Therefore, it will help us throughout the book.

There are several approaches in explaining probability. We will follow the axiomatic definition of probability in this book. Therefore, we will start with this topic. Next, we will cover random variables. We will also introduce probability density and distribution functions associated with random variables.

A.1.1 Axiomatic Definition of Probability

Axiomatic definition of probability constructs the overall probability theory based on three axioms taken de facto as true. Before introducing them, we should first consider some set concepts to be used while dealing with probability. We can form a set by random experiments such that outcome of this experiment define set elements. Here, randomness stands for the case such that the outcome for the experiment is not deterministic. In other words, any outcome within the set can happen. Sample space is the set of all outcomes of the random experiment. An event is a subset of the sample space in which we will define the probability on it. The event should satisfy certain conditions from a theoretical perspective. Since we will be dealing with practical application of probability theory in this book, we will assume these conditions are satisfied.

Before going further, let us consider a sample random experiment as flipping a coin. We know that outcome of this experiment can be either head (H) or tail (T).Therefore, we can define our sample space for this random experiment as $\Omega = \{H, T\}$. Since we will need subsets to define events, we can write $A = \{H\}$, $B =$

© The Author(s), under exclusive license to Springer Nature Switzerland AG 2025
C. Ünsalan et al., *Embedded Machine Learning with Microcontrollers*,
https://doi.org/10.1007/978-3-031-70912-8

$\{T\}$ as events. Axiomatic definition of probability assumes three conditions to assign probability on the events A and B as

1. $P(A) \geq 0$ and $P(B) \geq 0$
2. $P(\Omega) = 1$
3. $P(A \cup B) = P(A) + P(B)$ if $A \cap B = \emptyset$

The first assumption indicates that the probability is always greater than or equal to zero. The second assumption indicates that the random experiment outcome can have the maximum probability value of 1. The third axiom indicates that if the intersection of two events (as A and B) is an empty set, then the probability of their union is the sum of their probability. These three axioms can be used in constructing the complete probability theory.

Let's apply these three axioms to our random flipping a coin experiment. The first axiom tells that $P(A) \geq 0$ and $P(B) \geq 0$. The second axiom tells that $P(\Omega) = 1$. Hence, we will get either head or tail as the end of our random experiment. We know that the event $A \cap B$ cannot occur since either a head or a tail comes after flipping the coin, but not both. Hence, the third axiom tells us that $P(A \cup B) = P(A) + P(B)$. We know that $A \cup B = \Omega$ in our experiment. This leads to $P(A) + P(B) = 1$.

As the reader may observe, we did not talk about the values of $P(A)$ and $P(B)$. We only know that they should be between 0 and 1 based on the three axioms accepted. We will handle assigning probability to events in the following sections as we introduce random variables and probability density functions.

A.1.2   Conditional Probability and Bayes Theorem

We may have knowledge that a condition has already been satisfied during a random experiment. This poses a constraint on the outcome of the experiment. Hence, probability values should be modified accordingly. This will lead us to conditional probability.

Let's consider a simple example on conditional probability. Assume that our random experiment is rolling a dice. The events that we will assign probability for this random experiment are integer values from 1 to 6. Now, let's show how a previously occurred event or satisfied condition can be used in updating probability of another event. Assume that we would like obtain probability of the event of getting 3 after rolling the dice. We can represent this event as $A = \{3\}$. Assume further that we know the outcome of rolling the dice is even. In other words, we know that the outcome can only be 2, 4, or 6. We can represent this knowledge as another event $B = \{2, 4, 6\}$. This information changes the probability of the event A. We can represent it as the probability of A "given that" B has occurred. We can represent this conditional probability as $P(A|B)$. We know that the value 3 is odd. Hence, we have $P(A|B) = 0$.

In fact, the conditional probability for the two generic events is defined as $P(A|B) = \frac{P(A \cap B)}{P(B)}$. In our example, $A \cap B = \emptyset$. Therefore $P(A|B) = 0$. Now, what if we know that the output is odd? We can apply the same formula and obtain $P(A|B) = P(A)/P(B)$ since $A \subset B$. Since $0 < P(B) \leq 1$, we can say that $P(A|B) \geq P(A)$. Therefore, knowing that B has occurred increases probability of the event A.

Bayes theorem will be the fundamental building block of Bayes classifiers. Therefore, let's first introduce this theorem by using the events A and B. Bayes theorem indicates that

$$P(B|A) = \frac{P(A|B)}{P(B)} P(A) \tag{A.1}$$

where $P(B)$ is the prior probability, $P(A|B)$ is the likelihood, $P(A)$ is the evidence, and $P(B|A)$ is the posterior probability. Hence, Bayes theorem offers a way of updating the probability of event B, when a new evidence arrives. We will use these definitions while forming the Bayes classifier in Sect. 6.2.

A.1.3 Random Variables

What is more important than probability is defining a random variable from the machine learning perspective. The random variable is a function which projects events of a random experiment to real numbers. Let's consider the coin flipping random experiment. Assume that we assign number 1 to the event A and 0 to the event B. Hence, we form a correspondence between the events in our random experiment and real axis. As the reader might guess, we are free to choose any value in assigning real numbers.

Where does the "variable" term come from? We know that each event in the random experiment has an associated probability. Assigned numbers to events are also associated with these probability values. Therefore, we can assume that each assigned number in the real axis has an associated probability value. We can represent this association with the random variable represented by a capital letter. We can represent its value by a small letter. We can define a random variable based on our previous example as $P(X = 1) = P(A)$ and $P(X = 0) = P(B)$. As can be seen here, we can discard underlying events in our random experiment. Therefore, we can assume that assigned numbers have their probability values. To note here, other numbers in the real axis can also be considered in this setup. Assume that we would like to obtain $P(X = 3)$. Since there is no set associated with the real value 3, we will have $P(X = 3) = P(\emptyset) = 0$.

At this point, we should talk about discrete and continuous random variables. As the name implies, a discrete random variable can only take discrete values. This will be the case for all our embedded applications since we are dealing with digital numbers with limited resolution. On the other hand, a continuous random variable can take any real value in operation. Although this setup has some strict restrictions in realization, it is used most of the times in mathematical derivations. We will also follow the same approach in this book.

A.1.4 Probability Density and Distribution Functions

We know that random variable has associated probability values (for discrete random variables extensively). Instead of representing these probability values separately, we can form a function to describe them. Hence, will have the probability density function (pdf). We can define pdf for the discrete random variable X as

$$f_X(x) = P(X = x) \tag{A.2}$$

The reader should pay attention to the uppercase and lowercase letter usage while writing $f_X(x)$. Here X indicates that the pdf is associated with the random variable X. x is the value taken by the random variable. As an example, we can have $f_X(3) = P(X = 3)$. Hence $f_X(3)$ gives us probability of the random variable X at value 3.

The actual name for the definition in Eq. A.2 for discrete random variables is the probability mass function. Since we will use continuous and discrete random variables in mathematical derivations and implementation, we will only use the pdf naming from this point on. Moreover, the pdf definition in Eq. A.2 is not applicable to continuous random variables. The main reason is that there are infinite number of x values to consider for a continuous random variable. Hence, it is assumed that $P(X = x) = 0$ for the continuous random variable X. Therefore, a different approach is needed for obtaining the pdf of a continuous random variable. To do so, we first define the probability distribution function as

$$F_X(x) = P(x \le X) \tag{A.3}$$

Then, we can form the pdf as

$$f_X(x) = \frac{dF_X(x)}{dx} \tag{A.4}$$

We will model uncertainty in machine learning using random variables and their pdf. To be more specific, the pdf will be used in modeling. Before going further, we should summarize some fundamental properties of probability density functions. Since a probability value cannot be negative, the pdf cannot take a value less than 0. Since all the events map to real axis, the sum of corresponding probability values will be 1. This leads to $\sum_{x=-\infty}^{\infty} f_X(x) = 1$ for the discrete random variable X and $\int_{x=-\infty}^{\infty} f_X(x)dx = 1$ for the continuous random variable X.

There are several probability density functions in literature. The two well-known and extensively used ones for continuous random variables are the uniform and Gaussian pdfs. The uniform pdf (in standard form) is defined as

$$f_X(x) = \begin{cases} 1, & 0 \le x \le 1 \\ 0, & \text{Otherwise} \end{cases} \tag{A.5}$$

The uniform pdf limits the range of random variable. All values are equally probable within this range.

The Gaussian (normal) is the most well-known pdf. It can be used to model a fairly wide range of real-life phenomena. The other advantage of the Gaussian pdf is that its characteristic can be controlled by only two parameters as mean, μ, and standard deviation, σ. The Gaussian pdf is defined as

$$f_X(x) = \frac{1}{\sqrt{2\pi}\sigma_x} \exp\left(\frac{-1}{2}\left(\frac{x - \mu_x}{\sigma_x}\right)^2\right) \tag{A.6}$$

As can be seen in Eq. A.6, the Gaussian pdf is defined for the real axis. The highest value of the function is reached when $x = \mu_x$. The spread of the function is described by σ_x. If σ_x is small, then the spread is small. Hence, the Gaussian function concentrates around μ_x. If σ_x is large, then the Gaussian function will be wide. To note here, σ_x^2 is called variance.

A.1.5   Parameter Estimation

As explained Sect. 5.1.4, we can form sample pdf by constructing the histogram from samples (observations) of any random variable. Although this gives us freedom, there are shortcomings in this method as well. We know that selecting the bin size is one issue. The second issue is that the pdf is not represented in functional form. Parameter estimation based pdf formation helps us at this step. The idea here is as follows.

We can observe the value of the random variable X. We can assume each observation to be a random variable since we do not know the value of each observation before the observation. Since these originate from the same random variable, we take them to be identical. Assume that we have n such random variables as $X_1, \ldots, X_n$. Hence, we will have $f_{X_1}(x) = \cdots, f_{X_n}(x) = f_X(x)$ for $-\infty < x < \infty$. Please remember that we have discrete random variables here. In order to have a tractable problem, we should assume that the random variables $X_1, \ldots, X_n$ are independent. As a result, we will have independent and identically distributed (iid) random variables representing our observations.

In order to model the pdf of random variable X, we should pick a pdf with adjustable parameters. Then, we can use observations to estimate the value of parameters by using an estimator. At this point we should define what the estimator is. An estimator is the function of random variables that are used to estimate the parameters, but does not depend on the parameters itself. One well-known estimator used in applications is the maximum-likelihood estimator. It depends on the likelihood function of the random variables $X_1, \ldots, X_n$ as their joint pdf as

$$L(\theta) = f_{X_1, \ldots, X_n}(x_1, \ldots, x_n; \theta) \tag{A.7}$$

for the unknown parameter θ, with realizations $x_1, \ldots, x_n$. Using the iid assumption, we will have

$$L(\theta) = \prod_{i=1}^{n} f_X(x_i; \theta) \tag{A.8}$$

The maximum likelihood estimator aims to find the parameter θ that maximizes $L(\theta)$ for the given realizations (observations) $x_1, \ldots, x_n$.

We can model the pdf as Gaussian with two unknown parameters as mean μ and standard deviation σ. Then, we can use the maximum likelihood estimator to estimate the parameter μ from realizations (observations) $x_1, \ldots, x_n$ as

$$L(\mu, \sigma) = \prod_{i=1}^{N} \frac{1}{\sigma\sqrt{2\pi}} \exp\left(-\frac{(x_i - \mu)^2}{2\sigma^2}\right) \tag{A.9}$$

To simplify calculations, we can take logarithm of both sides. Then, Eq. A.8 becomes

$$\ln(L(\mu, \sigma)) = \sum_{i=1}^{N} \ln\left(\frac{1}{\sigma\sqrt{2\pi}} \exp\left(-\frac{(x_i - \mu)^2}{2\sigma^2}\right)\right)$$

$$= -N\ln(\sigma) - \frac{N}{2}\ln(2\pi) - \frac{1}{2\sigma^2} \sum_{i=1}^{N} \left(x_i^2 - 2\mu x_i + \mu^2\right) \tag{A.10}$$

By taking derivative of our objective function $\ln(L(\mu, \sigma))$ with respect to μ, we obtain

$$\frac{d\ln(L(\mu, \sigma))}{d\mu} = \frac{2}{2\sigma^2} \sum_{i=1}^{N} x_i - \frac{2N\mu}{2\sigma^2} = 0 \tag{A.11}$$

We set this derivative to zero as we are seeking the μ value which gives us the maximum likelihood. Simplifying equation A.11 leads to

$$\mu = \frac{1}{N} \sum_{i=1}^{N} x_i \tag{A.12}$$

We can repeat same process for standard deviation estimation. Thus, we take derivative of Eq. A.10 with respect to σ as

$$\frac{d\ln(L(\mu, \sigma))}{d\sigma} = -\frac{N}{\sigma} + \frac{2}{2\sigma^3} \sum_{i=1}^{N} (x_i - \mu)^2 = 0 \tag{A.13}$$

Table A.1 Pin usage table for CN4 header of the STM32 board

Pin	Port name	Usage area
1	PI3	Digital I/O, EXTI, TIM8 ETR, SPI2 MOSI, I2S2 SD, FMC D27, DCMI D10
2	PH6	Digital I/O, EXTI, TIM12 CH1, SPI5 SCK, I2C2 SMBA, FMC SDNE1, ETH RXD2, DCMI D8
3	PI0	Digital I/O, EXTI, TIM5 CH4, SPI2 NSS, LTDC G5, I2S2 WS, FMC D24, DCMI D13
4	PG7	Digital I/O, EXTI, USART6 CK, LTDC CLK, FMC INT, DCMI D13
5	PB4	Digital I/O, EXTI, TIM3 CH1, SYS JTRST, SPI3 MISO, SPI2 NSS, SPI1 MISO, I2S2 WS
6	PG6	Digital I/O, EXTI, LTDC R7, DCMI D12
7	PC6	Digital I/O, EXTI, USART6 TX, TIM8 CH1, TIM3 CH1, SDMMC1 D6, LTDC HSYNC, I2S2 MCK, DCMI D0
8	PC7	Digital I/O, EXTI, USART6 RX, TIM8 CH2, TIM3 CH2, SDMMC1 D7, LTDC G6, I2S3 MCK, DCMI D1

Hence, our σ estimation will be

$$\sigma = \sqrt{\frac{1}{N} \sum_{i=1}^{N} (x_i - \mu)^2} \tag{A.14}$$

A.2   STM32 Board Pin Usage Tables

Pins of the STM32 board are arranged in four headers, named CN4, CN5, CN6, and CN7. Usage areas of these pins are tabulated in Tables A.1, A.2, A.3, and A.4. Unused pins can be used for more than one purpose. To note here, we only summarized the usage areas of the pins to be considered in this book.

A.3   Changing the Default Toolchain of Mbed Studio

We will need to change the default toolchain of Mbed in several parts of the book. Therefore, we summarize the steps needed for this operation in this section. To note here, we covered the steps for the Windows operating system here. For other operating systems, please see [21].

Table A.2 Pin usage table for CN5 header of the STM32 board

Pin	Port name	Usage area
1	PA0	Digital I/O, EXTI, USART2 CTS, UART4 TX, TIM8 ETR, TIM5 CH1, TIM2 ETR, TIM2 CH1, SYS WKUP, SAI2 SDB, ETH CRS, ADC1 IN0, ADC2 IN0, ADC3 IN0
2	PF10	Digital I/O, EXTI, LTDC DE, DCMI D11, ADC3 IN8
3	PF9	Digital I/O, EXTI, TIM14 CH1, SPI5 MOSI, SAI1 FS B, QUADSPI BK1 IO1, DAC EXTI9, ADC3 IN7
4	PF8	Digital I/O, EXTI, UART7 RTS, UART DE, TIM13 CH1, SPI5 MISO, SAI1 SCKB, QUADSPI BK1 IO1, ADC3 IN6
5	PF7	Digital I/O, EXTI, UART7 TX, TIM11 CH1, SPI5 SCK, SAI1 MCLKB, QUADSPI BK1 IO2, ADC3 IN5
6	PF6	Digital I/O, EXTI, UART7 RX, TIM10 CH1, SPI5 NSS, SAI1 SDB, QUADSPI BK1 IO3, ADC IN4

Table A.3 Pin usage table for CN6 header of the STM32 board

Pin	Port name	Usage area
1	NC	Not connected
2	IOREF	Connected to 3V3
3	NRST	External reset
4	3V3	3 V input or output
5	5V	5 V input or output
6	GND	Ground voltage
7	GND	Ground voltage
8	VIN	Voltage input

In order to change the default toolchain of Mbed Studio to GCC_ARM, we should first locate its installed path. The GCC ARM compiler can usually be found in the folder "C:/Program Files (x86)/GNU Arm Embedded Toolchain/9 2020-q2-update/bin" folder. Please note that the folder name can be slightly different for some installations. Therefore, we strongly suggest the reader to check related folders if there no exact match. After finding the GCC ARM compiler path, we should go to the path C:\Users\username\AppData\Local\Mbed Studio in the Windows

Table A.4 Pin usage table for CN7 header of the STM32 board

Pin	Port name	Usage area
1	PB8	Digital I/O, EXTI, TIM4 CH3, TIM10 CH1, SDMMC1 D4, LTDC B6, I2C1 SCL, ETH TXD3, DCMI D6, CAN1 RX
2	PB9	Digital I/O, EXTI, TIM4 CH4, TIM11 CH1, SPI2 NSS, SDMMC1 D5, LTDC B7, I2S2 WS, I2C1 SDA, DCMI D7, DAC EXTI9, CAN1 TX
3	AVDD	Connected to Vref+
4	GND	Ground voltage
5	PI1	Digital I/O, LED1, EXTI, TIM8 BKIN2, SPI2 SCK, LTDC G6, I2S2 CK, FMC D25, DCMI D8
6	PB14	Digital I/O, EXTI, USB OTG HS DM, USART3 RTS, USART3 DE, TIM8 CH2N, TIM1 CH2N, TIM12 CH1, SPI2 MISO
7	PB15	Digital I/O, EXTI, USB OTG HS DP, TIM8 CH3N, TIM1 CH3N, TIM12 CH2, SPI2 MOSI, RTC REFIN, I2S2 SD
8	PA8	Digital I/O, EXTI, USB OTG FS SOF, USART1 CK, TIM8 BKIN2, TIM1 CH1, RCC MCO1, LTDC R6, I2C3 SCL
9	PA15	Digital I/O, EXTI, UART4 RTS, UART4 DE, TIM2 ETR, TIM2 CH1, SYS JTDI, SPI3 NSS, SPI1 NSS, I2S3 WS, I2S1 WS
10	PI2	Digital I/O, EXTI, TIM8 CH4, SPI2 MISO, LTDC G7, FMC D26, DCMI D9

operating system. There, we have to create a JSON file named as "external-tools.json." This file content should be as in Listing A.1.

Listing A.1 Mbed Studio JSON file content

```
{
    "bundled": {
        "gcc": "C:/Program Files (x86)/GNU Tools Arm Embedded/9
            2019-q4-major/bin"

    },
    "defaultToolchain": "GCC_ARM"
}
```

References

1. Alpaydın, E.: Introduction to Machine Learning, 4th edn. The MIT Press, Cambridge (2020)
2. Arm: https://www.arm.com/company/news/2021/02/arm-ecosystem-ships-record-6-billion-arm-based-chips-in-a-single-quarter. Accessed: 24 June 2024
3. Arm: https://os.mbed.com/platforms/ST-Discovery-F746NG/. Accessed: 24 June 2024
4. Arm: Arm Cortex M7 Processor Technical Reference Manual, revision r1p2 edn.
5. Cho, K., van Merrienboer, B., Gulcehre, C., Bahdanau, D., Bougares, F., Schwenk, H., Bengio, Y.: Learning phrase representations using RNN encoder-decoder for statistical machine translation (2014). arXiv:1406.1078
6. Dempster, A.P., Laird, N.M., Rubin, D.B.: Maximum likelihood from incomplete data via the EM algorithm. J. R. Stat. Soc. Ser. B (Methodol.) **39**(1), 1–22 (1977)
7. Duda, R.O., Hart, P.E., Stork, D.G.: Pattern Classification, 2nd edn. Wiley, Hoboken (2000)
8. Ester, M., Kriegel, H.P., Sander, J., Xu, X.: A density-based algorithm for discovering clusters in large spatial databases with noise. In: Proceedings of the Second International Conference on Knowledge Discovery and Data Mining, vol. 96, pp. 226–231 (1996)
9. He, K., Zhang, X., Ren, S., Sun, J.: Deep residual learning for image recognition. In: 2016 IEEE Conference on Computer Vision and Pattern Recognition (CVPR) (2015)
10. Hochreiter, S., Schmidhuber, J.: Long short-term memory. Neural Comput. **9**, 1735–80 (1997)
11. Hoke, B.: https://github.com/EmbeddedML/Datasets/blob/main/temperature_dataset.csv. Accessed: 24 June 2024
12. Howard, A.G., Zhu, M., Chen, B., Kalenichenko, D., Wang, W., Weyand, T., Andreetto, M., Adam, H.: MobileNets: Efficient convolutional neural networks for mobile vision applications (2017). arXiv:1704.04861
13. Howard, A., Sandler, M., Chu, G., Chen, L.C., Chen, B., Tan, M., Wang, W., Zhu, Y., Pang, R., Vasudevan, V., Le, Q.V., Adam, H.: Searching for MobileNetV3. In: Proceedings of the IEEE/CVF International Conference on Computer Vision (2019)
14. Hu, M.K.: Visual pattern recognition by moment invariants. IRE Trans. Inf. Theory **8**(2), 179–187 (1962)
15. Iandola, F.N., Han, S., Moskewicz, M.W., Ashraf, K., Dally, W.J., Keutzer, K.: SqueezeNet: AlexNet-level accuracy with 50x fewer parameters and <0.5MB model size (2016). arXiv:1602.07360
16. Jakobovski, Z.J.: https://github.com/Jakobovski/free-spoken-digit-dataset/archive/v1.0.10.tar.gz. Accessed: 24 June 2024
17. Kwapisz, J.R., Weiss, G.M., Moore, S.: Activity recognition using cell phone accelerometers. ACM SIGKDD Explor. Newslett. **12**, 74–82 (2011)
18. Lab, W.: https://www.cis.fordham.edu/wisdm/includes/datasets/latest/WISDM_ar_latest.tar.gz. Accessed: 24 June 2024
19. LeCun, Y.: http://yann.lecun.com/exdb/mnist/. Accessed: 24 June 2024
20. Malli, R.C.: https://github.com/rcmalli/keras-squeezenet. Accessed: 24 June 2024

21. Mbed: https://os.mbed.com/docs/mbed-studio/current/installing/switching-to-gcc.html. Accessed: 24 June 2024
22. NumPy: https://numpy.org/doc/stable/reference/random/generator.html#distributions. Accessed: 24 June 2024
23. OmniVision: OV5640 datasheet (2011)
24. PyTorch: https://pytorch.org/vision/stable/models.html#table-of-all-available-classification-weights. Accessed: 24 June 2024
25. Romeu-Guallart, P., Zamora-Martinez, F.: https://archive.ics.uci.edu/dataset/274/sml2010. Accessed: 24 June 2024
26. Rumelhart, D., Hinton, G., Williams, R.: Learning representations by back-propagating errors. Nature **323**, 533–536 (1986)
27. Sandler, M., Howard, A., Zhu, M., Zhmoginov, A., Chen, L.C.: MobileNetV2: Inverted residuals and linear bottlenecks. In: 2018 IEEE/CVF Conference on Computer Vision and Pattern Recognition (CVPR) (2019)
28. Scheck, T.: https://github.com/scheckmedia/keras-shufflenet. Accessed: 24 June 2024
29. Schubert, E., Sander, J., Ester, M., Kriegel, H.P., Xu, X.: DBSCAN revisited, revisited: why and how you should (still) use DBSCAN. ACM Trans. Database Syst. **42**(3), 1–21 (2017)
30. Scikit-Learn: https://scikit-learn.org/stable/modules/generated/sklearn.svm.SVC.html. Accessed: 24 June 2024
31. Scikit-Learn: https://scikit-learn.org/stable/modules/generated/sklearn.tree. DecisionTreeClassifier.html. Accessed: 24 June 2024
32. Sparks, P.: The route to a trillion devices. The outlook for IoT investment to 2035. Arm (2017)
33. STMicroelectronics: STM32F745XX STM32F746XX, docid027590 rev 4 edn. (2016)
34. STMicroelectronics: Application note Tutorial for MEMS microphones, an4426 edn. (2017)
35. STMicroelectronics: RM0385 Reference manual, STM32F75xxx and STM32F74xxx advanced Arm®-based 32-bit MCUs, rev 8 edn. (2018)
36. STMicroelectronics: Application note Interfacing PDM digital microphones using STM32 MCUs and MPUs, an5027 edn. (2019)
37. STMicroelectronics: User manual Camera module bundle for STM32 boards, um2779-rev 1 edn. (2020)
38. STMicroelectronics: User manual Discovery kit for STM32F7 Series with STM32F746NG MCU, um1907 edn. (2020)
39. STMicroelectronics: IMP34DT05 Datasheet MEMS audio sensor omnidirectional digital microphone for industrial application, ds12725-rev 4 edn. (2021)
40. STMicroelectronics: https://www.st.com/en/evaluation-tools/32f746gdiscovery.html. Accessed: 24 June 2024
41. STMicroelectronics: https://www.st.com/en/development-tools/stm32cubeide.html. Accessed: 24 June 2024
42. STMicroelectronics: https://github.com/STMicroelectronics/stm32ai-modelzoo/tree/main/ image_classification/src/models. Accessed: 24 June 2024
43. STMicroelectronics: https://github.com/STMicroelectronics/stm32ai-modelzoo/blob/main/ image_classification/src/models/resnetv1.py. Accessed: 24 June 2024
44. STMicroelectronics: https://github.com/STMicroelectronics/stm32ai-modelzoo/blob/main/ image_classification/src/models/st_efficientnet_lc_v1.py. Accessed: 24 June 2024
45. Tan, M., Le, Q.V.: EfficientNet: Rethinking model scaling for convolutional neural networks (2020). arXiv:1905.11946
46. TensorFlow: https://www.tensorflow.org/api_docs/python/tf/constant. Accessed: 24 June 2024
47. TensorFlow: https://www.tensorflow.org/api_docs/python/tf/Variable. Accessed: 24 June 2024
48. TensorFlow: https://www.tensorflow.org/api_docs/python/tf/dtypes. Accessed: 24 June 2024
49. TensorFlow: www.tensorflow.org/datasets/overview. Accessed: 24 June 2024
50. TensorFlow: www.tensorflow.org/tutorials/keras/save_and_load. Accessed: 24 June 2024
51. TensorFlow: https://www.tensorflow.org/api_docs/python/tf/keras/losses. Accessed: 24 June 2024

52. TensorFlow: https://www.tensorflow.org/api_docs/python/tf/optimizers. Accessed: 24 June 2024
53. TensorFlow: https://www.tensorflow.org/guide/basic_training_loops. Accessed: 24 June 2024
54. TensorFlow: https://www.tensorflow.org/api_docs/python/tf/lite/Interpreter. Accessed: 24 June 2024
55. TensorFlow: https://www.tensorflow.org/lite/guide/inference#load_and_run_a_model_in_python. Accessed: 24 June 2024
56. TensorFlow: https://www.tensorflow.org/model_optimization/guide/pruning. Accessed: 24 June 2024
57. TensorFlow: https://www.tensorflow.org/model_optimization/guide/clustering. Accessed: 24 June 2024
58. TensorFlow: https://github.com/tensorflow/tflite-micro/tree/main/tensorflow/lite/micro/tools. Accessed: 24 June 2024
59. TensorFlow: https://www.tensorflow.org/api_docs/python/tf/keras/layers/Conv2D. Accessed: 24 June 2024
60. Ünsalan, C., Gürhan, H.D., Yücel, M.E.: Programmable Microcontrollers: Applications on the MSP432 LaunchPad. McGraw-Hill, New York (2018)
61. Ünsalan, C., Gürhan, H.D., Yücel, M.E.: Embedded System Design with Arm Cortex-M Microcontrollers: Applications with C, C++, and MicroPython. Springer Nature, Berlin (2022)
62. Werbos, P.: Backpropagation through time: what it does and how to do it. Proc. IEEE **78**(10), 1550–1560 (1990)
63. Zhang, X., Zhou, X., Lin, M., Sun, J.: ShuffleNet: An extremely efficient convolutional neural network for mobile devices (2017). arXiv:1707.01083

Index

GPSR Compliance

The European Union's (EU) General Product Safety Regulation (GPSR) is a set of rules that requires consumer products to be safe and our obligations to ensure this.

If you have any concerns about our products, you can contact us on ProductSafety@springernature.com

In case Publisher is established outside the EU, the EU authorized representative is:

Springer Nature Customer Service Center GmbH
Europaplatz 3
69115 Heidelberg, Germany

The manufacturer's authorised representative in the EU is Springer
Nature Customer Service Centre GmbH, Europaplatz 3, 69115 Heidelberg,
Germany. If you have any concerns regarding our products, please
contact ProductSafety@springernature.com

Printed and bound by CPI Group (UK) Ltd, Croydon, CR0 4YY
29/04/2026
02099539-0001